Contemporary
Quality Function Deployment
for Product and Process Innovation

Towards Digital Transformation of Customer and Product Information in a New Knowledge-Based Approach

Contemporary Quality Function Deployment for Product and Process Innovation

Towards Digital Transformation of Customer and Product Information in a New Knowledge-Based Approach

Thomas Lager
Mälardalen University, Sweden

World Scientific
NEW JERSEY • LONDON • SINGAPORE • BEIJING • SHANGHAI • HONG KONG • TAIPEI • CHENNAI • TOKYO

Published by

World Scientific Publishing Co. Pte. Ltd.
5 Toh Tuck Link, Singapore 596224
USA office: 27 Warren Street, Suite 401-402, Hackensack, NJ 07601
UK office: 57 Shelton Street, Covent Garden, London WC2H 9HE

Library of Congress Cataloging-in-Publication Data
Names: Lager, Thomas, author.
Title: Contemporary quality function deployment for product and process innovation : towards digital transformation of customer and product information in a new knowledge-based approach / Thomas Lager (Mälardalen University, Sweden).
Description: New Jersey : World Scientific, [2019]
Identifiers: LCCN 2018056447 | ISBN 9789813279872 (hbk : alk. paper)
Subjects: LCSH: Quality function deployment. | Production management--Quality control. | Technological innovations.
Classification: LCC TS156 .L287 2019 | DDC 658.5/62--dc23
LC record available at https://lccn.loc.gov/2018056447

British Library Cataloguing-in-Publication Data
A catalogue record for this book is available from the British Library.

For any available supplementary material, please visit
https://www.worldscientific.com/worldscibooks/10.1142/11265#t=suppl

Desk Editors: Herbert Moses/Karimah Samsudin

Typeset by Stallion Press
Email: enquiries@stallionpress.com

To Gunilla, my best critic and true companion.
This book could not have been possible without
your unfailing support and assistance.
I am forever grateful.

There are some things which cannot be learned quickly, and time, which is all we have, must be paid heavily for their acquiring. They are the very simplest things, and because it takes a man's life to know them the little new that each man gets from life is very costly and the only heritage he has to leave.

Ernest Hemingway, *Death in the Afternoon*

Preface

Principal ideas and scope of the book

In today's highly competitive industrial climate, and in times of tight company resources and a continual need to improve company performance in all areas, the use of effective and efficient tools and development methodologies for R&D is one avenue to follow. In R&D management, methodologies to facilitate and improve the performance of various R&D activities come into fashion and, at times, go out again.

Sometimes, methodologies disappear because they are not good enough to withstand the tough industrial environment and its high demands on industrial usability. However, fortunately, today there are a large number of well-proven methodologies available. Thus, the question is not really whether they should be deployed or not, but rather *which* methodologies should be used and *why*, *when* and *how* they should be deployed.

Vintage Quality Function Deployment

> **This book is, foremost, an updated and extended handbook and guiding manual for the general use of the Quality Function Deployment (QFD) methodology for product and process innovation in the manufacturing industries.**

Two decades ago, the QFD methodology was still in its infancy in the Western world; there were few publications on the subject, and there was as yet practically no scientific evidence on company outcomes. Some of the available QFD books listed in the bibliography remain interesting to read, either in full or in selected parts, even if all of them are more than 20 years old.

However, since the industrial environment for product innovation is rapidly changing, supporting methodologies need to be adjusted and improved in order to be of good use for industry practitioners. Thus, this book is intended, first of all, to be an update on general QFD methodology use as such. It advocates that the industrial user should take advantage of four already well-proven benefits of using QFD in company product innovation in order to:

- **Create** a reliable and fact-based interface between the company and its customers — an innovation approach in the spirit of good old TQM.
- **Translate** the Voice of the Customer into a product attribute dimension — using the House of Quality.
- **Promote** company cross-functional product innovation — using QFD as an instrument to bridge organizational interfaces.
- **Address** the fuzzy front end of product innovation — a proactive use of QFD on the company's upfront product innovation activities.

In the deployment of the QFD methodology, there appears to have been some general misunderstanding about unproven benefits and the "cause and effect" of expected outcomes. Thus, we should begin by getting this sorted out once and for all. *Using a "retrievable and accumulated company knowledge base" coupled with "improved company communication," you will get "better decision support," assisting in the development of "better product specifications," resulting in "better products" and ultimately leading to "better overall customer satisfaction."*

Such scientifically proven outcomes from QFD industrial use should certainly incentivize product innovation professionals to introduce and make good use of the QFD methodology, focusing on its knowledge creation and organizational learning.

In search of a holistic perspective in product innovation

In today's product innovation, the problem is often not a lack of relevant information but rather an extreme overload of information — an inability to "see the forest for the trees." For this reason, it is generally difficult to see "the whole" and how individual pieces of information link to each other. Thus, there is a need for supporting innovation methodologies, and in particular for the matrix approach in the use of the QFD methodology.

The QFD matrices are outstanding instruments to grasp "the whole" in product innovation, and the gradual upgrade of the matrices over time will capture new patterns emerging from the new information. In the process industries, which generally produce non-assembled products of a more or less homogenous nature, the QFD methodology will directly address the total product (i.e., "the whole"). This truly important advantage in the use of the QFD methodology in the process industries has been grossly neglected in previous presentations.

Contemporary Quality Function Deployment

The book takes a groundbreaking knowledge-based approach in the use of the QFD methodology for the creation of Integrated Knowledge Platforms (IKP) supporting the product innovation work process and company organizational learning.

The Quality Function Deployment product innovation methodology, originally developed by the late Professor Yoji Akao, is a reliable and still highly trusted instrument for improving company product innovation performance. Nevertheless, this book takes a novel approach in the methodology's industrial application and use; as such, it aspires to offer a significant contribution to the available QFD reference literature. Moreover, the book seeks to introduce a new perspective on industrial QFD implementation and use. Three important messages for more efficient use of the QFD methodology are:

- ***Use** the QFD methodology as a supportive instrument to enhance your product innovation work process.*
- ***Discover** that the QFD matrices are outstanding instruments for capturing both explicit and tacit knowledge and for organizational learning.*
- ***Consolidate** the knowledge and matrix information into coherent and reusable IKPs.*

The key to future successful product innovation and renovation is an integration of knowledge along the total value/supply chain in a company end-to-end perspective. Thus, this book advocates that product innovation methodologies not only must reach out for the end-users but also must facilitate the integration of suppliers and supplied products or raw materials. Moreover, a new knowledge-based approach in the use of the QFD methodology for the creation of supportive IKPs for the product innovation work process is a key conceptual idea brought forward in this book.

Companies cannot afford to start from a "clean slate" in the future if they want to develop products in a cost-efficient manner and a reasonable timeframe. In order to rectify such a state of affairs, product innovation work processes and projects must start from easily retrievable and previously accumulated company platforms of knowledge. Adoption of the QFD tool and matrices in building such well-structured platforms related to individual products or product families is the heart of the matter and the "hidden gem" of successful future QFD methodology use.

Such an approach will help not only to take the "fuzziness" out of the "fuzzy front end" but also to take advantage of the diverse number of benefits one should be able to expect from a full use of the QFD methodology. It is argued in this book that the inefficient and costly use of the QFD methodology in "ad hoc" development projects should be abandoned and replaced by the use of the methodology in building coherent and reusable company IKPs for innovation. Such platforms are company frameworks and points of departure for faster product innovation; they are not part of the product innovation work process as such, but they serve subsequent project-based product innovation activities. This approach creates a number of advantages by making it possible to:

- ***Use** IKPs as launch-pads for the creation of product innovation projects.*
- ***Deploy** the IKPs for a dynamic integration of all matrix information, including the upstream & downstream matrices contained in the mpQFD system.*
- ***Utilize** IKPs to speed up the company's overall product innovation work process.*

- ***Revitalize*** *company use of the QFD phase progression systems in order to secure product manufacturability (processability) in innovation.*
- ***Institute*** *a new "end-to-end" product innovation work process perspective in the use of QFD in product innovation.*

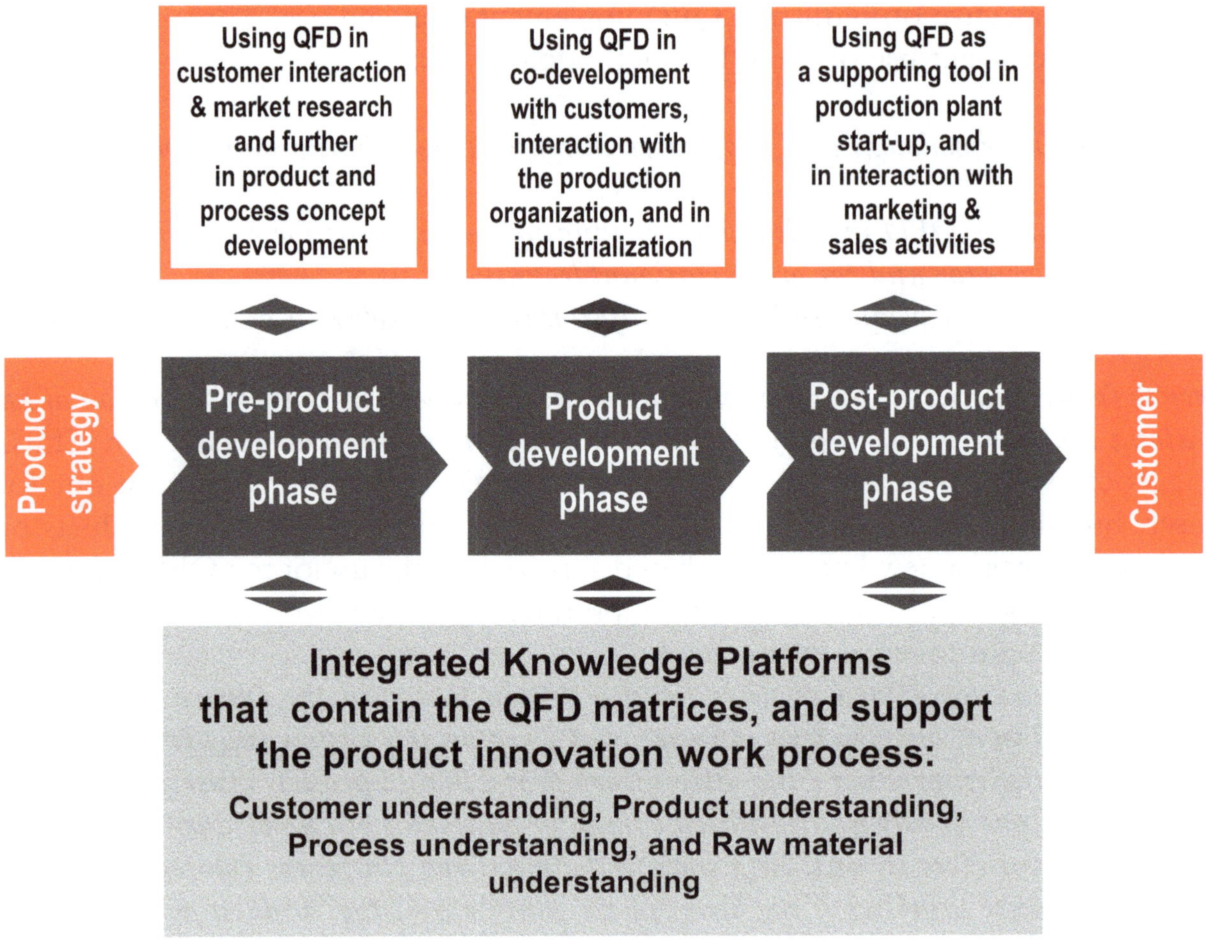

Similar to when Jack Welch stood on the barricades promoting Six Sigma ideas for improved overall company performance and securing its success in General Electric, company management and product innovators that come on strong for the application of "Contemporary QFD" with not too many ifs, ands, and buts are likely to reap the greatest rewards from QFD methodology use.

A process-industrial perspective on the use of the QFD methodology

The reader is introduced to the Advanced Multiple Progression QFD system (mpQFD), which is tailor-made for process-industrial product innovation applications and in use of its upstream & downstream matrices.

The QFD methodology was born and developed in Japan as a supporting tool for assembly-based industries, where products are manufactured and contain a very large number of components. In those industries, the most significant difficulty in using the QFD methodology is the large number of hierarchic levels of components and sub-components that constitute the overall product. Even if only a minor part of an assembled product is selected for a QFD project, this small system still contains a number of components. The ease of application of QFD to non-assembled products is obvious, since the House of Quality captures all Product Design Requirements (i.e., the total product).

Phase progression of Product Design Requirements into the production process using the classical model of the Four Phases of Matrices for assembled products is often a somewhat complicated procedure, and experience shows that few companies even attempt it. However, the phase progression into the process-industrial production systems is, comparatively, a much less complicated operation. ***The use of the upstream & downstream matrices in the mpQFD system thus represents "phase progression simplified."***

Thus, taking the process-industrial innovation characteristics as a point of departure, the good news is that the use of the QFD methodology is generally more advantageous in this cluster of industries than in other manufacturing industries.

Reading suggestions and targeted readership

Around 300 BC, the Greek mathematician Euclides wrote his famous but rather complicated book about geometry, Ta Stokeion. At that time, the Egyptian King Ptolemaios, who thought the book was too strenuous to read, asked for an easier version. Euclides curtly responded, "There is no royal road in geometry." Unfortunately, there is neither a royal road in learning the methodology of Quality Function Deployment. Professor Akao's advice on learning the QFD methodology was generally to run a QFD project — that is, "learning by doing". The present author certainly highly approves of such an approach; however, learning the QFD methodology is made even easier by the use of a good manual introducing the methodology and how to use it. This book is intended to serve this purpose. The total book content is divided into three sections and in the first section, containing Part I and Part II, the general use of the QFD methodology as a supportive instrument for the industrial product innovation work process is introduced.

In the second section, containing Part III and Part IV, the general methodological use of the Quality Function Deployment and the mpQFD system in particular are presented in detail.

In the third section, containing Part V and Part VI, new perspectives on the industrial application and implementation of the methodology and Integrated Knowledge Platforms are presented in a general overview.

Industry professionals

The majority of this book can be used in product and process innovation in all kinds of manufacturing industries, since the deployment of the *Voice of the Customer* and the *House of Quality* are identical in both the process industries and other manufacturing industries. However, Part IV of this book is dedicated specifically to QFD use in process-industrial product development. The phase progression methodology and the corresponding matrices in Part IV will surely also provide an out-the-box perspective for developers of assembled products. The illustrative case involves reporting from a process-industrial innovation environment, but the explanatory examples are certainly also useful in other manufacturing industries.

For newcomers and product developers that are not acquainted with the methodology or have never run a QFD project, the book offers a guiding framework and a handbook for the methodology's industrial use. While if it is recommended to take advantage of an experienced QFD facilitator in the first company QFD project, the purpose of the book is to provide readers with a good and solid understanding of how to use the methodology and how to implement it in an industrial setting. For the experienced QFD practitioner, the book will offer new perspectives on a more efficient and effective use of the methodology.

Industry professionals in the process industries

While the methodology is used in many sectors of the process industries, the methodological adaptation and presentations in textbooks for this family of industries are scarce, if not non-existent; this book is intended to fill this gap. For professionals in all sectors of the process industries, this book is a guiding manual on the use of the Advanced Multiple Progression QFD system (mpQFD), specifically adapted to a process-industrial context. In particular, the use of the upstream matrices is thoroughly reviewed and explained, underscoring the importance of achieving product "processability."

Academic scholars and students

For scholars in Innovation Management who are not already familiar with the QFD methodology in product innovation, the book will facilitate an improved understanding of the subject matter and provide new theoretical insights and knowledge about the methodology's use. For those familiar only with QFD use in other manufacturing industries, the application in a process-industrial environment offers a supplementary new knowledge area. Moreover, the use of the methodology in the creation of IKPs in a new knowledge-based approach in the use of the QFD methodology is a new road for further exploration. For students in all disciplines of the natural sciences, innovation management, product design and engineering, this book is an updated textbook on the QFD methodology as such. Each part of the book concludes with a number of questions that can be used as instruments in lecturing and for more advanced tutorials.

Following the footsteps of the late Professor Yoji Akao

The content of this book is not only based on the author's own research in Innovation Management but also relies extensively on the author's industrial experiences as a facilitator of the QFD methodology in various sectors of the process industries. In this capacity, the author has introduced the methodology in many companies in different sectors of manufacturing industries over a considerable span of time and has acted as a QFD facilitator in more than 15 large and complex product innovation projects.

As a visiting researcher invited by Professor Takao Fujiwara to Toyohashi University of Technology in Japan in the year 2000, the author also had the privilege of having the late Professor Yoji Akao as his mentor and prime supporter in the development of the *Multiple Progression QFD (mpQFD) system*, which has been specially adapted to process-industrial conditions for innovation. Their encouragement, suggestions, and assistance were decisive in the successful outcome of the development of the QFD system. Now, the author hopes to follow the late Professor Akao's footsteps by further developing the industrial use of the methodology and promoting the use of this tool to enhance company product innovation capabilities.

As a scholar in the management of innovation and technology, the author has not only had access to and digested relevant academic research and publications related to the QFD methodology but also has published extensively in reputed scientific journals and international conferences. The overall perspective on the industrial use of the QFD methodology has always been related to product innovation and the methodological use of QFD as a facilitating tool in the company product innovation work process. The combination of profound theoretical insight with operational industry best practice in the use of the QFD methodology is the hallmark of this book.

About the Author

Dr. Thomas Lager is an affiliated Professor at Mälardalen University, School of Innovation, Design and Engineering in Sweden. He is the General Manager of blinab "boutique" management consultancy in Sweden, in the area of Management of Innovation & Technology in the Process Industries.

He was previously a Professor and Chair in Innovation Management at University Mohammed VI Polytechnique in Morocco. He was an affiliated Professor at the Centre for Innovation, Technology and Entrepreneurship at Grenoble Ecole de Management in France. He was formerly an adjunct Professor and Director of the Centre for Management of Innovation and Technology in the Process Industry at Luleå University of Technology in Sweden. He holds an MS degree in Mining Engineering from the Royal Institute of Technology in Sweden. He has a PhD in Mineral Processing and a PhD in Business Administration and Economics from Luleå University of Technology. He has served as a section head at the National Industry Board in Sweden. Lager has also served 15 years in the Process Industry mainly in the capacities of Production Engineer in Sweden and Africa and R&D Manager in the Swedish mining industry.

Academic affiliation

Prof. Dr. Thomas Lager,
Mälardalen University,
School of Innovation, Design and Engineering,
P.O. Box 325, Smedjegatan 37, SE-631 05 Eskilstuna, Sweden
Web: www.mdh.se
E-mail: thomas.lager@mdh.se

Company affiliation

Prof. Dr. Thomas Lager,
Blinab,
B&L Innovation AB,
Rinkesta Säteri, Mejeriet, SE-63537 Ärla, Sweden
Web: www.blinab.com
E-mail: thomas.lager@blinab.com

Acknowledgements

A large number of people, in different ways and capacities, have contributed during the long period of my overall development of the QFD methodology. To all of these people, I would want to say that if you are not mentioned here, nevertheless please know that you have my sincere appreciation.

I would first like to thank Ms **Karimah Samsudin**, Editor at World Scientific Publishing Company, for her belief that a book on this subject would be an important contribution in the area of product innovation. Many thanks for your continuous support during the writing and production process.

Two individuals have been truly important in my development and further use of the QFD methodology. Professor Dr **Takao Fujiwara** who initially supported and arranged for my stay in Japan as a guest researcher and visiting Professor at Toyohashi University of Technology; many thanks for your assistance. Late Professor Dr **Yoji Akao**, a founder of the QFD methodology, who always supported my development of the Multiple Progression QFD system for process-industrial use and gradually became a sincere friend and a mentor. His philosophical approach and attitude have always been a beacon that has guided me in the industrial use of the methodology.

Five people mentioned below have been involved as readers of the first draft of the book manuscript, with the general idea that they should represent and provide expert advice from different sectors of QFD process-industrial use. I sincerely apologize for asking you all to digest such a raw, unstructured, and sometimes incomplete draft manuscript.

Ms **Emilia Liiri** (Senior project manager at Holmen Group; head office), a reader from the Forest Industries. As a project manager and user of the QFD methodology, including both the *House of Quality* and upstream matrices, I have had the privilege of working with you and assist in the capacity of an QFD expert. Now, I would like to acknowledge that your profound knowledge of the QFD methodology in industrial applications, your suggestions, pertinent questions, and general sound advice have surely made this book a much better and hopefully more readable. Thank you very much for an excellent contribution, Emilia!

Mr **Magnus Tottie** (Product Development Manager at the LKAB Group), a reader from the Mineral Industries. Magnus was heavily involved in LKAB's first QFD project, which was an early pioneering use of the QFD methodology in a process-industrial context, for which I functioned as the external facilitator. We afterwards coauthored a presentation of selected results in the R&D Management journal, and I have had the privilege to continue as a company advisor in several QFD projects, including the recent development of an advanced Product Matrix. As a speaking partner in my previous development of the methodology, and now as a supporter for this book project, I gratefully acknowledge your support.

Dr **Robert Svanberg** (former R&D Director at ArlaFoods), a reader from the Dairy Industries. In his previous capacity as an R&D Director in the largest beer manufacturing company in Sweden, his advice in the development and refinement of the illustrative case and associated matrices has been invaluable. Any errors related to the presented matrices and the conclusions are only a consequence of the author's inability to understand his good suggestions. As an experienced user of the QFD methodology and the mpQFD system in particular in many process-industrial applications, it has been a privilege to assist you as an expert facilitator, and now I also thank you for your important comments on the industrial use of the methodology.

Dr **Rachid Gamal** (head of Business Process Management at Nestlé), a reader from the Food and Drinks Industries and a brilliant former doctoral student of mine at Grenoble Ecole de Management. Over the extended period of the writing of this book, you have served as a sounding board, particularly with regard to questions related to the use of facilitating methodologies like QFD as supportive instruments for an improved product innovation work process. Thanks for stimulating discussions in Vevey and in Marrakech and for your never-ending kind encouragement.

Dr **Peter Samuelsson** (former Senior Vice President R&D at Outokumpu), a reader from the Steel Industries. Bringing Innovation Management into the board room and possessing a profound knowledge of the "nuts and bolts" of the production system in the Steel Industries, Peter has since long been a "comrade in arms" in my constant struggle to develop and provide improved tools and best practice for Innovation Management in R&D organizations. Your idea that product innovation projects should never start from "scratch" but rather from well-delineated platforms of previously accumulated knowledge has been further developed in this book into the concept of *Integrated Knowledge Platforms*; many thanks for stimulating discussions and comments on the industrial use of the QFD methodology.

I have worked on this book for a long time and owe many people; my special thanks to:

Mr **Göran Bäckblom** (former Vice President of Public Affairs at the LKAB Group). As a former R&D manager in a number of companies in the Mineral Industries, but not yet acquainted with the QFD methodology, you were asked to help with an "out-of-the-box" perspective on the book content. Your comments related to a need to better communicate the "central message" to product innovation professionals, and to clarify expected benefits from the industrial use of the methodology to Innovation Managers, are appreciated and acknowledged.

Mr **Joseph Craig** (founder and owner of the company IDEACore). I have had the honour to know and collaborate with Joe for over two decades in his capacity as a developer of software for innovation, and for the QFD methodology in particular. As owner of the company IDEACore and provider of the market leading QFD Designer software, I have had a continuous dialogue with him with regards to software improvements and use, and I have also truly enjoyed his hospitality and friendship during our meetings in Detroit. I now look forward to our joint efforts to revitalize and stimulate the industrial use of QFD methodology, joined in our strong belief that this methodology is the preferred instrument for enhanced future product innovation and renovation of excellence.

Mr **Shigeki Nitta**. Shigeki, you have not only assisted me in translating some of the beautiful Japanese Kanji figures but also have always shown a keen interest in my research and book writing. I now look forward to introducing this novel knowledge- and platform-based QFD approach and philosophy with you in the Japanese heavy industry.

Mr **Per Lindstedt** (founder and owner of the company Value Model Arusid AB). We did one of the first QFD projects in Sweden together. This was a good start for my long QFD journey. All the best to you, Per.

Last but not least, the assistance from Ms **Stephanie Clauson** has been invaluable in improving my poor English. You have succeeded in transforming my sometimes rather ugly prose into a more digestible manuscript. Many thanks!

Contents

Preface ix
About the Author xvii
Acknowledgements xix

Part I The Theory and Methodological Use of Quality Function Deployment in Industrial Product Development 1

Chapter 1 The Theory of QFD and Related Matrices, Rooms and Symbols — Getting Acquainted with the Fundamental Philosophy 5
Chapter 2 The Development and Manufacturing of Beer — An Introduction and Background for the Process-Industrial Illustrative Case 23
Chapter 3 Facilitating Methodologies for Enhanced Product Innovation 37
Chapter 4 Summing Up and Some Issues to Reflect Upon 43

Part II Process-Industrial Product Innovation and the Use of the QFD Methodology as a Facilitating Development Tool 45

Chapter 5 Contextual and Inherent Conditions for Innovation in the Process Industries 49
Chapter 6 The Process-Industrial Product Innovation Work Process 55
Chapter 7 A Review of QFD Use in the Process Industries — Lessons from the Past 65
Chapter 8 An Analysis of Necessary QFD System Capabilities for Enhanced Process-Industrial Product Innovation — Perspectives for the Future 71
Chapter 9 Summing Up and Some Issues to Reflect Upon 77

Part III **The "Voice of the Customer" and "House of Quality" — The Common Denominators and Matrices in All QFD Systems for All Manufacturing Industries** **81**

Chapter 10 The Voice of the Customer — The Facilitating Instrument for an Improved "Market & Customer Understanding" 87

Chapter 11 The House of Quality — The Facilitating Instrument for an Improved Product & Competitor Understanding 115

Chapter 12 Matrix Analysis and Target Setting for New or Improved Products 139

Chapter 13 Summing Up and Some Issues to Reflect Upon 147

Part IV **The Advanced Multiple Progression Quality Function Deployment System (mpQFD) — Introducing the Upstream & Downstream Matrices Adapted to Process-Industrial Conditions for Innovation** **151**

Chapter 14 The Process Matrix — The Facilitating Instrument for an Improved "Process Understanding" 157

Chapter 15 The Raw Material Matrices — The Facilitating Instrument for an Improved "Raw Material Understanding" 177

Chapter 16 The Product Matrices — The Facilitating Instruments for an Improved "Understanding of Product Functionalities" 195

Chapter 17 Multiple Progression Within and Outside Company Demarcations — Addressing the Total Supply/Value Chain in Product Innovation 213

Chapter 18 Summing Up and Some Issues to Reflect Upon 219

Part V **Consolidating Matrix Information into Integrated Knowledge Platforms (IKPs) — New Perspectives on the Industrial Use of the QFD Methodology** **223**

Chapter 19 Expected Company Outcomes from the Use of the QFD Development Methodology — Scientific Evidence from the Past 225

Chapter 20 A Review of the QFD Methodology in the Perspective of Company Knowledge Creation and Utilization — Discover the "Hidden Gem" of QFD Use in Product Innovation! 231

Chapter 21 Consolidating the Knowledge and QFD Matrix Information into Coherent and Reusable Integrated Knowledge Platforms (IKPs) 245

Chapter 22 Summing Up and Some Issues to Reflect Upon 255

Part VI Guidelines for Company Implementation of the Quality Function Deployment Methodology and Integrated Knowledge Platforms 259

Chapter 23 General Success Factors for Company Use of the QFD Methodology 261

Chapter 24 Implementation of the mpQFD System and IKPs in the Company R&D Organization — Best Practice Recommendations from a Seasoned QFD Facilitator 265

Chapter 25 Using QFD as the Company's Select Methodology to Reinforce "Inter- and Intra-Organizational Integration" in Product Innovation 281

Chapter 26 Summing Up and Selected Questions About Quality Function Deployment (QFD) and Related Integrated Knowledge Platforms (IKPs) 291

Appendix A 301

Appendix B 303

Appendix C 305

Appendix D 309

References 313

Index 323

Part I

The Theory and Methodological Use of Quality Function Deployment in Industrial Product Development

An entirely new approach to product development was introduced at Mitsubishi's Kobe Shipyard at the beginning of the 1970s.

品質機能展開

The methodology was called HIN-SHITSU KI-NOU TEN-KAI (see the Kanji figures above), which has been translated into English as quality function deployment (QFD). The exceptional outcomes of using this product development methodology in industrial innovation amount to a real success story.

In this first part of the book, you will be briefly introduced to the philosophy, general theory, and methodological use of QFD. You will then be familiarized with QFD's matrices, individual rooms and commonly used symbols, as well as alternative QFD systems. An introduction and a background to the illustrative case of beer development and manufacturing are then presented, and finally the general industrial use of facilitating methodologies and tools will be discussed.

Product innovation perspectives

Future product innovation of excellence and related cost-efficient production of new or improved products require effective systems for management of *Customer Requirements* and for monitoring customer satisfaction. Such systems must be adapted to the company's business environment and information needs from both a short- and a long-term perspective. They must be tailor-made to the company's stakeholder structure and should be perceived by its customers as an important part of its quality assurance system for future products and services. Well-functioning systems of this kind are very important tools for organizational learning and they represent a key competitive factor; therefore, copying them or stealing them should be difficult for other companies.

Customer demands on a product, incorporating the clarification of individual *Customer Requirements*, their hierarchical structure, importance ratings and an assessment of how well they are met in comparison with competing products, are necessary foundations for sound product innovation. Even if technical product solutions to a certain extent are generated within the company R&D department as a result of technology push, market pull and delivering customer satisfaction will always be paramount. Hence, even a revolutionary new product created within company limits will also finally meet its customers.

Company innovation, however, not only requires efficient and effective systems for the management of *Customer Requirements* and surveillance of customer satisfaction but also must certainly incorporate proper arrangements regarding how to translate such *Customer Requirements* into related *Product Design Requirements*, operational product specifications and Target Figures. Moreover, such *Product Design Requirements* must also be further progressed into company production systems in order to manufacture high-quality products at a low cost. To accomplish such objectives in company product innovation, it is necessary to take advantage of supporting tools and methodologies, and it is advocated that the QFD methodology can provide such a coherent knowledge base for the product innovation work process.

Using the QFD methodology to listen to the "Voice of the Customer"

Before starting the process of collecting customer requirements and data, it is necessary to identify who the customers are for the product one intends to develop. The first step in the QFD process is to have product developers and marketers to establish customers' demands and needs through direct contact and preferably via an interactive dialogue. When stating their requirements, customers generally do not present them in a well-structured form or in any particular order of importance. One must therefore ask customers to rate their various requirements on a scale of importance. When the time comes for a customer to make a purchasing decision, he or she will — either systematically or, more likely, instinctively — compare the performance of product attributes with similar products. A company should therefore also ask customers to rate the performance of the company product in comparison with the performance of its main competitors. All this information is contained in the *Voice of the Customer*, thus providing a holistic perspective on customer preferences.

Using the House of Quality to translate Customer Requirements into Product Design Requirements

Customer Requirements are often expressed in terms like "the car must give an easy ride", "the machine should be painted in a nice color" or "the drink must have a full-bodied flavour". One must therefore translate these qualitative *Customer Requirements* into technically quantifiable and measurable requirements. This was facilitated using the fundamental "QFD methodological matrix approach" and positioning *Product Design Requirements* at a right angle to the *Customer Requirements*, thereby defining a relationship matrix in which every possible relationship can be identified, visualized and assessed.

In the construction of the *House of Quality*, a new dimension of the *Customer Requirements* has now been created, in which the importance of the *Product Design Requirements* can be translated into *Technical Importance Ratings*. Here, one can now make a new competitive analysis, but this time in a technical dimension. After the *House of Quality* has been built, it is necessary to make a tour of inspection of the house and, in a comprehensive analysis, squeeze out the wealth of information that is available in the matrix in a combinatorial effort.

In the completed *House of Quality*, one can analyse and evaluate how good the company product is when compared to its competitors and how well it has satisfied the various customer requirements transposed into technical dimensions. The Target Figures that are developed from a platform of a QFD matrix are not ad hoc targets but are developed in the process of "interactive target setting", since the selected Target Figures can be traced back to the total matrix information.

An illustrative QFD case — the development and manufacturing of beer

In order to facilitate understanding of the QFD methodology, in this book you will be introduced to an illustrative case from the food industries: The development and manufacturing of beer. Products in this industry may have different degrees of complexity, but in general they are like most of the products from the process industries: always produced in highly complex production processes, often using a wide spectrum of well-defined high-quality raw materials. In this part of the book, you will be introduced to the development and manufacturing of beer, while the application of the QFD methodology and the matrices related to beer manufacturing will be presented in depth in Parts III and IV.

An investment perspective on the use of development methodologies

In times of tight company resources and a continual need to improve company performance in all areas, including R&D, the use of effective and efficient tools and development methodologies for R&D is one avenue to follow. It is not an easy task to estimate and budget the cost for development projects, and it is certainly not easier to estimate the cost of using new methodologies in advance. As financial outcomes from the use of methodology can be difficult to calculate, it is probably better and representative of a more dynamic mental model to regard the introduction and use of methodologies as an investment. For such assessments, calculation of returns may not be the best tool; rather, one should refer to the sound judgement of people who were involved in the use of the methodology.

Chapter 1

The Theory of QFD and Related Matrices, Rooms and Symbols — Getting Acquainted with the Fundamental Philosophy

The size and complexity of today's industrial organizations have made the distance between production, development, marketing and the customer so great that company product developers sometimes have difficulty in clearly hearing the "Voice of the Customer" and disseminating it further within the organization.

In today's world, the extensive internal and external collaborations, sophisticated product specifications and shorter product development cycles of the company make developmental work very demanding on the R&D department. It is no longer sufficient to develop a product that pleases the customer; it must also be better than the competing products in the global marketplace and must be produced using cost-efficient production processes. The QFD methodology can address all these aspects of today's product innovation; moreover, it also serves as a tool for "knowledge creation and utilization", which is a major theme and message of this book.

1.1 QFD as a facilitating methodology in company product innovation

There are three exceptional benefits that usually stand out in company use of the QFD methodology in product innovation.

(1) *Using QFD on a company's proactive upfront product innovation activities — addressing the fuzzy front end of product innovation*
One, and possibly the most important, reason for the success of the QFD methodology is that, in a Japanese fashion, it moves product innovation activities toward the front of the product development work process. Making greater efforts in the definition of a

new or improved product in the pre-development part of the work process will reduce later rework and "fire-fighting" when the product is being produced and introduced on the market. This is also one reason why the QFD methodology is strongly recommended for use especially in the early phases of product innovation. However, in the following parts of the book you will discover that the methodology could be even more beneficial to company product innovation if it is deployed before project-based development activities commence in the creation of *Integrated Knowledge Platforms* as an overall support for the product innovation process.

(2) ***Using QFD as an instrument to bridge organizational interfaces — promoting a company's cross-functional product innovation***
If we compare old-fashioned craftsman-type manufacturing with modern production of goods and services in large industrial corporations, we find that a dramatic change has taken place in the relationship between the producer and the customer. Craftsmen and small businessmen have a very close and direct relation with their customers, in which they play the roles of producer, product developer and marketer — a truly interdisciplinary function. Direct dialogue quickly establishes exactly what the customer wants and how much the customer is willing to pay. By contrast, product developers and production engineers in large corporations today unfortunately often have limited direct contact with or access to customers or consumers, whose wishes filter into the company via the marketing department, then through development and hopefully also into production (Tottie and Lager, 1995).

This process involves a serious risk of the customer's original requirements or wishes — or what in QFD terminology is called the *Voice of the Customer* — being lost, misinterpreted, or even distorted. When the frequent and spontaneous contact between the producer and the user of a product is broken, a system error or dysfunctionality of the organization is created, which is one reason for the use of methodologies like QFD as "organizational integrating instruments". The visual data-presentation format of the matrices is easy for engineers, innovators and marketers to use. In Chapter 25 of Part VI, you will find guidelines on how the QFD methodology can be used as an unparalleled instrument for bridging inter- and intra-organizational interfaces.

(3) ***An innovation approach in the spirit of total quality management (TQM) — creating a reliable and fact-based interface between the company and its customers***
Because the methodology is called "Quality Function Deployment" (QFD), there is often a misconception that this methodology is primarily a methodology in the TQM "tool-box". It is not! The QFD methodology is primarily a product innovation methodology, but it serves five important TQM strategic objectives: Customer focus, fact-based decisions, emphasis on work processes, continuous improvements and participation of all employees. A proper use of the QFD methodology supports all these strategic TQM goals, and one can thus say that the use of the QFD methodology leads to quality in company product innovation.

> Like beauty, quality is in the eye of the beholder. The beholder, in the case of product-development activities, should be the customer. Hence, any definition of quality should be supplied by the customer — which is what QFD ensures. Quality, which is in the eye of the customer, is communicated via the Voice of the Customer.
>
> (Eureka and Ryan, 1988)

The following chapters will present numerous industrial testimonials and scientific evidence supporting the excellent results of using the QFD methodology in company product innovation. Please note that the following part of the book comprises only an introductory presentation and a brief review of the methodology, which will be presented in full in Parts III and IV.

1.2 A brief history of the QFD methodology

It is probably no coincidence that the QFD methodology was first conceived and adopted in Japan, where great importance has always been attached to long-term strategic development and a "knowledge-based approach" in innovation, in contrast to the "quarterly management" philosophy practised by many Western companies.

An explanation for the success of QFD can be sought in the continual intensification of competition between companies' products and the ever-increasing demand for lower product development costs and shorter product innovation lead times. In Japan, where quality work has been carried out since the beginning of the 1950s, one important person in this quality movement at the end of the 1960s was Professor Yoji Akao, who is generally considered the "spiritual father" of the QFD methodology (Akao, 2003, Akao *et al.*, 1987), see Fig. 1.1. The methodology's breakthrough in the Japanese

Figure 1.1 The late Professor Yoji Akao, the spiritual father of the QFD methodology (Akao, 1997).

manufacturing industry is often ascribed to Toyota Auto Body, whose use of QFD successfully contributed to solving the company's problems with low-quality, rusting cars. The methodology was initially called Quality Deployment until, at the beginning of the 1970s, the product matrix was introduced at Mitsubishi Heavy Industries' Kobe Shipyard. Professor Furukawa assisted in the use and further development of the methodology and also acted as an important link in transferring it to the United States at the beginning of the 1980s.

The original system, developed by Professor Akao (1990), is still the dominant system used in Japan and was introduced in the United States in a slightly modified form by GOAL/QPC (King, 1987). The system includes a large number of matrices and is often referred to as the *Matrix of Matrices.* A simplified version, developed by Dr Fukuhara, received wide publicity when it was presented in the Harvard Business Review under the designation "House of Quality"; the ladder of four consecutive matrices is often referred to as the *Four Phases of Matrices* (Hauser and Clausing, 1988). The QFD methodology was first applied in United States in the automotive industry, where it was successfully introduced by the American Supplier Institute (ASI, 1989) and GOAL/QPC.

The two aforementioned matrix structures represent the QFD systems that are in practical use worldwide today in other manufacturing industries. Outside Japan, the methodology gained international acceptance during the 1980s. Its introduction in the European industry followed during the early 1990s, and the first Swedish and European QFD project in the process industries was reported in the mid-1990s (Tottie and Lager, 1995). A QFD system adapted to process-industrial product innovation was later introduced by Professor Lager (2005b) as multiple progression QFD (mpQFD).

1.3 Listening to the "Voice of the Customer"

When a new house is built, one of the most important matter to attend to is to ensure that the foundation is strong and of good quality. The foundation for the *House of Quality* in the QFD methodology is the *Voice of the Customer*, which is the most important part of a good *House of Quality.*

In Fig. 1.2, the different parts of the *Voice of the Customer* are illustrated, and the individual rooms have been numbered to facilitate the connection with the explanatory text in this chapter.

The first step in the QFD process is to have product developers and marketers establish customers' demands and needs through direct contact, preferably in an interactive dialogue. If this is not properly done, or if the traditional view that the people in the company "already know what the customer wants" prevails, there is a huge risk that the subsequent construction of matrices will be banal, meaningless or even misleading. Furthermore, the techniques used to establish *Customer Requirements* must naturally be

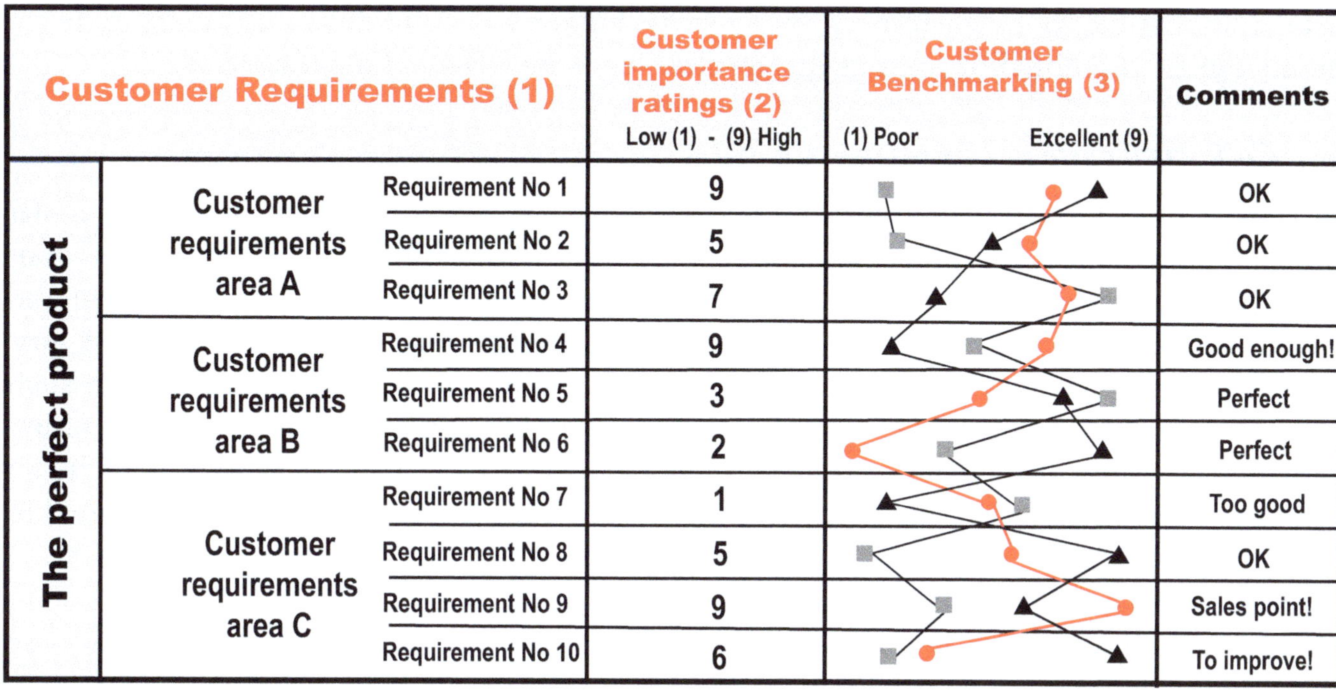

Figure 1.2 A simplified illustration of the *Voice of the Customer*. *Customer Importance Ratings* of individual *Customer Requirements* (1 = Low to 9 = High) and Benchmarking (1 = Poor to 9 = Excellent) are presented on ordinal scales. The red snake plot is the company's own product.

adapted to the structure of each individual company's clientele and the nature of its products.

(1) *Identifying customer requirements*

Before starting the process of collecting customer requirements and data, it is necessary to determine who the customers are for whom the product is proposed to be developed. Traditional segmentation of the market (by geographical location, size, etc.) can usefully be supplemented by further dimensions of customer structure, such as existing and potential customers, satisfied and dissatisfied customers, and so on. This will be further explained in Part III of the book. Moreover, in today's competitive climate, customers have come to expect something new excitements; they expect the supplier to come up with their own ideas and innovations (technology push). The use of the "Kano model" to identify such excitements will be presented later on in Part III.

When *Customer Requirements* have been acquired, the next step is to apply some kind of affinity criterion in order to sort customer requirements into homogeneous groups and gradually build up a hierarchical structure. While doing so, one should also try to reduce the number of requirements to manageable proportions without losing contact with what we call the *Voice of the Customer*. In labelling customer requirements, then, one should try insofar as possible to use customers' own words. Figure 1.2 shows how customers' requirements can be presented on various hierarchical levels in a *Voice of the Customer.*

(2) *Collecting customer importance ratings*

In stating their requirements, customers generally do not arrange them in any particular order of importance. A product development program can easily get off track at this point if the people in the company try to decide which customer requirements are important and which are not. This is a matter for the customer to decide. One should therefore ask customers to rate their various requirements on a scale of importance. The results from such activities are presented in Fig. 1.2, where each requirement has been assigned a rating showing how much importance the customer attaches to a particular requirement being satisfied. How this can be accomplished in many different ways will be further explained in Part III.

(3) *Performing a customer benchmarking*

When the time comes for the customer to make a purchasing decision, he or she will — either systematically or, more likely, instinctively — compare the features of a selected product with a competing product. The customer will in fact be making a competitive analysis, or a customer benchmarking, to use the currently popular term. It is thus important for a company to do the same thing before embarking on its development program. One should therefore also ask customers to rate the performance of the company product on a scale in comparison to those of the best competitors. Figure 1.2 shows how the results can be presented, using a snake plot. Note, however, that it has not yet been decided whether or to what extent attempts should be made to satisfy the various requirements.

A preliminary synthesis of the Voice of the Customer

In order to briefly introduce the preliminary use of the *Voice of the Customer*, the results from a simplified analysis are presented in Fig. 1.2. The beauty of this kind of analysis is the combination of the Importance Ratings and Benchmarking results for each *Customer Requirement*:

- Requirement No. 4's position is good and does not necessarily need to be improved further, since the company product position is already outstanding.
- Requirement No. 6's position is excellent, since being a good performer here is totally unimportant to the customer.
- Requirement No. 9 is usually called a "sales point", since this requirement is of utmost importance to the customer and the company's own product's position is excellent; must be exploited by Sales & Marketing!

- Requirement No. 10 is of a medium importance to the customer, and the poor position for the company product gives incentive for company product improvements.

1.4 The "House of Quality" — the common denominator and component for all QFD systems

A good house can be built in many shapes, in many sizes and with many different kinds of rooms, and a well-designed house is adapted to the tenant's needs. In a similar vein, the *House of Quality* should be "custom-made" to company's project needs, and the different kinds of rooms selected should be adapted to the product's specific and individual development requirements. It usually takes considerable amount of time and effort to erect a well-built house; in the same way, it is often a demanding exercise to build a well-functioning *House of Quality* with no "hidden faults".

In presenting the QFD methodology, it is important to decide upon a common language with clearly defined terms. In this book, the following definition from system theory has been used:

> A system can be defined as a set or assemblage of entities (elements or components) interrelated to each other and to the whole so as to achieve a common goal.
>
> (Neelamkavil, 1987)

The various matrices are thus also called "components", and the total set of matrices is called the "QFD system". This terminology emphasizes the interrelationship between the respective matrices but also reflects the classical picture from system theory of information input, the transformation concept for the total system and for individual components, and the final output from components and systems which is also achieved by the QFD methodology and individual matrices in phase progression. The individual parts of the matrices are (traditionally and also in this book) called "rooms", and you will find out that, as with "real" houses, the architecture of a *House of Quality* can be designed according to the user's needs and wishes.

The translation of *Customer Requirements* (usually called WHATs) into *Product Design Requirements* HOWs takes place in the *House of Quality* for all QFD systems. The terminology varies from one source to another, and the WHATs and HOWs in this book are, in Akao's publications, called "Demanded Quality" and "Quality Characteristics" (Akao, 1990), respectively. ASI calls them "Customer Wants" and "Design Requirements" (American Supplier Institute, 1989), and Hauser and Clausing (1988) call them "Customer Attributes" and "Engineering Characteristics".

In Fig. 1.3, the different parts of the *House of Quality* are illustrated, and the individual rooms have been numbered to facilitate the connection with the explanatory text in this chapter.

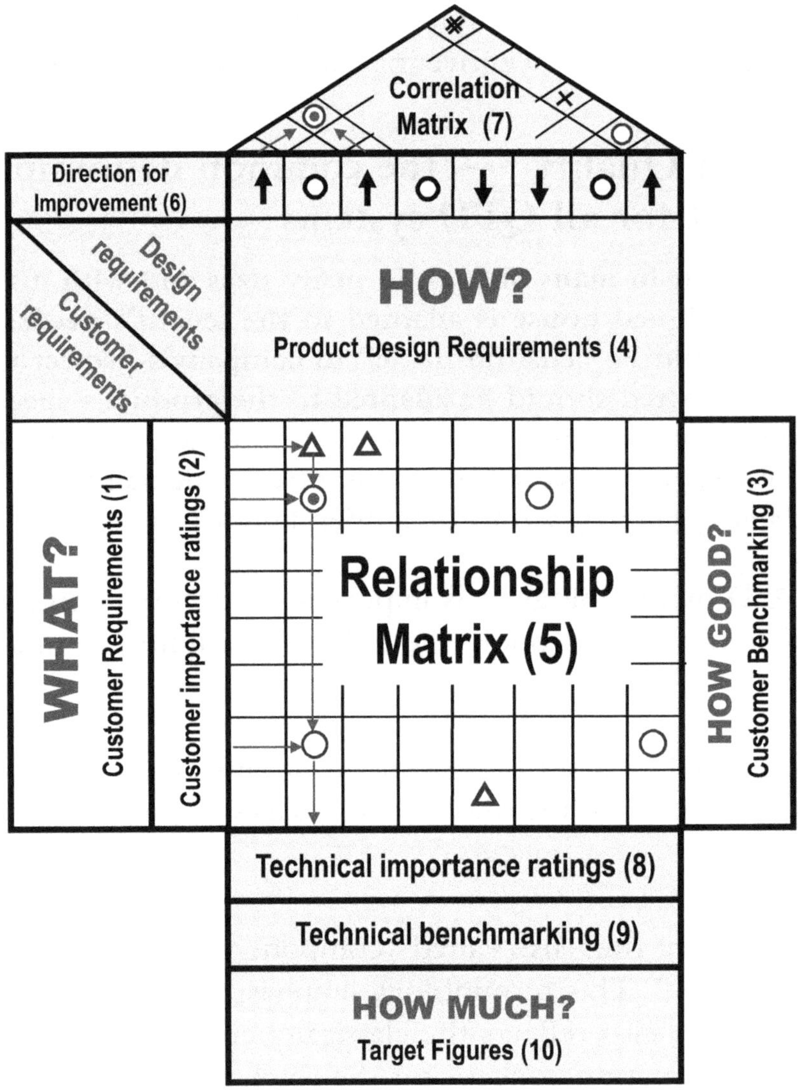

Figure 1.3 The *House of Quality*. Only some of the most commonly used rooms are included. The individual rooms in the *House of Quality* that are essential in the further discussion of the systems are those for Importance Ratings, Benchmarking (HOW GOOD?), Directions for Improvements, Relationship Matrix, Correlation Matrix, and Target Figures (HOW MUCH?).

(4) *From Customer Requirements to Product Design Requirements*

When designing a new house, one usually starts to dream how such a place should look and the preferred general architecture. After some time, one must define and transfer such ideas into a basic design. In a similar way, the *Voice of the Customer* must be translated into measurable *Product Design Requirements* that can guide future product innovation.

What one must try to do is to translate such qualitative requirements into more technically quantifiable terms. In the QFD process, one takes each *Customer Requirement* WHAT separately and asks how and by what technique it can be measured. One thus adds a matching list of *Product Design Requirements* HOWs to the list of *Customer Requirements*, as

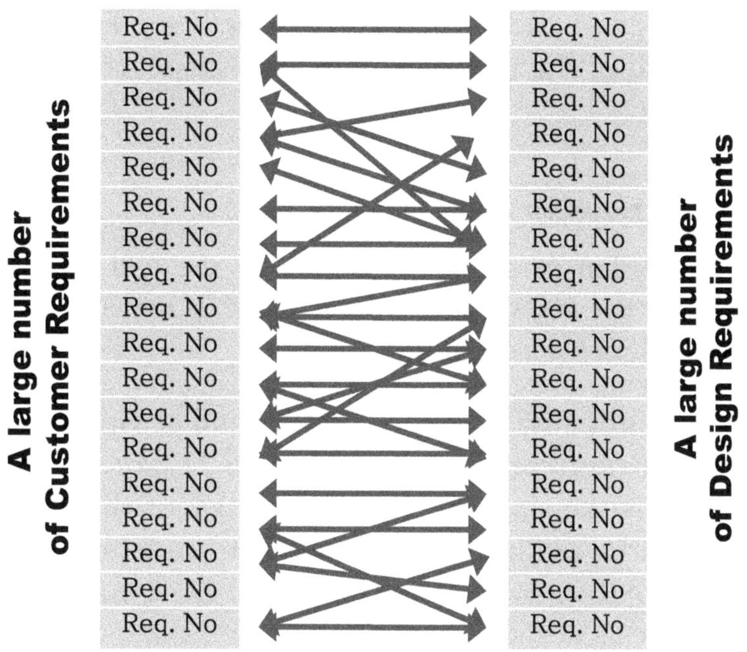

Figure 1.4 Multiple relations between *Customer Requirements* (WHATs) and *Product Design Requirements* (HOWs).

illustrated in Fig. 1.4. In company product development, and when single *Customer Requirements* are to be translated into *Product Design Requirements*, each requirement in the customer dimension often relates to multiple requirements in the product attribute dimension, as illustrated in Fig. 1.4.

In other manufacturing industries, the number of *Customer Requirements* WHATs, either on a single component or on the whole product in the process industries, usually fall in the range of 50–100 requirements. Assuming that the corresponding number of *Product Design Requirements* HOWs fall in a range of 20–40 requirements, the number of potential requirement relationships will fall in a range of 1000–4000. Assuming a matrix "fill" (relations) of about 50% results in 500–2000 relations to review. It is quite easy to comprehend that such a large number of relations would be impossible to keep track on without some sort of methodological tool.

However, it is important to recognize not only whether there is a relationship but also how strong such a relationship is, as well as to identify the evidence and the underlying mechanism behind each relation. If lacking a proper methodological tool, each product developer must try to keep this vast amount of information as a kind of mental map, which one can easily see is not a very reliable platform for development. For a team of developers, the situation is even more complicated, since individuals usually have different opinions regarding whether there is a relation and, if so, how strong it is.

This problem illustrated in Fig. 1.4 was solved with the fundamental QFD methodological matrix approach of positioning the *Product Design Requirements* HOWs at a right angle to the *Customer Requirements* WHATs and thereby defining a relationship matrix in which every possible relationship can be identified and assessed, see Fig. 1.3. The Matrix system was thus born, and, because of the matrix's house-like shape, this matrix was later denominated the *House of Quality.*

(5) *Constructing the Relationship Matrix*

When an owner takes possession of a new house, he or she will receive the keys to open it. Combining the WHATs and the HOWs into the relationship matrix is the key that unlocks the overall use and structure of the *House of Quality.*

Plotting all the HOWs at right angles to the WHATs and setting up a matrix in which the relation of each *Customer Requirement* to each *Product Design Requirement* is represented by a separate relation area is the starting point in the use of the QFD methodology. One can afterwards use this matrix, as illustrated in Fig. 1.3, to determine whether there is any relation between a *Customer Requirement* and a *Product Design Requirement* and, if so, how strong the relation is.

In the relationship matrix in the *House of Quality*, the following three symbols are usually used:

◎ **A strong relation:** Marked with a double ring (often in different colours) and in numerical calculations assigned with the value of 9. This symbol indicates: Here there is definitely a strong relation.

○ **A medium strong relation:** Marked with a single ring and in numerical calculations assigned with the value of 3. This symbol indicates: There is most likely a relation here.

△ **A weak relation:** Marked with a triangle and in numerical calculations assigned with the value 1. This symbol indicates: It is possible that there is a relation here.

No relation: An empty relation, which is in numerical calculations assigned with the value 0. The absence of one of the above symbols indicates: There is most likely no relation at all here.

An interesting possible relation: Marked with a question mark. An area in which there is no existing knowledge but that is of definite interest to explore further.

The symbols listed above, which usually work very well as visual symbols in practical matrix work by the QFD teams, are also used in Japanese scratch sheets in horse racing; generally, in normal business applications, they indicate:

- **Double ring:** Excellent.
- **Single ring:** Good.
- **Triangle:** Tolerable or neutral.
- **Cross:** Bad; not used in the relationship matrix but in the Correlation Matrix.

(6) *Selecting directions for improvement*

One generally discovers many details in a new house that need to be improved. Likewise, there may be several *Product Design Requirements* that need to be improved in a new product. In this room, on top of the relationship matrix, the direction for improvements for each *Product Design Requirement* will be identified, to later support the construction of the correlation matrix (the "roof").

When all *Product Design Requirements* have been identified and used for the construction of the relationship matrix, it is essential to establish the desired directions for improvement of each individual *Product Design Requirement* for a new or improved product. The three possible alternatives to select from are:

↑ **The more the better:** Illustrated with an arrow pointing upwards.

↓ **The less the better:** Illustrated with an arrow pointing downwards.

○ **Target value:** Illustrated with a ring.

This room must be completed before the "roof" of the *House of Quality* can be made. Naturally, in the development of this room one will find out interesting characteristics of the individual *Product Design Requirements*, and how to develop this room will be explained further in Part III. However, in selecting between the different alternatives, one must assess each *Product Design Requirement* "in isolation" from the influence on other *Product Design Requirements*; their correlations will thus afterwards be taken care of in the "roof" of the *House of Quality* in Fig. 1.3.

(7) *Designing the Correlation Matrix*

A well-built house has a roof that protects it from heavy rain and snow; to do so, the roof is often constructed with steep sloping planes. On top of the *House of Quality*, you will find a "roof" of a similar shape. The purpose of this roof is to provide a room where the relations between individual *Product Design Requirements* can be explored and assessed.

When one wants to improve a *Product Design Requirement* and thus changes the intensity value of a selected HOW on a product, such a change may often affect other HOWs. Increasing the hardness of a material, for example, often reduces its toughness. It is thus important to investigate and establish such correlations and, if such correlations exist, whether changes in individual *Design Requirements* reinforce or counteract each other. Thus, one must construct a roof over the relationship matrix, which is usually called the *Correlation Matrix*. In the Correlation Matrix, the following symbols are commonly used:

◎ **A double ring:** A strong positive relation.
○ **A single ring:** A weak positive relation.

- ✱ **A double cross:** A strong negative relation.
- × **A single cross:** A weak negative relation.

Thus, one establishes which *Design Requirements* are "friendly" and which are "hostile" to one another. The Correlation Matrix is a very important building block in the *House of Quality*, because a well-optimized product is often the result of a number of carefully considered compromises between individual *Product Design Requirements.* In other manufacturing industries producing assembled products, the vast numbers of multiple components usually do not interact substantially or could be selected in order not to interact. In the process industries, which usually produce more or less homogenous and non-assembled products, this matrix room is consequently of a much greater importance in product innovation. Part III of the book will explain in further detail how to develop and use the Correlation Matrix correctly.

(8) *Calculating Technical Importance Ratings*

Houses are often constructed with a basement, and one often finds a spacious basement to be useful for many different purposes. The *House of Quality* is generally constructed with many different kinds of floors; in its "basement", one of the compulsory rooms is the room where *Customer Importance Ratings* are recalculated into Technical Importance Ratings.

In the construction of a *House of Quality*, a new dimension of *Customer Requirements* has been created, in which the importance ratings of *Product Design requirements* can be translated into *Technical Importance Ratings.* In this step, one multiplies all relevant relations for each column of HOWs with its *Customer Importance Rating*, and in the total sum for each HOW one gets the translation into *Technical Importance Ratings.* This provides one with an instrument which tells approximately how important the various *Product Design Requirements* are on the basis of customers' original ratings. This is one of the QFD methodology's most important features and represents a significant benefit of the *House of Quality*, as it facilitates the further prioritization and target setting for individual product attributes.

(9) *Performing a Technical Benchmarking*

A whole new technical dimension of *Customer Requirements* has been created and one can now make a new competitive analysis, in this technical dimension. One can thus run a benchmarking of each measurable product attribute in the technical dimension HOW, which gives an opportunity to compare the performance of the company product with competitors' products. Alongside the customer competitive assessment room, we have a new room in the house, which is called *Technical Benchmarking.* In contrast to the importance of customer's involvement in doing a *Customer Benchmarking*, this *Technical Benchmarking* is an entirely internal affair in which the company product and competitive product properties are examined and assessed for each *Product Design Requirement*. In the process industries, this type of benchmarking is generally conducted in the company's own laboratory, and the procedures will be further detailed in Part III.

Conducting a Matrix Analysis

The first thing the owner of a new house often likes to do is to walk around in the house and make an inspection all the rooms and facilities that have been included in the design, in order to decide how to use them in the best way. After the *House of Quality* has been built, it is now time to make a similar inspection and, in a comprehensive analysis, squeeze out the wealth of information that is available in a combinatorial effort.

Surprisingly, the Matrix Analysis is an often neglected activity in presentations of the use of the QFD methodology in textbooks and articles. The reason for this unfortunate omission may be an unfamiliarity with the methodology as such or the exhaustion of the QFD project team after the construction of the *House of Quality* and the completion of all rooms. However, capturing the "low-hanging fruit" of the methodology takes place not only during the learning phase of matrix development but also in the thorough discussions and analysis of the wealth of information within the *House of Quality* through a holistic perspective on all matrix rooms. In Parts III and IV of this book, this important learning experience in using the matrices in product development will be thoroughly discussed and an extensive guide on how to carry out a Matrix Analysis for all the individual QFD matrices will be presented. As one industry representative exclaimed during this process: "Now I suddenly understand that the QFD methodology is really not about developing matrices, but about how you should use the matrices in product innovation".

(10) *Target setting for phase progression and concept development*

When the owner of a new house finally moves in, taking possession of the house and starting to live in it, this is the desired target for the whole endeavour. In a similar way, once all of the above steps have been completed, it is time for the project team to start using the *House of Quality*. The heart of the QFD methodology is thus not the matrices as such but rather the use of the matrices in product and process innovation. The project team can now, "under the same roof", operationalise its knowledge into a platform for the future development of new product concepts.

We have now finished building the *House of Quality* and can analyse and evaluate how good the company product is compared to competitors' products and how well the company product has satisfied the various customer requirements that we have transposed into technical dimensions. However, it is important to note that it has still not been decided whether or to what extent we are going to improve (or reduce) the intensity of product attributes, and one has certainly not yet developed a new product!

However, one has developed a "fact-based" platform for further development of a new or improved product, and can set Target Figures for a new product, which are contained in the room in the bottom of the matrix presented in Fig. 1.3. The target-setting process will be further discussed in Part III; in brief, such targets can be looked upon as a "sieve" that can be deployed to sift future product and related process concepts. The targets that are developed using a platform of QFD matrices are thus not ad hoc targets but, rather, are developed in the process of "interactive target setting", since the selected targets can be traced back from the total matrix information.

1.5 Progressing Product Design Requirements into the production system

The QFD methodology today is unfortunately often presented in a rather superficial manner, using simplified descriptions of the *House of Quality* and finishing with a presentation of how customer demands are being progressed into production using four consecutive matrices. However, looking into the industrial use of the methodology, one finds that few companies have ever tried the phase progression into deeper levels, and thus most companies have not used and exploited the methodology to its full extent.

On the basis of the *House of Quality*, one can set *Design Requirements*, *Technical Importance Ratings*, and *Target Figures*, not only for a new or improved product but also as a guide for its necessary related production system capabilities. One can thus use the HOWs and their target figures from the *House of Quality* as WHATs in the upstream matrices to project the *Voice of the Customer* deeper into the company's internal production environment. There are two QFD system architectures that are commonly used today in assembly-based industry and one system architecture for the use of QFD in the process industries.

1.5.1 *The Matrix of Matrices — a system designed by Professor Akao in Japan and further promoted in the United States by GOAL/QPC*

The Matrix of Matrices system should be considered an advanced system for experienced QFD users. The original system (see Fig. 1.5) presented by Akao (1990) contains in principle 16 matrices (components) and includes the following four areas (columns): Quality Deployment, Technology Deployment, Cost Deployment, and Reliability Deployment.

Akao recommends starting with the first column. The system is presented in a slightly different fashion by GOAL/QPC, as concept development is included and there are 30

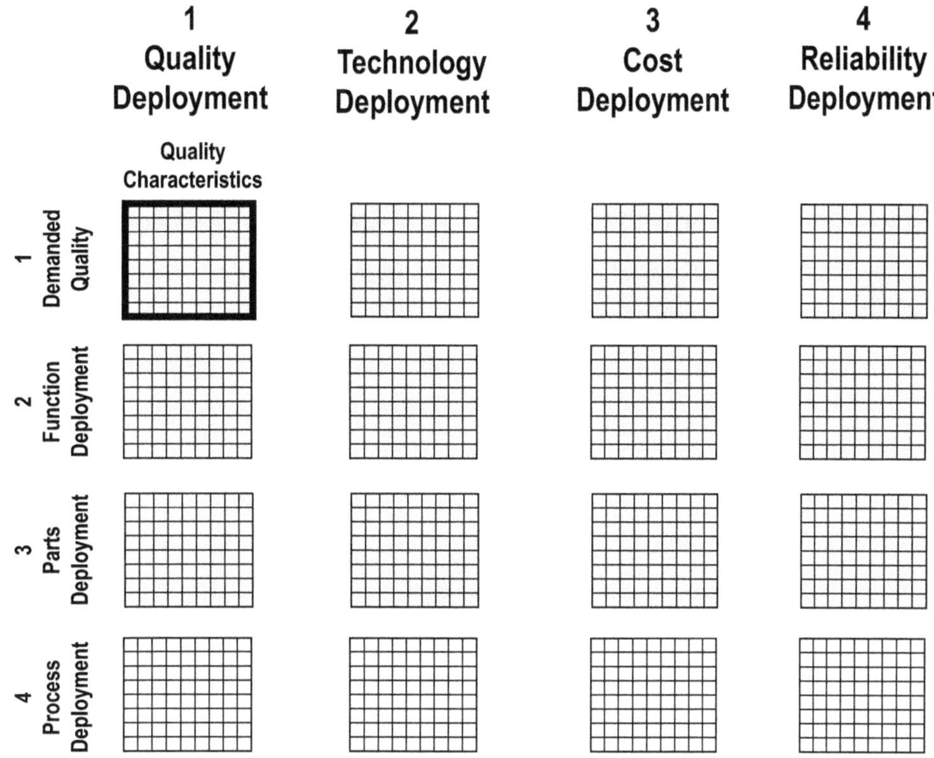

Figure 1.5 The "Matrix of Matrices", a simplified presentation of the system and its components, after Akao (1990).

matrices in the system (King, 1987). To simplify matters, the individual areas of the matrices have not been named in Fig. 1.5. The number of matrices in the system may appear intimidating, but we can take comfort in Akao's advice to use as many matrices as necessary for each application.

1.5.2 *The four phases of matrices — a system designed by Dr Fukuhara in Japan and further promoted in the United States by the American Supplier Institute*

Starting with the development of the *House of Quality*, the *Customer Requirements* are translated into *Design Requirements* in this system. These *Design Requirements* can serve as WHATs in a consecutive matrix to express the demands on *Part Characteristics*, and they can then be further progressed to the *Process Planning* and *Production Planning* matrices (see Fig. 1.6). This transformation is often called phase progression. Thus, using the four matrices, the customer WANTs are progressed from product development and design into manufacturing. It is important to recognize that not only the individual product properties but also their associated calculated importance ratings and, of utmost importance, the selected Target Figures are progressed to the following matrix. Opinions differ as to how many of the product properties should be included in this progression.

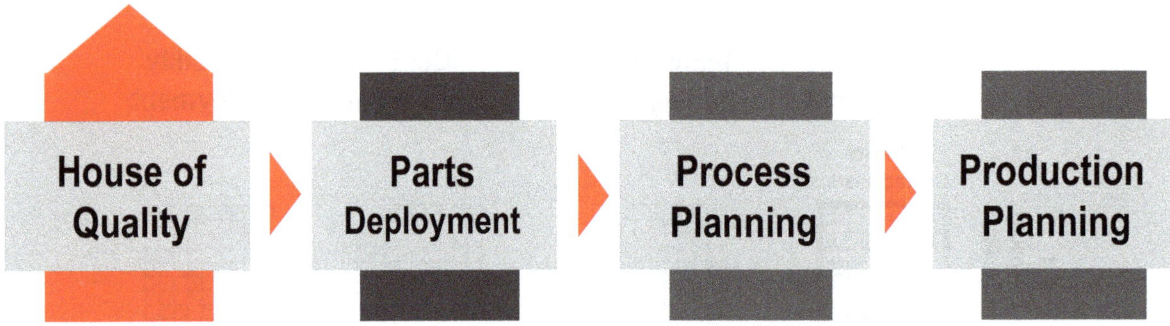

Figure 1.6 The four phases of matrices; a simplified presentation of the system and its components, after (ASI, 1992).

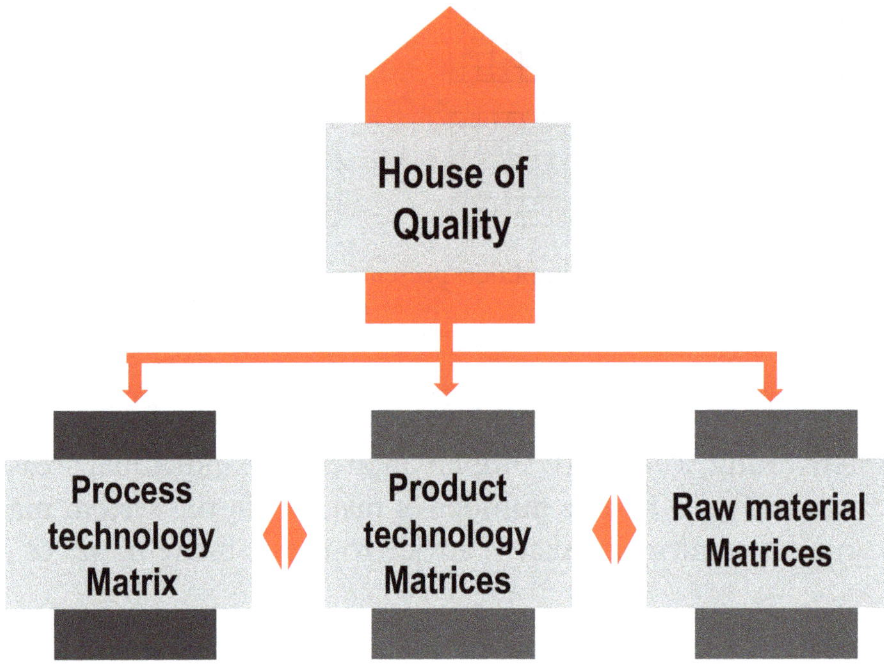

Figure 1.7 The mpQFD system for process-industrial use (Lager 2005b).

ASI and Professor Akao advocate that the properties should be subjected to a rigorous priority-ranking process to focus the matrix development on new, important, and difficult product properties.

1.5.3 *Multiple progression QFD (mpQFD) for process-industrial applications, designed by Professor Lager*

This fairly new QFD system was developed by Thomas Lager (under the mentorship of Professor Akao) in order to provide a QFD system especially adapted for process-industrial applications, see Fig. 1.7. The four-stage progression system for other manufacturing industries is not applicable to the process industries, because the latter's products are not assembled components. It is argued in this book that in the process industries the products and related production processes are firmly interlocked, and thus product and process innovation must always go hand in hand. Consequently, *Customer Requirements* regarding the product must first be translated into *Product Design Requirement*s and then must be further progressed into the production system as requirements of process capabilities and raw material properties.

The multiple progression QFD system (mpQFD), is thus specifically designed to fit process-industrial needs and use (Lager 2005b). The content in the *Process Matrix* and *Raw-material Matrix* must be separated, since only the raw material characteristics can be further progressed as demands on supplied products. The product technology matrix relates selected *Product Design Requirements* to *Explanatory Product Characteristics* and underlying mechanisms for the creation of functionalities. The system was presented and published in 2005 and has proved to be a useful QFD system for many different sectors of the process industries (see e.g., Lager and Kjell, 2007; Tottie *et al.*, 2016).

Chapter 2

The Development and Manufacturing of Beer — An Introduction and Background for the Process-Industrial Illustrative Case

Methodologies and the QFD methodology, in particular, tend to be rather difficult to comprehend using only theoretical descriptions and models. The understanding of complicated methodological issues thus often benefits from simplified or artificial examples (QFD on a mousetrap) or the use of simplified illustrative industrial cases.

Throughout this book, the "development and manufacturing of beer" has been selected as the illustrative case for how to use the QFD methodology in product innovation. The criteria for the case selection were:

- A product and a related production process that a reader easily can identify as a process-industrial product and production process.
- A product that is the final outcome of a long supply/value chain that starts with a number of primary (first-tier) raw materials.
- A case that well illustrates how the mpQFD system can be used in progressing *Customer Requirements* into the production system.
- A product and process about which the reader not only may be familiar with but also may be interested to learn more about.

The selected case originated some years ago when the author introduced the QFD methodology in the largest company manufacturing beer and soft drinks in Sweden. For the purpose of this work, the project results have been partly disguised, somewhat simplified and sometimes purposely altered in order to exemplify and highlight different aspects of the use of the QFD methodology in product development.

Throughout this book, this case, shown in Fig. 2.1, will serve to illustrate the use of the mpQFD system for the development of new or improved products. In this chapter,

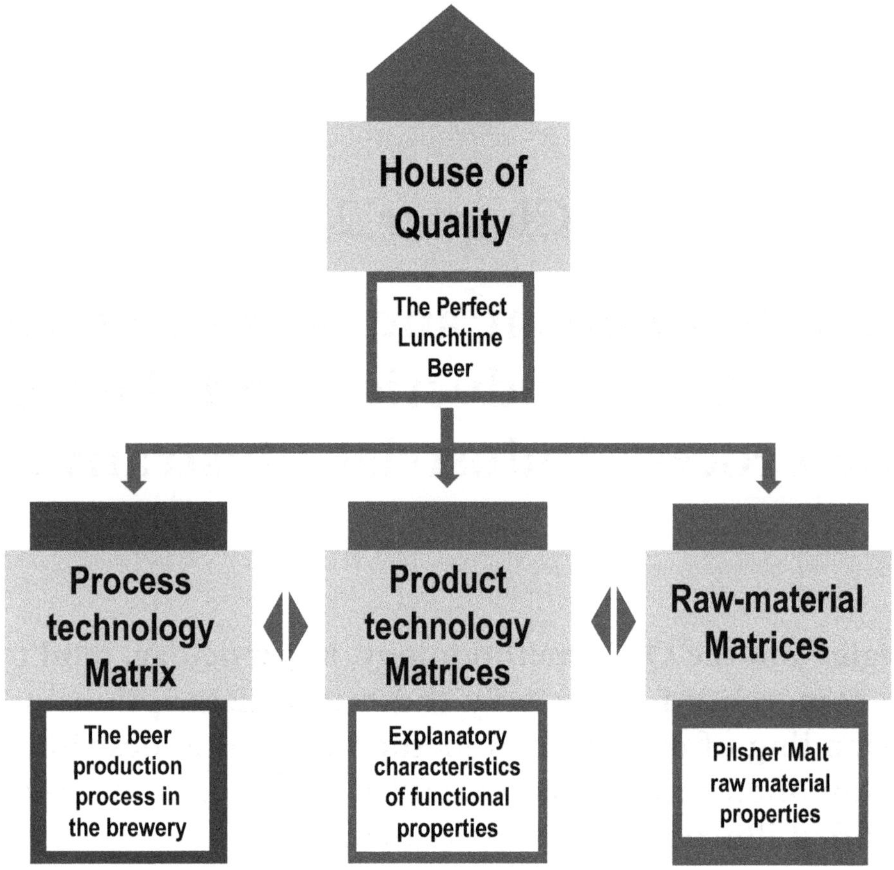

Figure 2.1 The use of the mpQFD for the development and manufacturing of beer.

you will initially learn some fundamentals of the development and manufacturing of beer. However, the application of the QFD methodology and the use of the mpQFD system for the development and manufacturing of beer will be discussed in Parts III and IV of the book. With reference to Fig. 2.1, the *Voice of the Customer* and the *House of Quality* for the perfect lunchtime beer will thus be presented and discussed in Part III, while the Process Matrix, the Raw-material Matrix, and the Product Matrices will be presented and discussed in Part IV.

The production system for beer manufacturing is a very complex biochemical process. Since product innovation in the process industries in reality is the development of a new process, you will learn in Part IV of the book how the QFD methodology in particular can benefit process-industrial product innovation.

Since large matrices generally are difficult to read and comprehend when reduced to book size, and since their complexities can interfere with learning, you will find two different versions of this case in the book. The "full versions" of the *Raw-material Matrices* are presented in Appendices A and B of the book. The "full version" of the *House of Quality* is presented in Appendix C and the "full version" of the *Process Matrix* is presented in Appendix D. However, it should be noted that the "full versions" of these matrices sometimes originated from even larger matrices developed in the industrial project. In the introduction of the mpQFD system in Parts III and IV, the "full versions" are sometimes referred to when illustrating certain complex and difficult methodological issues. However, in the general introduction of the QFD methodology in the following chapters, you will only be presented with small "mini versions" derived from the "full matrices".

In those "mini versions" only a few *Customer Requirements* have been selected from the *Voice of the Customer*, and only a few related *Product Design Requirements* have been selected for the development of a "mini version" of a *House of Quality*. In a similar vein, all other matrices have been reduced in size following a similar principle.

The case has been selected in order to illustrate the methodological use of QFD and the mpQFD system in particular; however, please note that it is not intended to be used as a manual for the development of beer and beer manufacturing. A brew-master or a professional "home-brewer" may spot minor or even major deficiencies in the relationship matrices and metrics. However, this will only serve the overall purpose and message of this book that the QFD methodology is an excellent tool for knowledge creation and utilization.

To reap the fruit of this illustrative case, it is necessary for the reader to possess a basic knowledge of beer quality characteristics and manufacturing. In the translation of *Customer Requirements* into *Design Requirements* and for the benchmarking processes, it is thus important to have a fundamental knowledge about beer styles and acknowledged beer quality parameters. In the phase progression of the *Product Design Requirements* into the related *Process and Raw-material Matrices*, it is also necessary to have a basic knowledge of the raw materials required for beer production and the rather complex biochemical brewing process. While the following sections may seem a bit technical, irrelevant or even boring to you, the information they contain will later contribute to providing an improved understanding of the methodology.

2.1 From farmers to finished products — revisiting part of the supply/value chain in beer manufacturing

From raw materials to supplied products to a consumer or industrial customer, the chain is often fairly long in many sectors of the process industries. The properties and quality of raw materials are often decisive for the selection of production technologies; moreover, they also often influence the properties of the final products. In beer production, one of the most important raw materials is the malt that is produced from barley. The supply/value chain from barley farmers to consumers is long, and the farmers' cultivation of barley will affect the quality of malt produced at the malteries that will later be used in the breweries. Moreover, the quality of barley will affect not only the malt product and brewing process at the breweries but also the quality of the finished product, as illustrated in Fig. 2.2.

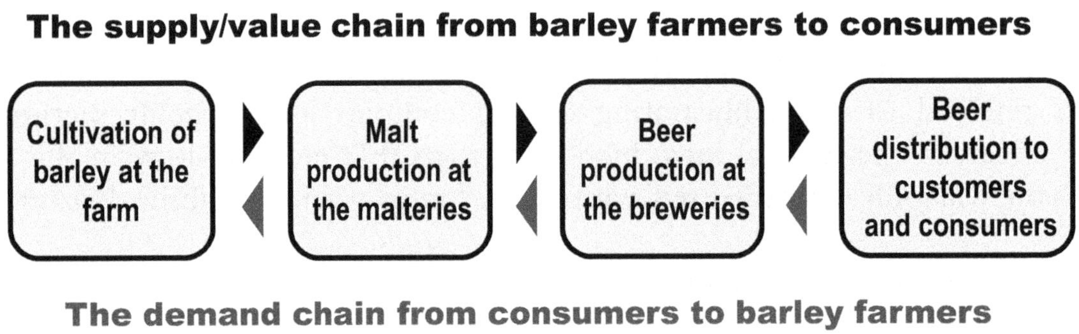

Figure 2.2 *The Consumer Requirements* for the beer product must be progressed backwards into the chain of suppliers, finally ending up as demands on the barley product delivered by the farmers to the malteries.

This is a common characteristic not only in the food and beverage industries but also in many sectors of the cluster of process industries. For companies in different positions within such long and often complex supply/value chains, a good understanding of these interrelations and their interfaces is a success factor in product innovation. Product innovation methodologies in a process-industrial context must thus be adapted to manage this complexity. In Part IV, you will be familiarized how the mpQFD system can help you to do this.

2.1.1 *Cultivation of barley by the farmers*

The raw material for the malting process is barley, and good barley quality is decisive for the production of high-quality malts. Referring to Fig. 2.2, the cultivation of barley for malting must produce grain that adheres to specific quality standards, under strict farming practices. Barley is a short-season, early-maturing crop that is tolerant to drought, alkali and saline soils. Spring barley matures within 60–90 days, making it a desirable crop with regard to water and nutrient efficiency use. The *Raw-material Design Requirements* HOWs for barley for malt production include, for example:

- **Physical properties:** Kernel size, Kernel roundness, husk thickness, moisture, etc.
- **Chemical and microbiological properties:** Protein content, enzyme activity, etc.
- **Functional properties:** Germination energy, Kernel dormancy, etc.

2.1.2 *Malt production at the malteries*

Malting is in reality the first step of the brewing process, but while some breweries have integrated malting facilities, the production of malt is usually carried out by external

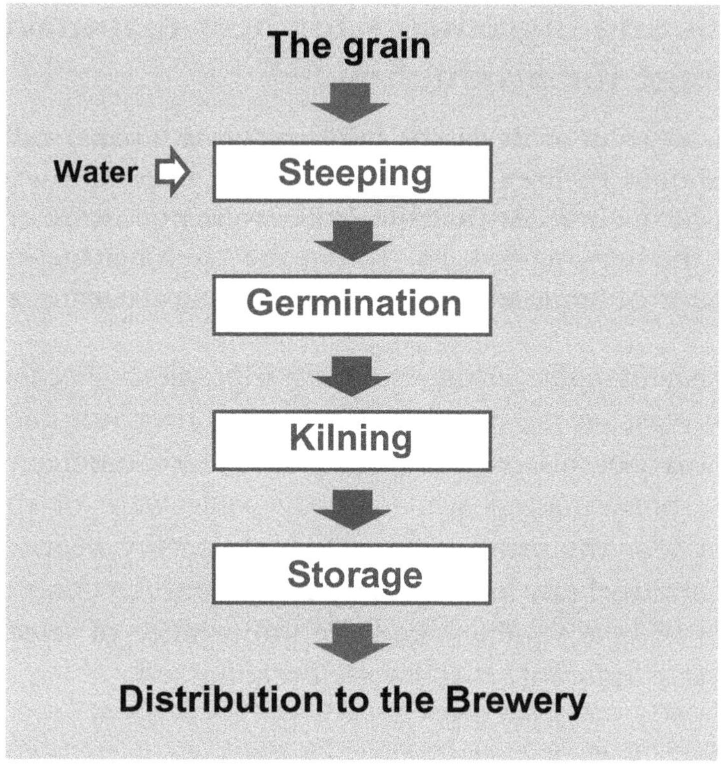

Figure 2.3 A simplified presentation of the four unit processes that make up the malting process at the malteries.

suppliers. In the malting process, barley and sometimes other cereal grains are transformed into malt through chemical and biochemical processes, as further illustrated in Fig. 2.3. Malting is thus a controlled germination of barley. The first step of malting, called steeping, is carried out in tanks during periods of different lengths and moisture. When the moisture content is increased, the germination starts and produces enzymes that are essential for modifying the grains in the production processes in the brewery. During the kilning process, the malt becomes friable and can be easily crushed in the brew-house.

From a customer perspective, the consumer of the beer will set the demands on the beer product — the "Voice of the Customer". The *Customer Requirements* must be translated into a technical dimension (*Design Requirements*) that not only will guide beer product developers but also will help the brew-master to select the proper process conditions and production technology in the brew-house, as well as the suitable raw materials. These requirements on raw materials will then guide the production process at the malteries and the cultivation of barley by the farmers.

2.2 Raw materials and ingredients for beer manufacturing — setting the stage for production

On the 23 April 1516, a group of Bayrische brewers wrote a paper called "Reinheitsgebot", prescribing that beer should be brewed using only malt, hops and water in order to secure the production quality of the beer. At that time, they were not aware of the existence of yeast and its importance for the brewing process. Today, the "Reinheitsgebot" is not valid even in Germany, but the concept of "non-additive brewing" is experiencing a renaissance.

In the early days of beer manufacturing, the only ingredients used were water, malt and hops. The influence of yeast on the brewing process was then not known, since this microbiological substance was not discovered until the end of nineteenth century by Louis Pasteur. However, the brew masters were already well aware of the importance of the "brewing environment" for the production of a high-quality beer. Today's brew masters have an abundance of optional raw materials and ingredients to choose from in the production of a large number of beer varieties, and the importance of selecting the proper yeast strain for a desired process outcome is today well recognized.

2.2.1 *Water*

The quality of the water used was initially the sole determinant in what style of beer it was possible to produce. The reason for this is that the chemical characteristics of water have a significant influence not only on the brewing process as such but also on the finished product. Today, water quality can easily be manipulated and properly adjusted to fit almost any kind of desired brewing process and style of beer. In addition to the importance of using water that tastes good and is microbiologically sound, the hardness (content of calcium and magnesium ions) of the water is the most important characteristic for a good brewing liquor. Calcium ions not only inhibit colour formation during the boil and facilitate protein coagulation but also influence the beer fermentation process.

2.2.2 *Malt*

While hops are important for the seasoning of beers, the malt is the fermentable material that often is considered to give a beer its "body and soul". At the malteries, the steeped barley is transformed into malt during the "kilning process"; the higher the kilning temperature, the darker the malt. Pilsner malt is generally used in brewing light coloured beer, while Wiener malt is used for darker beers, and Munchener malt is used for even darker products. At higher kilning temperatures, Caramel malt, Chocolate malt, and the even darker Black malt are produced. Normally, a mix of different types of malt is used in the brewing process. In Part IV, malt will be used to illustrate the development of a *Raw-material Matrix*. You will then discover that around over 20 different measurable properties of the malt must be considered by the brew-master in the selection of a proper malt.

2.2.3 *Adjuncts*

Adjuncts are un-malted grains that supplement the malted barley, primarily in order to cut costs but sometimes also to improve beer properties like head and flavour. They can be added as solids in advance of the mashing process if they are starchy, direct to the wort kettle as syrups to increase brew-house capacity, or as post-fermentation ingredients. Corn is often used in the making of pale lager beers, and the use of rice can lighten the "beer body" and add sweetness to the product. Flavourings of many different kinds can also be used as additives depending on the beer style, and spices like cinnamon, coriander, pepper and orange can optionally be added in the fermentation process. Different kinds of vegetables and fruits have long since been used as well, and today these are sometimes added in the form of flavour extracts or fruit juice directly to the finished product.

2.2.4 *Hops*

Hops are used in the brewing process to balance the sweetness from the malt and to give the beer a more bitter flavour. Only the cones of the female plant are used, and the flavour and aroma come from the resins and oils in the lupulin glands. The hop character is also to a large extent dependent on where the hops have been cultivated. Hops can be added to the wort-boiling process as fresh hops, pellets or liquids, and the brewing processes for the addition of hops varies. Bitter hops are usually added at the outset of the wort-boiling process, while aromatic hops are normally added later, since the aroma otherwise will be lost during the wort-boiling process. The alpha-resins (humulone, cohumulone and adhumulone) are the most important flavour characteristics of hops and are mainly responsible for the bitterness of the beer. The essential oils are responsible for the aroma of the hops; the most important of these are humulene, myrcene, farnesene and caryophyllene.

2.2.5 *Yeast*

In the early days of beer production, "wild yeasts" were the organisms that caused the fermentation of beer, and the fermentation process could thus be characterized as "spontaneous". At that time, and also today in the production of ale, the yeast rises to the top of the fermentation vessel in a process called "top fermentation". In 1881, Emil Hansen of the Carlsberg Brewery in Denmark was the first to develop a yeast strain that secured a "bottom fermentation" — that is, a fermentation process in which the yeast sinks to the bottom of the fermentation vessel. When the aerated (or O_2 injected) chilled wort begins to ferment, the yeast transforms the sugar into alcohol and carbon dioxide in a complex biochemical production process. Each yeast strain has its own particular flavour signature and produces beer with a unique ester flavour profile. Today, every brewery uses its own strain of yeast, and when out-licensing a beer brand it is usually only the yeast strain that is supplied to the foreign brewery.

2.3 A guided tour of the Brew-house — visiting a complex biochemical process

> The varieties of barley used; the place and the season in which they were grown; the procedure of malting, especially at the kilning stage; the number of different malts employed; the method of their infusion or decoction; the use of less traditional grains or other brewing sugars as "adjuncts" to the main mash; the way in which the solution of fermentable sugars is run off; the character of water; the length of the boil in the brew kettle; the way in which the kettle is heated; the variety of hops added; the stages at which they are used; the type of yeast; the temperature and duration of fermentation; the further maturation and conditioning; the type of filtration and pasteurization; all of these elements shape the beer. Among them, and especially in the temperature and duration of various processes, the permutations are infinite.
>
> **(Jackson, 1991)**

The above quotation illustrates the fact that the development and manufacturing of beer is a very complex production process including several process steps, each with a number of unit operations. A simplified overview of the beer manufacturing process is presented in Fig. 2.4. In the following sections, each step of the process will be explained in further detail.

Today, when the malt is bought from an external supplier, the brewing process starts when the malt is crushed together with water and fed into the "mash-tun" (usually two vessels) in the brew-house. After the mashing process step, the so-called "grist" is filtered and removed before the mash moves on to the wort-boiling process. At this stage, hops are added and the brewing process continues, sometimes in many separate process steps.

Next, the hops are removed, and the clarified wort is passed through a whirlpool to separate unwanted proteins. The wort is then passed through a cooler, then yeast is added, and the fermentation process step follows in a separate vessel. After fermentation, the wort is matured and conditioned in tanks. The wort is usually filtered and pasteurized before being passed on to pressurized holding tanks (Fix, 1999).

2.3.1 *Step 1 — the mashing process*

Before the malt is fed into the mash-tun, it must be crushed to expose the starchy endosperm (kernel) to the brewing liquor. During the mashing process, the malt is mixed with hot water in order to activate the enzymes in the malt and transform the starch into sugar (during this process, the colour of the wort gradually also gets darker). The "mashing scheme" is the temperature/time profile of the mashing process, and there are numerous different mashing schemes for the brew-master to choose from (the holding times at different temperatures are called "rests").

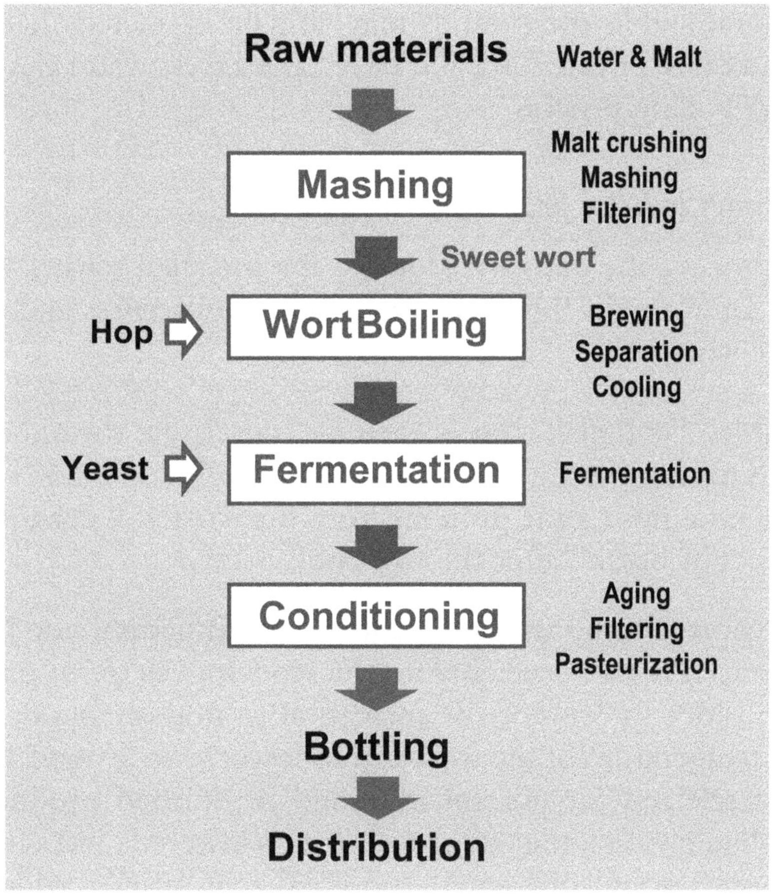

Figure 2.4 A simplified presentation of the beer production process at the brewery.

The temperature during the first phase of the mashing process is usually around 35–40°C (this phase is often called acid rest). In this initial period of grain liquefaction, the beta glucans are broken down. When the temperature reaches 60–65°C (the most important phase of the mashing process), the alpha-amylase enzyme transforms the starch into maltose sugar. However, in the production of low-alcohol beers, the temperature is immediately raised to 70° in order to minimize the production of maltose sugar. In the production of lager beers, a high-temperature, short-time mashing scheme usually has rests at 60° and 70°, with a total mashing time of only around one hour.

The "mashing thickness" refers to the concentration of grains in the mash and is measured in litres of water per kilograms of grain (an average mashing thickness is around 2.5). The pH of the mash is another important process parameter, and the enzyme activity is favourable around a pH of 5.3. A lower pH is acceptable, but a higher pH not only sharply decreases the enzyme activity but also can negatively influence the malt flavour of the finished product. After mashing, the residual grains are separated and, in order to improve extract recovery, the residuals are washed (sparged) with hot water (74°). However, during this process, some undesirable compounds may also be extracted that may influence the flavour negatively and cause both staling and haze. The excess water can then be evaporated in the following wort-boiling process.

2.3.2 *Step 2 — the wort boiling process*

In the wort boiling process, the hops are added to the so-called "sweet-wort", and the wort is vigorously boiled for around 70 min in order to give bitterness and aroma to the wort. The objectives of the wort boiling include:

- Increase the sugar level and make the wort sterile (deactivate the remaining enzymes that would be harmful to the wort).
- Dissolve bitter compounds (some proteins from the wort will coagulate with wort acids and precipitate as "hot break" after the wort boil).

The "hot break" is removed using whirlpool equipment. If natural hop flowers are used and not hop extracts, a screen is also used. Before the wort moves on to fermentation, it must be chilled quickly in order to prevent the formation of undesirable bacteria. In a gradual cooling process, the temperature of the wort is thus lowered to around 10–17°C dependent on the type of yeast selected for use and the kind of finished product to be produced. During the wort-boiling process, the oxygen content is lowered, and since the fermentation process requires a large amount of oxygen, more oxygen must be added after cooling the wort to a suitable level.

2.3.3 *Step 3 — the fermentation process*

After cooling and O_2 addition, the wort continues to the fermentation process. Fermentation is generally acknowledged as one of the most important, as well as one of the most difficult and complex, parts of the brewing process. During fermentation, the yeast "eats" the sugar and produces ethyl alcohol (C_2H_5OH) and carbon dioxide (CO_2). At an alcohol content around 8%, the fermentation process is retarded and at an alcohol content of more than 14% the yeast is killed (thus making this the maximum alcohol content of naturally produced beer). During the fermentation process, maltose sugar (the extract) is gradually consumed, but the dextrin is left, and the "degree of fermentation" indicates how well the fermentation process has performed. The fermentation process is retarded by lowering the temperature, but sometimes part of the extract (less than 1%) is preserved in order to secure a "post fermentation" later on.

In "top-fermented" beer (fermentation at a temperature around 10°C), the yeast cells do not split up and are "floated" by the CO_2 gas to the top of the vessel. During "bottom fermentation" (at a temperature around 20°C), the yeast cells split up and are "flocculated", thus sinking to the bottom of the vessel. During fermentation, the fermentable sugar is transformed into alcohol, carbon-dioxide and flavouring substances; however, some undesirable substances must afterward be diverted, along with excess carbon dioxide.

2.3.4 *Step 4 — the conditioning process*

After fermentation, it is of vital importance that no oxygen is allowed to be in contact with the beer, since oxygen is the main cause of "stale" beer. The conditioning process includes ageing, filtering and pasteurization before the beer moves on to bottling and distribution. This is an important part of the brewing process, and the quality, flavour and aroma of the beer will now be greatly improved. When the beer is transferred from the fermentation vessel at a temperature around 3–6°C, it contains a maximum of around 0.2% oxygen, which must be increased to around 0.5% in the finished product. The increased level of oxygen can be achieved by post-fermentation in a closed vessel when the pressure is controlled by a valve (bunging pressure), but, if necessary, additional CO_2 can also be administered.

Since the wort not only contained maltose sugar but also other fermentable sugars that can produce undesirable compounds, these must be removed. The maturing process during storage is dependent on temperature and time (generally between 2 and 4 weeks) in order to produce a good-quality beer. During the final part of the maturing process, the beer is chilled close to the freezing point in order to precipitate undesirable protein compounds and avoid chill haze. Most of the undesirable particles sink to the bottom of the vessel during storage, but there are still microscopic particles in the beer that must be removed in a filtering process. There are many alternative processes, but today filtration in membrane filters is getting more and more common. After filtration, the beer is usually pasteurized in order to remove undesirable microorganisms that otherwise can harm the quality of the beer, which will increase the shelf life of the product. Today, pasteurization is generally carried out at a temperature of around 70° in a plate pasteurization vessel before the beer goes to bottling.

2.3.5 *Step 5 — the packaging (bottling) process*

Food products are contained, protected and carried in packages. The original function of packaging was to protect food from external contaminants and ensure its preservation. Today, however, food packaging not only must use novel barrier technologies to better manage products' shelf life and avoid any migration of hazardous substances into the food but also provide convenience to the consumer, such as ease of handling and use. As with all products from the food and beverage industries, the packaging of beer is thus a vital part of the product. Long ago, beer was delivered in pitchers or other kinds of vessels, but today it is bottled or canned if not delivered to bars and pubs as draught beer. In the bottling process, cleanliness is essential in all steps. After the labelling process, the bottles are packaged for distribution.

2.4 Beer styles and beer quality characteristics

> Beer can be taken simply as a refreshing drink, or as a balm at the end of the day. The search for the perfect pint should last a lifetime. In the meantime, there is a classic style of beer for every mood and every moment. Each can be judged only according to the characteristics of its style. To say that a Pilsener is better than a Stout is like arguing that a Chardonnay is more worthwhile than a sherry, or apples more important than oranges.
>
> (Jackson, 1991)

2.4.1 *An introduction to the family of beer styles*

One could certainly first recognize the "spontaneously fermented" beers as a distinct group of beers. Within this group are beers produced in Belgium, such as Lambic, Gueuze and Kriek, the last of these having a cherry flavour. One usually distinguishes between "top-fermented" and "bottom-fermented" beers, and in the previous section you have been briefly been introduced to their different production process characteristics (Svensson, 1995).

Within the group of "top-fermented" beers, one finds the Ales (e.g., Pale Ale and Brown Ale), Stout, Porter and the German Weissbier, the last of these being produced from wheat as a raw material. Among the "bottom-fermented" types of beers, which today certainly are the most common types, one finds Lager, Pilsner, Dortmunder, Vienna style and Munich Pale style. Of course, this is a very simplified overview intended only to serve as an introduction to the wealth of different kinds of beer styles and brands on the European market. Different beer styles are often presented in glasses of characteristic shapes and sizes, as illustrated in Fig. 2.5.

Figure 2.5 Different kinds of beer are sometimes served in glasses with a specific design related to the product.

2.4.2 *Beer as a companion to food*

In some countries, the natural companion to food is wine, with different kinds of wine intended to pair with certain types of food. In such contexts, beer is more often consumed in other social contexts. On the other hand, in countries where beer is considered the natural companion to food, the styles of beer are often generally adapted to the kind of food that is served. Some general advice to consider when selecting beer as a mealtime drink is as follows:

- Spicy food served in a hot climate often goes well with well-chilled (7°), light-bodied, and fruity kinds of beers like Corona (Mexico) or Budweiser (USA).
- Food like "pub grub" served in the UK (e.g., roast beef and cheddar cheese), goes well with fruity and aromatic ales like Bass Pale Ale or Bitter (UK), served at 13°.
- Food that is mildly spiced but heavy and is served in a colder climate requires a more aromatic, hoppy kind of beer like a Pilsner Urquell (Czech), served at a temperature of around 9°.
- Asian food of a sweet-and-sour character goes well with light fruity lager beer styles like Asahi beer (Japan).
- Seafood like oysters or salmon goes well with a dry stout or porter like Guinness (Ireland), served at around 15°.

Thus, different kinds of food may be best paired with different kinds of beer products. Nonetheless, a global beer manufacturer must try to supply a "good compromise" for a general lunchtime beer. Finding such a "good product compromise" is often difficult, but is a most important task in product innovation if too many product varieties are to be avoided.

2.4.3 *Beer quality characteristics and defects*

Each style of beer and each beer brand produced at different locations have different individual characteristics. Each brewery has a brew-master with access to a well-equipped laboratory for measuring a wide array of beer characteristics, and large breweries also have an expert "sensory panel". The sensory panel, if properly educated, must also be recognized as a "measuring instrument" of a very high quality. The following metrics for measuring beer quality are often relevant:

- **Appearance:** Colour, cloudiness, head (frothability and froth stability).
- **Aroma (bouquet):** Hoppiness, maltiness, fruitiness.
- **Palate (flavour):** Body, bitterness, sweetness, complexity, finish.
- **Miscellaneous:** Style of beer, defects.

There are a number of defects that may influence the quality of beer; of these, it is generally acknowledged that "staling" is the most common defect in packaged beer. Since effective additives for the prevention of staling tend to have undesirable side effects and neutral additives rarely seem to work, the best option for brewers is to understand and avoid the fundamental mechanisms that cause staling in order to achieve an acceptable (albeit finite) shelf life. Some degree of clarity is generally also considered to be a desirable property even for darker beers, and thus "haze" or cloudiness is considered a defect for many beers. The most important type of beer haze is "chill haze", which may appear when a beer is well chilled. However, some beer types like Weissbier are unfiltered and thus are expected to be somewhat "cloudy".

2.4.4 *How to taste and judge beer*

Beer often has a less distinctive taste than wine, and it can also include a large number of sensory components. It is therefore not an easy task to judge and recognize different kinds of beer products and individual producers. For this reason, it is recommended that beginners should start with "open testing", in which the producer and the beer brand are openly displayed, tested and discussed. Another kind of testing, which is probably the most common, is "half-blind testing". In this type of testing, the different beers are presented by producer, type and characteristic, and afterwards the test panel will try to identify which kind of beer is in each glass (in this kind of testing, one should select beers that are similar in colour). The most difficult kind of testing is "blind testing", for which it is recommended to select a theme, such as "different types of beer", "different products from a single manufacturer", or "country style".

Chapter 3

Facilitating Methodologies for Enhanced Product Innovation

From a resource allocation perspective, is pouring more money into R&D the best way to achieve greater innovation? Looking at innovation through the lens of effectiveness and efficiency, the answer to this question is certainly no. The effectiveness (doing the right things) and efficiency (doing those things well) of using the innovation resources are the hidden variables that should incentivize companies to use the best practices and tools in management of innovation.

In times of tight company resources and a continual need to improve company performance in all areas, including R&D, the use of effective and efficient tools and development methodologies for R&D is one avenue to follow. But is the price a company pays to learn, introduce and use a methodology worth the benefits it will share in various forms (e.g., product improvements, better work processes, etc.)? Sometimes new methodologies have hardly been introduced in a company before new tools are presented and advocated by academics, consultants or pioneering industry professionals. It is thus important not only to have a clear company strategy for the selection and implementation of methodologies and tools for innovation but also to secure organizational solutions to make company methodology use sustainable. There are certainly pros and cons for the use of development methodologies, but the frequent industrial use of different types of methodologies and the associated company resources they demand make assuring the productivity of development methodologies a growing company concern for R&D.

3.1 An investment approach to the introduction and use of development methodologies

Financial calculation of the profitability of using a methodology may be difficult if not impossible, but using an investment model as a mental roadmap will give the methodology

a more fair chance to survive. However, the ultimate issue is not only whether a methodology can improve various aspects of product development but, rather, whether the resources committed yield sufficient returns to make the investment an attractive option and alternative for allocating limited company resources.

Methodologies cannot be assessed on their own merits; they can only be assessed in relation to the activities they are meant to facilitate or improve. In the case of methodologies that are supposed to facilitate and improve R&D work processes, and product development in particular, the assessment of the product innovation results is consequently the ultimate measure of their industrial usability.

3.1.1 *Estimating the cost and outcomes of methodologies*

It is not an easy task to estimate and budget the cost for development projects, and it is certainly not easier to estimate in advance the cost of using new methodologies. It is often costly to introduce a methodology on a select product development project for the first time in a company. However, using the same methodology again in a new product development project can be expected to cost less when the team already is familiar with it. If the methodology is used repeatedly (e.g., in the development of new product generations), the cost of using the methodology will be even less, because it will be well known and the intellectual infrastructure will be available.

Assuming that methodologies are introduced and used in an efficient manner, what are the expected outcomes? This is often called performance. However, it should be noted that what one should measure is not the effectiveness of a methodology as such but rather the improvement in the activity that it intends to facilitate. The share of the output that can be attributed to the use of a methodology can thus be called "methodological effectiveness".

3.1.2 *Measuring the outcome from product development*

A single product development project does not deliver its final outcome within the R&D function. A cross-functional R&D work process is interacting with marketing and production functions, and the final outcome of the total product development activity cannot be measured until the product has been produced and sold on the market. The financial return must thus often be calculated over an extended period of time. One must also recognize that the use of a methodology may have effects at many different levels in a company; in addition to assisting with the design of better products (R&D level), it may also improve total company communication between functions (company level), give better product specifications (project level) and result in generally improved customer understanding.

Successfully using a methodology in an individual development project is a different matter from using it successfully in a company's product development process, or even for the company as a whole. Outputs and outcomes often occur at all these levels, depending on the objectives. If, for example, a company has decided to make use of the QFD methodology to improve company work processes or cross-functional communication, the measurement of the results from a single project use of QFD is certainly not enough for such an assessment.

3.1.3 *An investment perspective on methodologies*

Thus far, we have only discussed the calculation of input of resources to use a methodology and the measurement of the share of the output that the use of a methodology is supposed to deliver. However, it is not until those measures are combined that the company can start to truly assess a methodology's usefulness. Productivity is defined as the relation between input and output for an activity or process. When using financial measures for input and output, one reaches a final destination in productivity calculations: Profitability. However, in this calculation a time factor must also be considered. If the time until one obtains the results from the use of a methodology is included, classic investment models are applicable.

The difficulty of calculating the return from invested resources is related not only to the aforementioned difficulty of estimating the share of the financial return from each product development project but also to the difficulty of estimating how many projects the cost of introducing the methodology should be shared with. It is obviously not correct to write off the whole cost of learning and introducing a methodology on a single project. It is fairly easy to account for the costs of using a methodology after a project is finished, but not in advance. Thus, it is very difficult to assess the financial share of the final outcome from the use of a methodology. This fact may cause companies to unjustly regard the introduction and use of a methodology as a cost item, while disregarding the subsequent revenue.

Although financial outcomes are difficult to calculate, it is probably better, and representative of a more dynamic mental model, to regard the introduction and use of methodologies as an investment. Following this line of thought, in a company's final assessment of a methodology, it is not sufficient to state types of outcomes and the importance of the results they produce; rather, one must also try to assess whether the activity was worth the effort. For this assessment, calculations of returns may not be the best tool; instead, one should consider the sound judgement of the people who were involved in the activity and the use of the methodology.

3.2 Integrating multiple methodologies in company work processes for innovation

It has been argued that sufficient adoption of appropriate tools in innovation management could significantly improve company performance (Nijssen and Frambach, 2000; Nijssen and

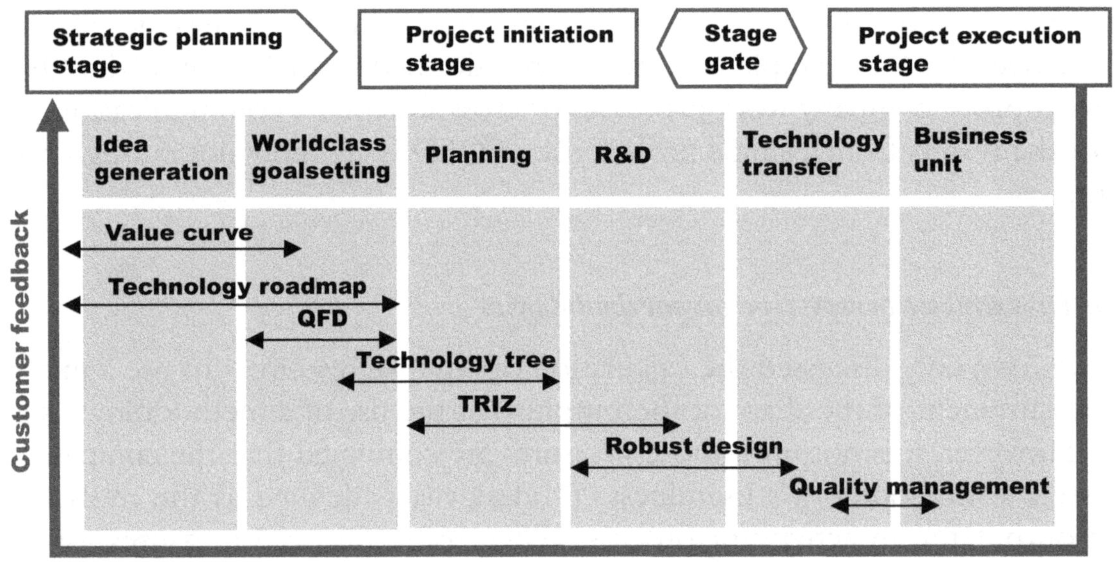

Figure 3.1 Deploying a portfolio of integrated development methodologies at Samsung (Park and Gil, 2006).

Lieshout, 1995). However, it has also been reported that the low penetration of tools and techniques is mainly due to a lack of awareness of their existence, as well as difficulties in their implementation once the tools are known (De Waal and Knott, 2013).

The QFD methodology is today often fully integrated with other methodologies (e.g., TQM, Design for Six Sigma). In addition, it is sometimes combined with other supplementary methodologies (e.g., Conjoint Analysis, Reengineering, Reverse Engineering, etc.) rather than used as a "stand-alone" development tool. In Fig. 3.1, Samsung Research Centre's use of multiple methodological tools for development can be seen to illustrate this situation well. The QFD methodology is used by this company as its primary tool for "World Class Goal Setting" for new products in the Project Initiation Stage. Goal setting (also called "target setting" for new products and production technology) is often, and rightly, considered to be one of the most difficult and critical activities in innovation. Goal setting involves setting Target figures that are necessary to gain a future competitive advantage but still possible to reach in efficient development work. Later on, this book will advocate that the QFD methodology should be used not only in this pre-development phase but in the total product innovation work process in an end-to-end perspective. Moreover, it will be further argued that in order to speed up the product innovation projects, the QFD methodology should be used to build *Integrated Knowledge Platforms* as supportive platforms for the company product development work process, well in advance of project-based product development.

The QFD methodology must today be regarded as a well-proven methodology for innovation and as such often combined with other methodologies like Design for Six Sigma. However, all methodologies must continually be assessed and further developed, and it is the belief of this author that, if improved and made more user-friendly, the QFD methodology has the potential to become the most important tool for companies' future product innovation.

In this chapter, it is advocated that facilitating methodologies like QFD can be important instruments for enhanced product innovation in all manufacturing industries. In the next part of the book, you will be introduced to the QFD methodology as the selected methodological tool and supporting instrument for the product innovation work process and for process-industrial innovation in particular.

Chapter 4

Summing Up and Some Issues to Reflect Upon

You have now become familiarized with the QFD product innovation methodology, one of the most widely used methodologies in product innovation. You have further been acquainted with the *Voice of the Customer* and the *House of Quality* as the common denominators of all QFD systems. You have learned that *Product Design Requirements* must be progressed as requirements on production system capabilities, in order to secure a cost-efficient product manufacturing process. However, it has also been pointed out that the process-industrial innovation environment requires adapted methodologies, which will be further discussed in the next part of the book.

You have also learned some fundamentals of beer production and beer products. The complex supply/value chain, which is common in many sectors of the process industries, has been highlighted, along with its importance for sound product innovation. The products produced by different companies in different sectors of the process industries are often considered "not very advanced" (an opinion not held by this author), but that the related production system often is very technically advanced is generally agreed upon. You have now learned from the introductory case presentation that the manufacturing of beer is a complex biochemical process and that not only the raw materials but also the production process influences the characteristics and quality of the final product. It is thus argued that it is not only important to understand customer preferences in the development of new or improved products but also that the upstream QFD matrices are paramount instruments for an improved understanding of how such products can be manufactured in the related production process.

It is advocated here that supporting development methodologies should not be presented as imposing additional costs in product innovation but rather as tools to improve the efficiency of the product innovation work process, thus yielding multiple returns on such an investment. It has further been shown that a spectrum of different product development methodologies often must be used in an R&D portfolio of tools and best practices.

Finally, it must be pointed out that, in addition to using the information from the individual matrices in product innovation, the richest fruits of the QFD process are often harvested at the matrix-building stage. It is therefore important not to get "hooked" on the matrix concept as such; one should not force the pace by trying to get the matrix completed quickly at the expense of failing to answer the difficult questions that arise in the course of the work, as it is often the answers to those questions that contain the seeds of new ideas and product concepts.

After reading this part of the book about QFD theoretical and methodology use, it is now important that you try to relate *t*his information to the perspective of your company product innovation behaviour and philosophy. In order to facilitate this activity, please reflect on the following questions:

- If you have heard about the product innovation methodology QFD but have not yet tried it in your company, is this because of a general skepticism against three-letter Japanese development methodologies, or is there any other reason?
- How do your product development teams interact with customers and consumers in order to understand their present and future product needs, wishes and desires?
- If you share the belief advocated here that cross-functional product innovation working practices are essential for successful product innovation and you have not practiced QFD, how do you alternatively arrange and promote such a development approach in your company?
- Do you keep track of competitive companies' product properties, specifications and functional properties? If so, what kind of system do you practice, and how systematically do you compare competing products with your company's products?
- Do you believe that it is necessary for Target Figures and product specifications to be measurable with internationally recognized test methods or, alternatively, with internally well-defined techniques?
- Are you presently using some sort of "target setting" best practice in the development of new products? If so, how do you relate product specifications of new products to known customer preferences?
- In the illustrative case of beer development and manufacturing, the importance of understanding the intimate relations between raw materials, production processes and finished products is highlighted. Do you find a similar situation in your company?
- It is pointed out that the development of a new beer product in fact is the development of a new process; the process embodies the product. How does this statement relate to your company's product innovation context?
- Is your company presently using any product innovation methodology, or do you believe that the "art of product innovation" is something that can only be "learned by doing"?
- R&D and product development expenditures are in reality not cost items, since the company surely expects returns from new product innovation of a considerably higher figure. Do you share the belief that the same is true for the use of product development methodologies?

Part II

Process-Industrial Product Innovation and the Use of the QFD Methodology as a Facilitating Development Tool

It is not to be said that prior knowledge in management of R&D and innovation in general does not apply to the process industries, but rather that best practice and methods from other domains may very well be useful for the family of process industries as well. However, the idiosyncrasies of process companies are likely to influence the conduct of R&D and innovation and call for adapted methodologies, tools and an actionable and improved knowledge base for R&D management and product innovation in the process industries.

(Lager *et al.*, 2013)

Writing a book about management of product innovation and QFD use, and the use of the QFD in the process industries in particular, requires careful consideration of the environment for process-industrial product innovation. The reason for this is that one must recognize the characteristic features of different kinds of industries that will influence not only innovation and innovation-related activities but also the proper use of development methodologies. In all industrial undertakings today, it is necessary to use supporting methodologies and tools in order to facilitate and improve the performance and quality of a spectrum of company activities, including R&D and product innovation. There are fortunately today an abundant number of such methodologies available, so the question is not really whether they should be deployed or not, but rather WHICH methodologies should be used and WHY, WHEN and HOW they should be deployed.

Acknowledging and ascertaining the unique conditions for R&D and product innovation in the process industries

In the following chapter, it is initially concluded that the nature of the production systems in the process industries significantly differs from that of other manufacturing industries; however, different sectors within the process industries share a large number of similar characteristics. Moreover, the intimate couplings between raw materials, production systems and products constitute unique "contextual innovation conditions" that must be taken much more into account in companies' product and process innovation and methodology use. Furthermore, the laboratory and pilot plant experimental environment and the interdependencies of product and process innovation constitute "inherent innovation conditions" that underscore the shared idiosyncrasies of innovation in this "family" of industries. As a consequence, management tools, methodologies and best practices from other manufacturing industries are of less interest for use in the process industries, while tools and methodologies adapted to the process-industrial cluster of industries are likely to be applicable to many sectors of the process industries (Lager, 2017a).

Focusing on the product innovation work process for QFD implementation and use

This book focuses on the use of the QFD methodology as a supportive tool for the process-industrial product innovation work process, and formal work processes will initially be discussed in the spirit of Edwards Deming: *If you can't describe what you are doing as a process, you don't know what you are doing.* Using the commonly deployed "Stage Gate" decision system as a point of departure, a simplified three-stage product innovation work process is initially depicted in Chapter 6, and the important relational aspect with the work process of associated process technology development is afterwards discussed. You will discover that this work process is of a substantially different character compared with similar work processes in other manufacturing industries — a fact that naturally places specific demands on the QFD system.

Lessons learned from process-industrial QFD use and necessary QFD system capabilities for this "family" of industries

Reviewing the extant literature on the industrial use of the QFD methodology shows that it has been primarily been deployed in other manufacturing industries, in spite of its multiple advantages in process-industrial use. You will learn from interesting previous industrial experiences of QFD use in different sectors of the process industries, resulting in the conclusion that a QFD system for use in the process industries must have the following capabilities:

- Must recognize that products from the process industries are not assembled components.
- Must address the fact that products, production processes and raw materials are very closely interrelated.
- Must have the ability to relate product functional design requirements to inherent product characteristics.
- Must be able to cope with interlinked chains of both external and internal customers and suppliers.
- Must facilitate the development of product and process concepts.

Chapter 5

Contextual and Inherent Conditions for Innovation in the Process Industries

A study of the 2,000 highest ranked companies by level of research and development (R&D) investment worldwide revealed that about 30% of those companies belonged to the process-industrial cluster. However, all sectors in that study, apart from pharmaceuticals and biotech industries, had an average R&D intensity of less than 5% and could therefore be included in the group of industries often categorized as low and medium technology industries (LMT).

(Lager, 2010)

The cluster of industries generally called the "process industries" spans multiple industrial sectors, constitutes a substantial part of the entire manufacturing industry and is generally considered to include petrochemicals and chemicals, food and beverages, mining and metals, mineral and materials, generic pharmaceuticals, pulp and paper, steel and utilities.

5.1 What is this thing called the process industries?

The process industries are a part of all manufacturing industries, using raw materials (ingredients) to manufacture non-assembled products in an indirect transformational production process often dependent on time. The material flow in production plants is often of a divergent v-type, and the unit processes are connected in a more or less continuous flow pattern.

(Lager, 2017a)

The above "intentional definition" (Lager 2017a) was selected for this book, but it must initially be recognized that this class of industries is not homogeneous; rather, it consists of a number of industrial sectors that are more or less similar in their individual characteristics.

A number of industrial sectors and industries have been selected (Lager, 2010) from all manufacturing industries presented in the statistical classification of economic activities in the European community (NACE, 2006). The following industrial sectors are thus suggested for inclusion in the cluster of process industries (associated NACE codes are presented in parenthesis):

- *Chemical & petrochemical industries* (chemicals, rubber, coatings, industrial gases) (20, 22)
- *Food & beverage industries* (10, 11)
- *Forest industries* (pulp & paper) (17)
- *Mineral & material industries* (minerals, cement, glass, ceramics) (08, 23)
- *Mining & metal industries* (05, 06, 07, 24)
- *Pharmaceutical industries* (incl. biotech industries and generic pharmaceuticals) (21)
- *Steel industries* (24.1, 24.2, 24.3)
- *Utilities* (electricity & gas, water, sewerage, waste collection and recycling) (35, 36, 37, 38)

5.2 Contextual production conditions of non-assembled products

> **The different sectors of the process industries share a large number of characteristics related to their production systems, but on the other hand, the nature of the production systems in the process industries significantly differs from the nature of the production systems in other manufacturing industries.**
>
> **(Lager 2017a)**

One of the principal differences between companies in the process industries and those in other manufacturing industries is that the products supplied to companies in the process industries, and often those delivered from them, are materials or ingredients rather than components or assembled products (Flapper *et al.*, 2002; Frishammar *et al.*, 2012). In Fig. 5.1, a simplified structural model of a company's production system in the process industries illustrates the system of material transformation from supplied raw materials into finished products to be marketed to external customers or consumers (Lager *et al.*, 2017a; Storm *et al.*, 2010).

Using the structural model in Fig. 5.1, a number of contextual conditions for innovation in the process industries have been selected (Lager, 2017a).

5.2.1 *Incoming material characteristics*

- *Types of incoming materials.* A key difference between companies in the process industries and those in other manufacturing industries is that the products supplied to them, and often also those delivered from them, are materials or ingredients rather than components (Frishammar *et al.*, 2012).

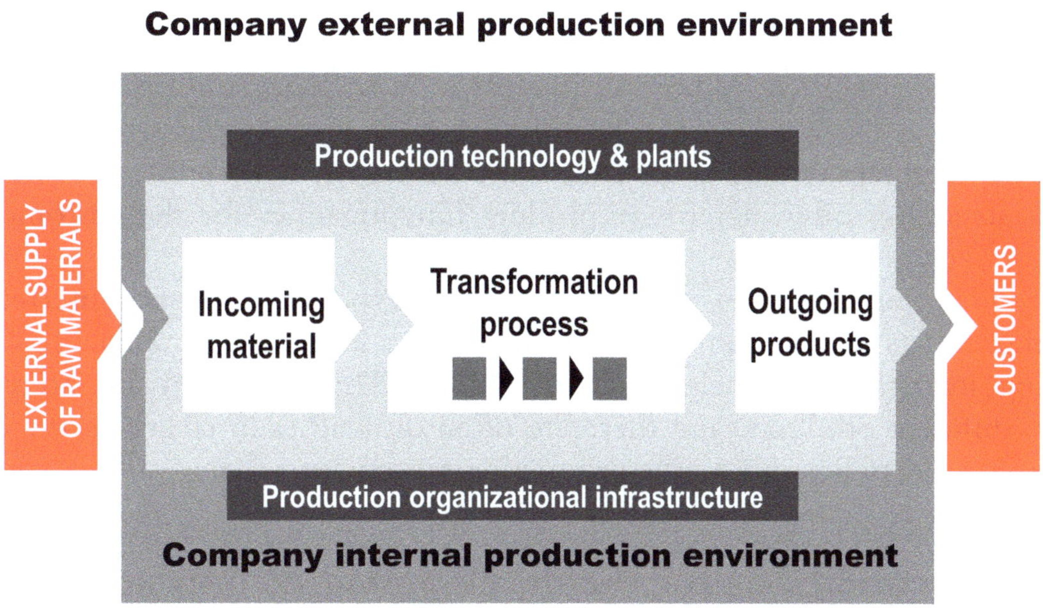

Figure 5.1 A simplified structural model of the production system in the process industries (Lager *et al.*, 2017c; Storm *et al.*, 2010).

- *Changeability of incoming materials.* Some sectors of the process industries, such as the food industries, use raw materials and produce products with limited shelf lives (Taylor *et al.*, 1981), which may have a strong influence on production system characteristics in these industries as opposed to other manufacturing industries.
- *Incoming material influence on finished products.* The characteristics of incoming materials not only determine the choice of unit operations and the design of the total production system but also may directly influence the properties of finished products (Samuelsson *et al.*, 2016).

5.2.2 *Transformation process characteristics*

- *Production system structural characteristics.* General structural characteristics of the transformation process in the process industries include batch processing, semi-continuous processing and fully continuous processing, but not "job-shop".
- *Material characteristics.* Operations in the process industries usually start with a small number of raw materials, from which hundreds or even thousands of final products are made (King, 2009), representing a V-type flow pattern (Chakravorty, 2000).
- *Material transformation characteristics.* In other manufacturing industries, the raw materials are "reconfigured" to become a finished product, but the materials remain more or less the same throughout all subsequent manufacturing activities. However, in the process industries, raw materials undergo an inherent transformational change, becoming an identifiably different material as a finished product (Floyd, 2010).

5.2.3 *Outgoing product characteristics*

- *Types of outgoing products.* Almost all process industries use and produce non-discrete materials, and, as previously noted, products manufactured in the process industries are not assembled components but are more or less homogeneous entities. Moreover, their inherent structures often determine product functionalities for the customers (Lager, 2010).
- *Product interrelationships in product portfolios.* Product portfolio interrelationships between products are sometimes high in the process industries compared to other manufacturing industries (Samuelsson *et al.*, 2016). There is often an interconnection between different products, and there are often difficulties in changing the content of total product portfolios due to the production structures.

5.3 Inherent conditions for the development of non-assembled products

> **The particular product and process innovation environment common among sectors belonging to the cluster of process industries requires unique experimental facilities and development approaches of a clearly different kind rather than what is common in other manufacturing industries.**
>
> **(Lager, 2017a)**

The environment for pre-product development activities and early experiments in the process industries is the laboratory (Lager, 2000), whereas in other manufacturing industries pre-development occurs in a design office. Development work is fundamentally experimental, and addressing up-scaling problems is a major development task. This is because new product properties are often added, while existing ones are lost, when an emerging concept moves from the laboratory to a pilot plant and then to full-scale production (Frishammar *et al.*, 2014). The difficulty involved in the scale-up of a biotech process from test quantities to commercial manufacturing volumes has been well recognized in previous studies (Linton and Walsh, 2007).

Development of functional product prototypes in other manufacturing industries is replaced by test runs in pilot plants or full-scale production in the process industries, where test batches for customers are made and adequate process conditions are specified (Lager, 2000; Pisano, 1996, 1997). Pilot and demonstration plants thus represent bridges between basic knowledge generation and technological breakthroughs on the one hand and industrial application and commercial adoption on the other (Frishammar *et al.*, 2014).

In product innovation in other manufacturing industries, it is important to consider that a product should be easy to manufacture (Boothroyd *et al.*, 1994); this is referred to as product manufacturability. In a similar vein for non-assembled products, it is also important to consider their "processability" (Lager, 2005b). A strong interrelationship between product and process innovation is thus often necessary for good innovation performance in the process industries (Lager, 2002). This was early recognized by Marx (1867), who noted that: "the process disappears in the product". For this reason, company product innovation and renovation often call for process innovation and renovation (Hirsch-Kreinsen, 2008). Reichstein and Salter (2006) thus advocated that they should be viewed as "brothers" rather than as "distant cousins".

This interdependency between product and process development and the notion that the products are intimately coupled with the production system were underscored by Tristan Roussell (2012), who boldly stated that: "in the biotech industry the process is the product". He then intuitively recognized that product development is always related to the development of a new process, and sometimes vice versa. The importance of this interrelationship between product and process development has since been reviewed and well clarified by (Hullova *et al.* 2016; Hullova *et al.* 2019). It has also been noted that their complementarity is likely to be dependent on the institutional context (Ballot *et al.*, 2015).

Chapter 6

The Process-Industrial Product Innovation Work Process

Product development is a process that always must start and finish together with the customers. The drivers for the process are the customers' needs and their willingness to pay for the product or service a company can develop and produce. In corporate annual reports and on company websites, "customer focus" in innovation is usually strongly emphasized and advertised; in reality, however, in their day-to-day operations companies often run into trouble explaining how this is being accomplished, or at least how it ought to be accomplished.

6.1 Introducing the formal work processes for innovation

If you can't describe what you are doing as a process, you don't know what you're doing.

W. Edwards Deming

Many activities in industry can be looked upon as "work processes", but a work process must have both a customer and a supplier as well as the existence of feedback loops. Melan (1992) characterized four different types of transformations occurring in different types of work processes:

- Physical — Change of state, converting raw or semi-finished material (e.g., assembling an automobile).
- Locational — Involving movement of objects or material from one place to another and storage (e.g., moving and warehousing).
- Transactional — Involving a value-implied exchange (e.g., banking or stockbrokerage).
- Informational — Data are the input, and the output consists of data converted into meaningful information (e.g., financial reports).

Other characteristics of a formal work process are that an "owner" of the process should be recognized and that the process should have clear boundaries, or "interfaces", with connecting work processes. However, the term "development process" is illusive, because it gives the impression of a completely ordered, linear and logical process that is controlled step by step. In contrast, those who have been directly involved in industrial development activities can testify that the development process is often of a very complex and sometimes even chaotic nature. However, early in 1974, Utterback presented a simple three-phase model for the development process; interestingly enough, this fundamental structure is still sound (Utterback, 1974):

- **Generation of an idea**, involving synthesis of diverse (usually existing, as opposed to original) information, including information about market and technology.
- **Problem solving**, including specific technical goals and designing alternative solutions to meet them.
- **Implementation**, consisting of manufacturing, engineering, tooling, plant start-up, and market launch required to bring an original solution or invention to its first use or market introduction.

6.2 The formal product innovation work process — lessons from past

> **A number of firms still rely upon product development artisans, people who just seem to know what needs to be done. The firm empowers them to do the job, then works hard to stay out of their way... Development groups in companies without formal development processes, or that do not follow their firm's process, fly by the seat of their pants or do what we always have done and hope that none of the details get lost along the way.**
>
> **(Griffin, 1992)**

Booz *et al.* (1982) introduced a simple linear model of product development with the following steps: idea, preliminary assessment, concept, development, testing, trial and launch. At that time, many companies may not have formalized their product development efforts — a fact that should not a priori be confused with the fact that many companies already had established routines and knowledge about how to develop new products. Research in this area by Cooper and Kleinschmidt (1986, 1988) gave a picture of both the industrial use of more formal work processes and how the work process could be designed and possibly improved. Recognizing the "product development process" as a number of activities separated by "decision points", the Stage Gate model made the development process more distinct. In order to improve an ad hoc process for product development, it could have been necessary to overstress the formality of such a new work process to induce an organization to change. The original linear Stage Gate model probably would have served this purpose well.

Later research by Cooper (1994) and other research in this area (Bower and Keogh, 1996; Kline, 1985; Schroeder *et al.*, 1986) gave a more complex picture of the "product development process" and indicated that future work processes should be more flexible and adapted to the individual characteristics of different development projects. Descriptive research and case studies sometimes presented a picture of a very chaotic development process. This observed fact, however, does not necessarily preclude the possibility that this chaotic process could be made less chaotic and that the output from development processes could be higher if they had had a more formal structure. Cobbenhagen *et al.* (1990) presented their important early views on future development processes as follows:

> Innovation capability is built up in a learning organization. Control of innovation processes therefore calls for a shift from formal planning (feed forward), towards an investment in feedback; stressing the aspect of product development as a knowledge building activity and organizational learning.

The importance of R&D and its work processes as "knowledge-building activities" is a theme that will be central in the further presentations and discussions in this book, conceptualized as *Integrated Knowledge Platforms*. How many structures should the product development process have, and how formalized should the process be? The good and simple answer proposed by Norling (1997) is that it depends on the type of project that is going on and the type of organization the activity is being performed in.

Reengineering of the R&D work processes later on turned out to be a very powerful tool for improving the R&D work process (see, e.g., Brockhoff *et al.*, 1997; Hammer, 1990; Liebeskind, 1998), but a restructuring and improvement of R&D work processes must always start with better definitions and descriptions of the development systems and their components. When companies in the process industries start to formalize work processes for product development, it is important to learn from industrial experience and academic research about the "product development process" in other manufacturing industries, as well as to identify the specific needs of the development environment for the process industries. The Stage Gate (sometimes called Idea-to-Launch) process advocated and advertised by Cooper must today be regarded as a "de-facto standard" for a formal product development process.

> A Stage-Gate process is a conceptual and operational map for moving new product projects from idea to launch and beyond — a blueprint for managing the new product development (NPD) process to improve effectiveness and efficiency.
>
> (Cooper, 2008)

The latest version of the system includes the following stages: Discovery (ideation), Scooping, Build business case, Development, Testing & validation, Launch and Post-launch review. In his late review of the Stage-Gate system, Cooper emphasize the importance of gathering the *Voice of the Customer* and undertaking a competitive analysis as well as the importance of the fuzzy front-end part of the system. Those activities include ideation, scoping the project, defining the product and building the business case. The Stage Gate system has been studied in a large sample of users and has proven to be an instrument that top-performing companies have used often and well (Cooper, 2012). However, the process does not stand unquestioned and has attracted a number of criticisms, summarized by Cooper himself:

> It is accused of being too linear, too rigid, and too planned to handle more innovative or dynamic projects. It is not adaptive enough and does not encourage experimentation. It is not context-based and its gates are too structured or too financially based and the system is too controlling and bureaucratic, loaded with paperwork, checklists and too much value added work.
>
> (Cooper, 2014)

The criticism of the Stage Gate system may naturally relate to the nature of the system as such, but it is more likely to stem from companies' inability to properly implement it. Cooper also articulated the fact that some people had misread, misapplied and abused an otherwise excellent system. Cooper's vision of the next generation Stage Gate system includes:

- Adaptable and flexible gates;
- Fast track versions of the system for low-risk and minor projects;
- Overlapping stages;
- Emphasis on the fuzzy front end.

6.2.1 *A simplified three-phase model of an innovation work process to be used in this book*

In reviewing the Stage Gate system, another weakness that can be noted is that the most important activity, "industrialization", is not properly visualized. Today, it is not sufficient to develop a product with competitive performance characteristics; such a product must also be manufactured in a cost-efficient production system. For this reason, "Design for Manufacturability" and the early involvement of a production system development or adaptation are of primary importance, especially in the process industries. This was also noted in the early model proposed by Booz *et al.*, but not in the simple three-phase Utterback model. Following in the footsteps of Utterback's three-phase model, the work processes in this book will be outlined as a simplified three-phase work process, as illustrated in Fig. 6.1.

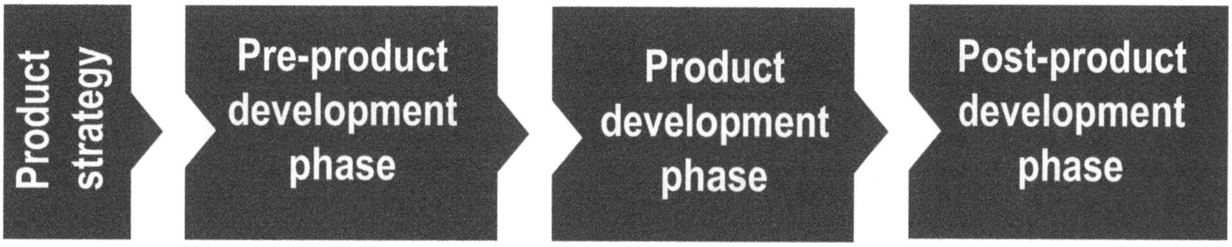

Figure 6.1 A simplified three phase model of the development work process.

6.3 The process-industrial experimental innovation environment

One striking difference between innovation in the process industries and in other manufacturing industries is the environment for experimental work, including laboratory and pilot plant testing and trials in demonstration plants.

In Fig. 6.2, the different development steps in the process-industrial experimental environment have been outlined in a simplified model. The test work sometimes include only one or two of these steps, sometimes in a different order and with different iterative loops. Each development step serves a different purpose, and short-cuts can be taken; laboratory results, for example, can be scaled up and applied directly in a production plant, while simulation can replace pilot plant testing (Lager, 2000). This is often today, as an estimate of technology maturity, expressed in terms of Technology Readiness Level (TRL).

6.3.1 *Laboratory testing*

Laboratory testing is often considered the starting point for product innovation and includes experimental work in the laboratory, either in batch testing or in more continuous laboratory testing. Most of the experimental work is often done in a laboratory environment because of the ease of experimentation and low costs, but the laboratory testing may have different objectives depending on the character of the development work. Development methodologies can certainly be improved in laboratory testing; one important step is the introduction of methodologies for experimental design and multivariate experimental evaluation, as well as other optimization and simulation methodologies.

6.3.2 *Simulation*

The use of simulation has already replaced test work in some applications and will, in many respects, revolutionize process-industrial innovation in the future. Simulation can be used, following all the steps presented in Fig. 6.2, as a supplementary activity to testing in

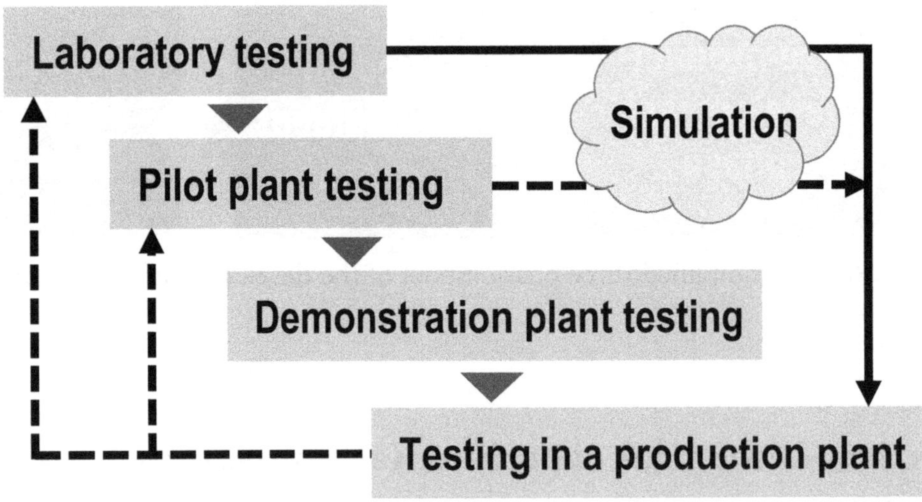

Figure 6.2 Different kinds of test environments in the process industries. Innovation does not necessarily involve all the steps shown, nor are they necessarily performed in the order shown. Several iterative loops are common. Simulation can be a supplementary activity or can replace practical test work. After Lager (2000).

laboratories, pilot plants and operating plants. Well-developed simulation programmes also have a strong potential to be used in training programmes for people both before and after the start-up of new process technologies and new production plants. The performance of simulation and simulation programmes, however, depends to a large extent on a fundamental understanding of the inherent process technologies. The advancement of simulation technology today is at such a stage that it is not advisable to start any major new installation of process technology without a previous or simultaneous installation of proper equipment and programmes for simulation (Contreras *et al.*, 2005; Harding and Popplewell, 2000; Laganier, 1996; Shelden and Dunn, 2001; Young *et al.*, 2001). Today, the term "Digital twins" can refer to advanced simulation of physical technologies or even large production plants.

6.3.3 *Pilot plant testing*

Moving from the laboratory stage to the pilot plant stage is an important step to improving the understanding of a production process, and the large cost increase associated with pilot plant testing makes the balance between pilot plant testing and laboratory testing an important issue. One reason for moving the development process into the pilot plant is to test the full process as a closed system and to determine the influence of circulating loads on the process results. Another reason may be to produce test samples in larger quantities for further testing at the customer's premises. For an in-depth presentation of pilot planting, design and construction, see Palluzi (1992, 1997).

6.3.4 *Trials in a demonstration plant*

Trials in a demonstration plant are often very expensive and are generally carried out in a major development project when the process is of a completely innovative character and existing pilot plant facilities cannot be used (Frishammar *et al.*, 2014). The primary objective of using a demonstration plant is not to use it as an alternative to the pilot plant but, rather, to run trials that complement previous pilot plant testing. These objectives may be:

- To test wear on the process equipment.
- To develop process control strategy and systems.
- To scale up the process to something between the pilot plant and the production plant.
- To produce larger samples of products for subsequent testing in the customer's process.

6.3.5 *Production plant tests*

Nowadays, many tests are conducted directly in the production environment, resulting in fast feedback and reliable test results; nonetheless, finding a correct balance between the goals of fast process development in plant tests and minimum production disturbance remains an important issue. Production plant trials may be undertaken to test new equipment, new process conditions, new process configurations, and so on. The role of the production plant as a source of innovation is of great importance, and it may also serve other purposes in future integrated development work involving the production plant operational crew and the development teams. The use of the production plant for more development activities and improvement work may have future organizational implications and lead to new development models. There are also two fundamentally different cases:

- The process is to be implemented in an existing production environment or existing production plant.
- The process will be implemented in a completely new plant that must be built before production starts, a fact that made part of the unique organizational solutions possible at Chaparral Steel (Leonard-Barton, 1992).

Different situations also exist according to whether the new process is to be integrated into an existing production structure or operated as a stand-alone process.

6.4 Current process-industrial innovation work processes

While product and process development in the process industries can sometimes be considered as two sides of the same coin (since product development often involves development of new technology), the lack of distinction often creates confusion both in the operational performance of industrial R&D and in academic research.

6.4.1 *Product and process development — two separate but interacting work processes*

To distinguish between the two terms product and process development, "product development" is defined in this book as follows:

> Product development is driven by a desire to improve the properties and performance of finished products, even if the nature of the practical development work is [the] development of process technology in a laboratory.
>
> (Lager, 2000)

Objectives for product development may include improving product properties, improving product quality (uniformity of composition), or creating more environmentally-friendly products. The product development work process in this book will be presented as shown in Fig. 6.3. Correspondingly, "process development" is defined as follows:

> Process development is driven by internal production objectives. Such objectives may be reduction of production costs, higher production yields, improvement of production volumes, environment-friendly production, etc.
>
> (Lager, 2000)

In a product development project, the "product development" activity and work process is the customer for the development of new process technology. For this reason, there is a need for a stronger integration of the development of new or improved products and the

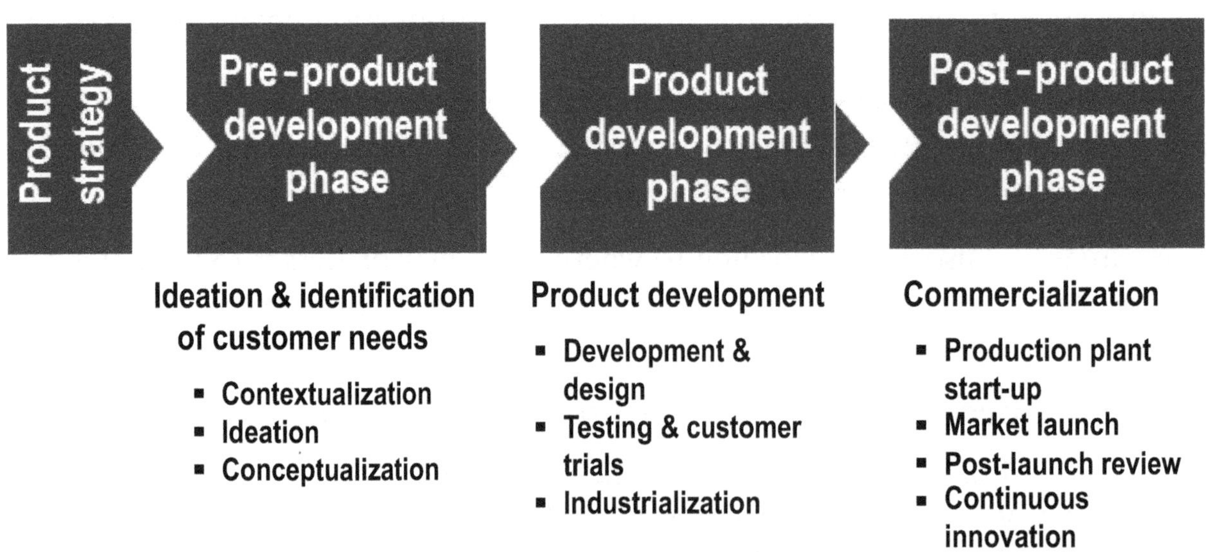

Figure 6.3 A simplified model of the product development work process used in this book.

development of new or improved production process technology. In the perspective of an improved product development work process for product innovation in the process industries, a conceptual model for the product development work process is outlined in Fig. 6.4.

It is advocated in this book that interaction with customers and consumers is one cornerstone for successful product development. However, since price/performance is a prime concern of customers, neglecting the production aspect in product development is a serious mistake. Thus, product development must understand not only what product attributes the customer values most but also how such attributes can be created in a cost-efficient production system.

Ease of manufacturing of new or improved products is nothing new in other manufacturing industries, and for decades the mantra of "design for manufacturability" has been well recognized and acknowledged in industrial life (Boothroyd *et al.*, 1994). Product developers of excellence today certainly acknowledge the need to manufacture products in cost-efficient production processes also in the process industries, but they usually do not recognize the importance of a very early integration between the work process for the development of new or improved products and the development of related and necessary new or improved production process technology. It has previously been argued in this part of the book that "the process embodies the product", and the consequence of this is thus that an improved product development work process must be interlocked with a modified process development work process.

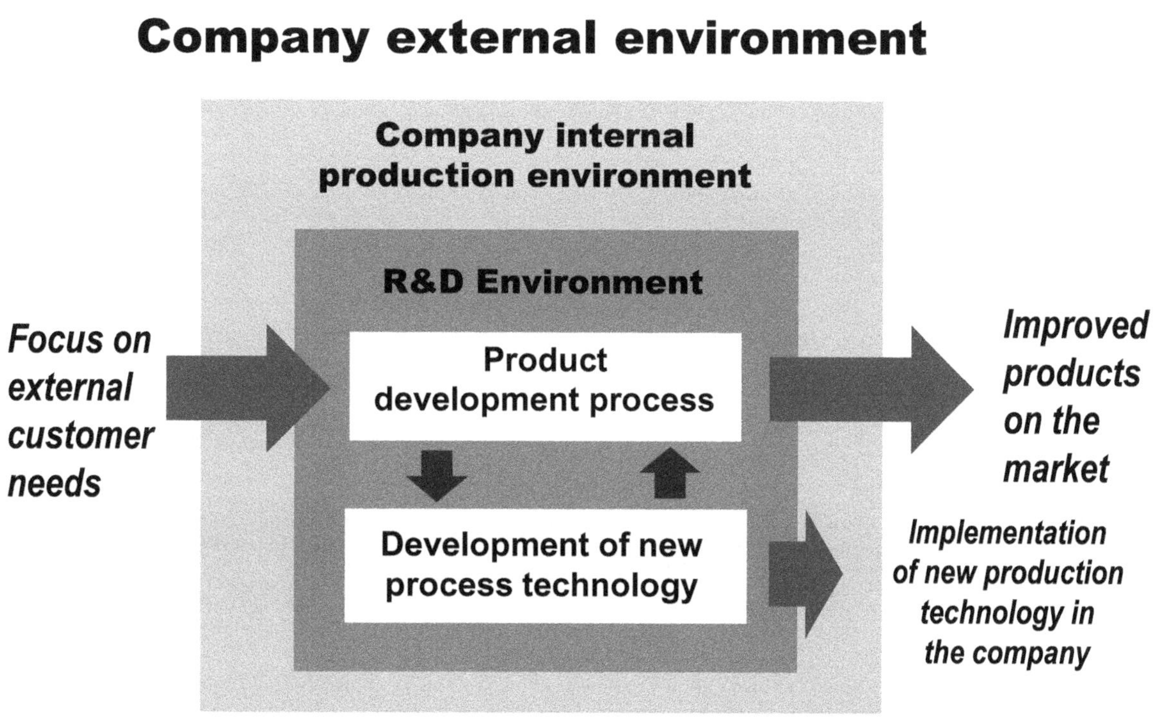

Figure 6.4 A revised structural model of the product development work process in the process industries that will be used as a fundamental platform for the use of the QFD methodology.

As shown in Fig. 6.4, the development of process technology must be transformed into a sub-process of the product development process. However, the pre-development phase is not the identification of production needs (process development), but rather the identification of the need for new or improved production technology identified in the product development work process and in Fig. 6.4, the modified interaction is illustrated. Methodologies that support product innovation in the process industries must thus recognize this fact and facilitate such an integration, and this integration must start early — preferably even before the pre-development phase.

Chapter 7

A Review of QFD Use in the Process Industries — Lessons from the Past

In the steel industry, the product concept is almost closed (compared with manufactured products where different parts can perform the same function), so the opportunities to balance cost and quality come only from the choice of alternative process paths, raw materials and the method of production.

(de Oliveira *et al.*, 1996)

The subject of the QFD methodology's application in the process industries is sparsely and unfairly treated in the reference literature. In the presentation of the Four Phases of Matrices system by Day (1993), the use of QFD for process industry application is treated in five pages and is mainly a reference to ASI QFD models (American Supplier Institute, 1989). In the extensive textbook on QFD by Cohen (1995), the subject of the process-industrial use of QFD is not discussed at all. The best presentation of the Four Phases of Matrices system for the process industries is to be found in the ASI workshop manuals (American Supplier Institute, 1989), where the methodology's adaptation for use in the process industries is presented on only a few pages, using the frequently cited case of the manufacturing of chocolate cake.

The presentation of the Matrix of Matrices system by Akao (1990), however, is of more interest to advanced users of QFD methodology than to beginners. In the early book by King (1987), presenting a slightly modified version of the Matrix of Matrices system, QFD use in process industry applications is not mentioned. In some other textbooks, the topic is likewise not treated at all (Bossert, 1991; Eureka and Ryan, 1988).

Reviewing the QFD literature shows that there does not exist a thorough literature review in the subject area of QFD use in the process industries. The following review thus not only serves as a foundation for this book but also can serve the purpose of a collection and reference list for QFD users in various sectors of the process industries. The references are often somewhat old, since they report from early trials and use of QFD, when the methodology was introduced in the industry. However, the lessons learned remain valid and useful for industry practitioners. The following review of process industry experiences has been grouped into four categories: Mining and metal industries, Food and beverage industries, Forest industries, Chemical and petrochemical industries.

7.1 Experiences from the Mineral and Metal Industries

> **The use of QFD helps to strengthen the product development process. The overview description of the product ensures that nothing is forgotten and it becomes possible to get away from single-parameter thinking and strive for optimum compromise at all times. The matrix itself will in fact supply impulses to seek new knowledge, because it clearly reveals gaps in knowledge such as customer requirements that have not been measured. This makes the matrix an aid to setting research priorities.**
>
> **(Tottie and Lager, 1995)**

The use of QFD to improve the quality of raw materials for ALCOA smelting plants is reported in an early paper by Williams (1992). The use of the methodology was focused on the importance of improvements and did not develop a classical total hierarchical structure of customer demands. One can observe that the customer is a company internal customer and is also far removed from the ultimate end user of alumina products. It was reported that deeper levels of QFD matrices were being used, but they were not presented.

Some methodological issues are discussed in a paper by Tottie and Lager (1995) presenting the use of QFD in the development of advanced iron ore pellets for the steel industries. In the process of listening to the customer, it was recognized that, although interviews were conducted with purchasers, blast furnace managers, logistics managers and process engineers, they only tried to interpret the voice of the real customer, which was identified as the blast furnace process itself. The engineers, with their expert knowledge, tended to express their demands in measurable technical properties of the product, representing a voice that cannot be ignored. One important benefit of the *House of Quality* was reported to be that it helps to get away from single-parameter thinking.

The use of QFD for ceramic bricks is briefly presented by Pertl (1992), but the presentation is more about other quality tools, though a simplified *House of Quality* is shown. A rather simplistic picture of the use of QFD, also in the ceramic industry, is given by Corma (1993).

The use of QFD at Bethlehem Steel Corporation is presented in an internal periodical publication by Mongeon (1996). The methodology was generally perceived to be a powerful development tool, but the company's experience suggested that more emphasis should have been placed on its use in the production process. Several successful projects were reported, and the conclusion reached was that the methodology can be of great assistance in the development and start-up of new production processes. The use of QFD in the

Brazilian Belgo-Mineira Steel Company was reported by de Oliveira *et al.* (1996). In this case, the Akao system was used, and the results from the QFD exercise were reported to be very rewarding. The project, related to the development of rods and bars for car suspension springs, included, most interestingly, not only the development of a *House of Quality* but also related separate raw material and process matrices and a complementary matrix for "Metallographic Structure". The use of QFD in Alumina Extrusion at Westinghouse Electric Corporation is presented by Calo and Kramer (1992):

> The efforts centered about a team including commodity customer, extrusion press supplier and extrusion die fabricator sub-supplier which represented a mix of speciality engineering expertise in performance/process modelling, extrusion press practices and die fabrication.... Relationships between customer requirements and product specifications identified the key specifications and led to the development of separate matrices for die design and press practices.

7.2 Experiences from the Food and Beverage Industries

> **In a Product Matrix, 21 new potentially measurable product properties were presented as WHATs in the matrix together with 13 selected important product properties from the HOWs in the House of Quality. It was identified that the product properties and related test methods that are necessary or can benefit product development, and the understanding of how a better product could be designed, are not always the same as the product metrics that are used for good process and quality control.**
>
> **(Lager and Kjell, 2007)**

Some papers about QFD in the Food industry (Duxbury, 1991; Hofmeister, 1991; Charteris, 1993) were published at the beginning of the 1990s. The first paper simply points out the possible advantages that could be obtained with QFD in a fairly superficial manner. The second, a book chapter by Hofmeister, presents only the ASI methodology, including some simplified examples. The paper by Charteris presents a vast number of complementary methodologies and statistical tools and mentions very briefly that the author participated in a QFD project on a dairy product with favourable results.

Both of the latter two articles take up the issue of deploying QFD on the associated package but do not bring any original methodological ideas forward. In two papers, (Bech *et al.* 1997) present the use of QFD on two food products, farmed smoked eel and frozen green peas. These papers present a somewhat unorthodox use of QFD, and one could question whether their *House of Quality* should really be called QFD, as it features unconventional symbols and negative relations in the relationship matrix. The use of complementary methodologies like experimental design and sophisticated statistical analysis nevertheless make interesting reading. Finally, two papers present the use of QFD in the chocolate industry and in the meat industry (Viaene and Januszewska, 1999; Guedes *et al.*, 1999).

7.3 Experiences from the Forest Industries

> **This paper is my testimonial to the power of QFD. Being the initial QFD "driver" for this project has been one of the most rewarding experiences of my career. Will I use QFD again? YES! I've already started another project. What advice would I give to someone who is thinking about using QFD; JUST DO IT.**
>
> **(Stitt and York, 1993)**

There are four articles dating from the early 1990s dealing with the use of QFD in Forest Industry applications. One article describes the introduction and use of QFD at Weyerhaeuser Company on oriented strand board (Hanson, 1993), and the other three cover various applications in the Kimberly-Clark Corporation (Scheurell, 1992, 1993; Stitt and York, 1993). The first paper by Hanson makes for very rewarding reading, as it focuses less on the matrices as such than the use of the methodology in three successful applications. As such, it stresses the need for proper training in QFD, the identification of different customer groups, the need to focus on the end user, the importance of distinguishing demands on a product from demands on a service, and the development of more customer-oriented test methods. Their intention was to progress further with deeper levels of matrices.

The two papers by Scheurell present the use of QFD in a start-up of a new product, using new technology and a new process to build a new mill with new equipment and train new operators (Scheurell, 1992). The first paper captures the team-building, objectives and industrial reality to such an extent that it could well be used in training. The team's feeling that the *House of Quality* was not being fully utilized encouraged them to develop a number of phase 2/3 matrices and complementary No. 4 matrices (the downstream matrices). They arrived at the philosophical conclusion that the core of QFD is not the matrices but rather what one does with them.

The follow-up paper (Scheurell, 1993) presented the results from the start-up of a new mill. One important finding is that if one includes start-up problems in the definition of the product development cycle, using the deeper levels of matrices in QFD, there can be a substantial improvement in time-to-market. Deploying the deeper levels of matrices makes QFD meet Statistical Process Control (SPC) in a natural way that was considered meaningful to operators in answering the question of what are the important product specifications to control. Scheurell presents very successful project results, including the performance of the product on the market.

The paper by Stitt and York (1993) describes an experience of using QFD to develop an improved, customer-preferred product using existing processes in an established facility. In spite of previous marginally successful QFD projects at the mill, the paper presents the successful use of QFD by further developing the phase 2/3 and 4 matrices (the downstream matrices). The importance of the process control part in matrix 4 is stressed, as is the usefulness of QFD in bridging organizational and plant interfaces. Stitt and York conclude that

QFD can be used in many applications, such as new products and processes, new products with existing processes, product and process improvements, and cost reduction programmes.

7.4 Experiences from the Chemical and Petrochemical Industries

> **The raw material by itself does not have full function; it is only an intermediate product in a process that will eventually allow the material to exhibit its function.**
>
> **(Oyaizu, 1990)**

In the book by Akao (1990), the usability of QFD in process industry applications is given one chapter, and the specific problems are well covered and illustrated by two industrial cases. The first case presented by Oyaizu (1990) concerns the production of synthetic rubber at Nippon Zeon Company. In the introduction, Oyaizu recognizes the difference between the process industries and other manufacturing industries as follows:

> In machine and assembly industries a demanded quality for a total system can be deployed into successively smaller subsystems. … In process industries, however, a mechanism that is necessary to realize some demanded quality cannot function independently of other necessary mechanisms and functions.

This underscores the importance of the development and use of the correlation matrix (the roof) of the *House of Quality*. The need to develop a complementary matrix to relate quality characteristics further to underlying structural properties is recognized, and the corresponding matrix is called "mechanism deployment" — an interesting notion which will be discussed further in Part IV in the development of the *Product Matrix*. The second case presented by Kubota (1990) refers to the use of QFD at Nippon Carbon. The use of QFD is highlighted as part of the overall quality system, but it is recognized that, in long production chains, there are intermediate products whose properties must be related to process parameters for a full process understanding.

In a later paper by Oyaizu (1994), the problem of capturing the latent needs of the customers for new materials and developing them into manifested needs is discussed. This theme is further elaborated in a later publication (Ootaki *et al.*, 1996). McLaughlin and Stratman (1997) propose using the S-curve concept together with QFD. The QFD methodology is presented and exemplified with a case from the Chemical industry. The use of QFD, not only to translate customer demands into product attributes but also to relate those attributes to process capabilities, is strongly recommended. The progression of data into deeper levels with QFD matrices is recommended in combination with the analysis of technology using the S-curve methodology.

Chapter 8

An Analysis of Necessary QFD System Capabilities for Enhanced Process-Industrial Product Innovation — Perspectives for the Future

It would not be unreasonable to speculate that the sparse reporting of the use of QFD in the process industries, and more specifically its use beyond the initial House of Quality, could be attributed not only to companies' lack of interest in or awareness of the methodology as such but possibly also to a lack of proper methodological guidance and the availability of a user-friendly QFD system adapted for use in the process industries.

In the management of R&D, methodologies and tools to facilitate and improve the performance of various R&D activities come and go. Sometimes methodologies disappear because they were not good enough to withstand the tough industrial environment and its high demands on industrial usability. Sometimes methodologies are introduced in companies with good results but do receive further attention or a "company champion" to keep them alive when new promising tools emerge and are promoted by company R&D management. However, Quality Function Deployment is now, after about 30 years of industrial use, a well proven and mature development methodology.

Not only is QFD a methodology that is frequently used in industrial development processes (Cristiano *et al.*, 2000), but also new development concepts like Design for Six Sigma rest on the foundation of QFD (Tennant, 2002). It is of interest not only to develop new methodologies and best practices but also to improve and further refine existing methodologies. This type of improvement must start with an assessment of the performance of the existing methodology, as well as an assessment of demands for improvements and adaptation of methodologies to future industrial needs.

The QFD methodology was born and developed in Japan primarily as a supporting tool for assembly based industries where products are manufactured and usually contain a very large number of components. In those industries, the most significant difficulty in using the QFD methodology is the large number of hierarchic levels of components and subcomponents that constitute the overall product (e.g., from overall *Customer Requirements* on the car to individual requirements on components in its suspension).

However, in the large cluster of industries we usually denominate the process industries, products are not created through an assembly of components, since the produced products are more or less homogenous and are further intimately related to the production process. For this reason, the commonly used and previously presented QFD systems for other manufacturing industries are not well-adapted for use in the process industries. This state of affairs is not uncommon, and most of today's available best practices, tools and methodologies are not designed for process-industrial use in spite of the importance of this very large group of manufacturing industries. Novitsky (1983) expressed this situation as: "Process Industries — where are you?"

The necessary QFD system capabilities for enhanced process-industrial product innovation will be analyzed and brought forward in this chapter. This review will also offer guidance for a further understanding and use of the mpQFD methodology as specifically adapted to process-industrial use, which will be comprehensively presented in the following Part IV. Even if the process-industrial use of the QFD methodology is not entirely forgotten in textbooks, scientific articles and other publications, the deployment of the QFD methodology in the process industries is grossly neglected. This is unfortunate, since the production of homogeneous products in the process industries simplifies the use of the QFD methodology compared with its use in assembly based industries. However, the specially adapted QFD system for process-industrial use, the mpQFD system, has still not yet received proper attention in a textbook and guide for industry implementation and use. This book is intended to serve this purpose.

8.1 A QFD system that recognizes that products from the process industries are not assembled components

The system must be adapted to the types of products that are produced in the process industries, which are not assembled components but rather more or less homogeneous products.

The lack of system → component → part hierarchical levels (e.g., car → door → door handle) in the product itself makes the development of a *House of Quality* easy for the kind of homogeneous products produced in the process industries, and since the products are not assembled components, there is no need for a multi-level QFD system. The original Four Phases of Matrices system is thus not applicable, since after the product we do not have parts but instead a long production process chain, sometimes with raw materials from another supplier.

The lack of an assembled component structure of the products also relates differently to the common recommendation to progress only those characteristics that are new, important or difficult into the deeper levels of the QFD system. First of all, it is recommended in this book that it is essential to include all product requirements in the *House of Quality*, since one can argue that it is important to know both what is and what is not important when developing products. Focusing on the improvement of important product properties while possibly reducing or removing unimportant product properties in a cost-cutting approach is thus strongly recommended.

Second, one product property in a more or less homogeneous product can seldom be changed without affecting another product property (either positively or negatively). Consequently, in contrast to changing part properties in assembled products, as in the automotive industry, there is a need for a QFD system that can handle interactive multiple product characteristics well. Company development of *Correlation matrices* in the *House of Quality* are thus much more important in process-industrial product innovation compared to product innovation in other manufacturing industries.

8.2 A QFD system that addresses the fact that the products, production processes and raw materials are closely interrelated

Not only does the finished product make demands on raw materials, but also the production process takes the liberty of making its own demands on the raw materials going into it; the latter type of demands can be called "processability requirements".

The upstream & downstream matrices (components) beyond the *House of Quality* must be well-adapted to the production environment in the process industries. This includes production of intermediate products within the internal process chains, the influence of production process design parameters on product properties, and the inherent demands from the production process on incoming raw materials (processability requirements). Not only does the production process in the process industries have customer–supplier interfaces on a company and production plant level, but also the company internal supply chains are often a long consecutive chain of internally produced intermediate product properties. The QFD system must be able to handle this complexity, and the intermediate products are often measured by a number of sophisticated test methods (Kubota, 1990).

It should also be noted that it is not only the finished product that makes demands on the raw materials. The production process makes its own demands on the raw materials going into it, which can be called "processability requirements". However, since such demands only make sense in relation to raw materials, this gives an incentive to split production process parameters and raw material properties into two separate matrices. The demands on raw material can then be progressed in an infinite chain from the end user and back to the raw material, which is not possible with the process parameters. Splitting up the process and raw material parameters will also make the individual matrices smaller and easier to handle (solving a problem with too large matrices reported in the literature).

The convincing results from the Forest Industry with regard to using the process control matrices suggest that they must be included in the QFD system (Stitt and York, 1993). Because they do not constitute a system or an overall model for process control, they should be integrated into the process matrices, and not giving the misleading idea that you are developing a control model with the QFD tool. The needs presented above are not addressed in the Four Phases of Matrices system. The use of separate process and raw material matrices is not formally presented in the Matrix of Matrices system; nevertheless, industrial users of this system have utilized this possibility (de Oliveira *et al.*, 1996).

8.3 A QFD system with an ability to relate product functional design requirements to inherent product characteristics

Products from the process industries that appear low tech to an unexperienced onlooker often have a high material complexity, but their product functionalities caused by such inherent material structures and compositions are "not on display".

Although products from the process industries appear superficially homogeneous, there are different degrees of homogeneity in different types of products, as has been previously discussed. The understanding of inherent product properties' relationship to product specification and metrics is of utmost importance for the product developer, but often not for the customer. The complementary matrices developed using a modified Matrix of Matrices system give good information about products from the process industries that are structurally complex (Oyaizu, 1990). This is also evident from the papers from the Food industries, where sensory properties of interest to the customer are related to the substances that create them (which are of lesser interest to the customer's perception of taste). The use of such a complementary *Product Matrix* has not yet formally been introduced in any QFD system but is a desirable capability of a QFD system adapted to process-industrial use, which will be presented in Part IV.

8.4 A QFD system that can cope with interlinked chains of both external and internal customers and suppliers

> **Successful development of products and processes depends to a large extent on the understanding and experience of operating in the chain-like structures of companies in the process industries.**
>
> **(Lager and Blanco, 2010)**

The importance of supply-chain collaborations and considerations for improved innovation performance has thus been stressed in many studies (Cantista and Tylecote, 2008; Sahay, 2003; Soosay *et al.*, 2008). A QFD system must be able to cope with the long and often complex production chain of multiple customers and suppliers to companies, who sometimes are located in the middle of this chain with no direct contact with the end user (see Fig. 8.1).

One approach to this problem is to see each company as both a supplier and a customer in this chain. The user demands on the functionality of the supplier's product in the customer's production environment are translated in the supplier's *House of Quality* into measurable *Design Requirements.* This approach gives a development perspective for the supplier

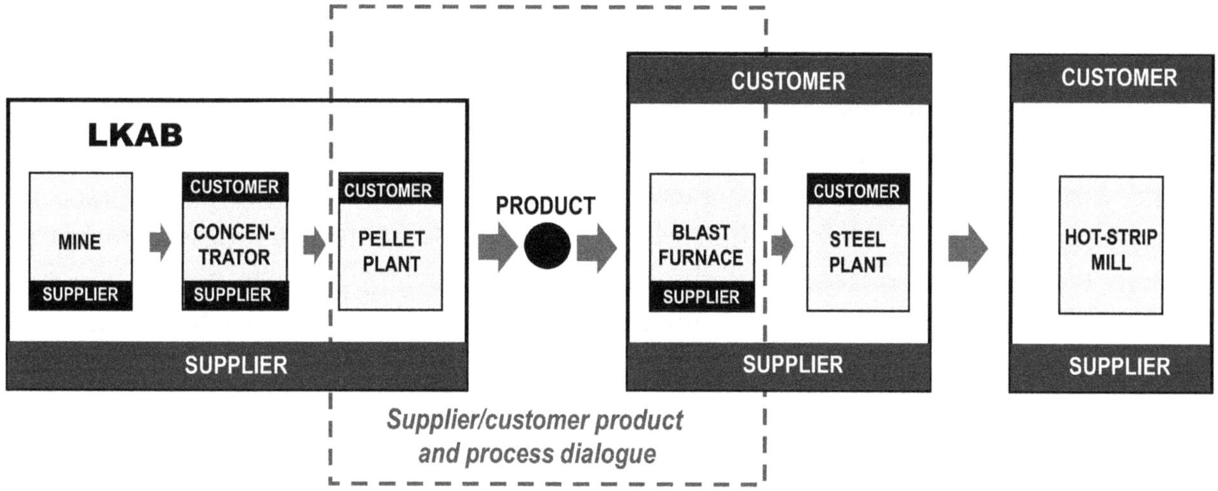

Figure 8.1 The process supply/value chain for a product supplied by the LKAB company (Tottie and Lager, 1995).

in his desire to dig deeper and understand the causes of the demanded product specifications from the customer and user of his product.

The alternative approach is to regard all companies, except the supplier to the end user, simply as actors in a long and complex production chain and to view the product specifications (hopefully reflecting the demands from the customer) as design requirements to be progressed backwards as far as possible in the chain of customers and suppliers. This approach requires a development of deeper underlying matrix levels for the chain of separate production units in the total production system.

There is strong evidence from the literature supporting the importance of going beyond the *House of Quality* using the upstream & downstream matrices. Previously available QFD systems for other manufacturing industries have not addressed this problem or the need for progressing customer requirements backwards into multiple chains of customers and suppliers. An adapted QFD system should be able to cope with both of the approaches presented above at the discretion of the user.

8.5 A QFD system that facilitates the development of product and process concepts

Concept selection is the Rubicon in the design process. ... The first stage of the design process identifies the requirements of the customers. From these customer requirements a list of product specifications are developed. The specifications are a list of functions that the product must provide and is given in a solution neutral form. Concept development is the next stage in the design process and involves establishing a conforming set of sub-systems.

(King and Sivvaloganathan, 1999)

A QFD system for the process industries must pay more attention to the intimate coupling between products and the production process and consequently between product and process development work processes. There is often a misconception among product developers in industry embarking on their first QFD project that when the *House of Quality* has been built, a new product has been created. Unfortunately, this is not the case, since this point marks only the beginning of the development work. In the pre-development phase, as shown in Fig. 6.3, the outcome of the pre-product development is not only a product concept; rather, in process-industrial innovation, a related process concept must also be developed.

The original Matrix of Matrices system does not make allowance for the development and selection of new product concepts; however, concept development is added in the slightly modified version by King. The problem with the absence of a "concept development phase" is even more pronounced in the Four Phases of Matrices system. The lack of "concept development" and a following "product development phase" in the matrix flight of stairs gives the false impression that one only has to press the button for the development of the four consecutive matrices in order to create a new product.

Chapter 9

Summing Up and Some Issues to Reflect Upon

You have now been acquainted with the unique process-industrial environment for product innovation and with the fact that a QFD system must be adapted in order to succeed under such process-industrial conditions. A general product innovation work process and a simplified process-industrial product innovation work process have been outlined. You have gained a fundamental understanding of the structural parts of the product innovation work process, which is most important for a proper use of the QFD methodology in general. The overall theme of this book is how the QFD methodology should support such a work process in order to make product innovation more efficient.

Experiences from the industrial use of the QFD methodology have been reviewed, and the general conclusion is that the methodology certainly is suitable for the process industries as well as for other manufacturing industries. Taking the process-industrial innovation characteristics as a point of departure, the good news is that the use of the QFD methodology is generally more advantageous in this cluster of industries than in other manufacturing industries.

The complexity and difficulty of the product functional analysis of assembled products consisting of intricate hierarchical system levels of multiple components, like primary functions and sub-functions, are striking. This is sometimes illustrated in the saying that "an airplane is 100,000 components flying in formation." Even on a high part level (e.g., the door of a car), each product contains a large number of components, and each component contains a number of sub-components. The application and use of the QFD methodology thus makes careful *preplanning* a vital activity in order to select and target the QFD projects. This is not necessary for non-assembled products.

Even if a minor part of an assembled product is selected for a QFD project (e.g., the system for operating the front window of a car door), this small system still contains a number of components. In the use of the QFD methodology, it is then necessary to combine the component matrices. The ease of application of QFD to non-assembled products is obvious, since the *House of Quality* captures the total *Product Design Requirements.*

Phase progression of *Product Design Requirements* into the production process using the classical model of the Four Phases of Matrices for assembled products is often a most complicated procedure, and experience shows that few companies even attempt it. However, the phase progression into the process-industrial production systems is, from a comparative perspective, a much less complicated operation. The use of the downstream matrices in the mpQFD system is thus "phase progression simplified."

The major difficulty in product innovation with non-assembled products is associated with changing individual design requirements that (for better or worse) influence and interact with others. This is an important aspect of process-industrial innovation that fortunately is well handled in the deployment of the QFD methodology using the roof of the *House of Quality*. For this reason, the development of the correlation roof in the *House of Quality* is of major importance for non-assembled products compared to assembled products, in which the individual components do not directly affect other components.

After reading this part of the book, it is now important that you try to relate this information to the perspective of your company's product innovation behaviour and philosophy. In order to facilitate this activity, please reflect on the following questions:

- If your company belongs to the family of process industries, please reflect on how common the presented "contextual conditions" for product innovation are among different sectors and for your company in particular. How do these characteristics influence successful product innovation behaviour?
- If your company belongs to other manufacturing industries, please reflect on how the presented "contextual conditions" for product innovation differ from the process industries and for your company in particular. How do these characteristics influence successful product innovation behaviour?
- If your company belongs to the family of process industries, please reflect on how common the presented "inherent conditions" for product innovation are among different sectors and for your company in particular. How does this influence successful product innovation behaviour?
- If your company belongs to other manufacturing industries, please reflect on how the presented "inherent conditions" for product innovation in the process industries differ from other manufacturing industries and for your company in particular. How does this influence successful product innovation behaviour?
- Is your company using a formal product development work process, or do you believe such formalization will stifle the creativity and productivity of the development teams?
- If your company uses a Stage Gate model for product development or a process of a similar kind, do you agree that such a process is mainly a model for decision making and is not really a comprehensive model of a product innovation work process?
- Reviewing the simplified three-stage product innovation work process presented in Fig. 6.1, how applicable is it to your company?

- Do you agree on the revised structural model presented in Fig. 6.4 and the proposed integration between product innovation and the development of associated related process technology?
- If your company belongs to the cluster of process industries, when you review process-industrial experiences from the use of the QFD methodology, do you know how the QFD methodology is practiced in your industrial sector?
- If your company belongs to the cluster of other manufacturing industries, when you review process-industrial experiences from the use of the QFD methodology, do you know how the QFD methodology has been practiced among your prime competitors?
- Were you previously aware of the very large differences between the development of non-assembled and assembled products, and do you agree that development methodologies must be adapted to such different industrial environments for innovation?
- Considering advanced product functionalities, how well do you understand how such functionalities are created and how the inherent product properties (structures) relate to such product functionalities?
- Is it important for your company's product innovation to relate well to the total internal and external supply/value chain? If so, what is your company's position in this value chain? How many internal production interfaces must be considered in your company value chain?
- In your company's product innovation, do you develop interactive product concepts in the pre-development part of the work process? If you belong to the family of process industries, do you additionally develop process concepts that are well connected to the product concepts?

Part III

The "Voice of the Customer" and "House of Quality" — The Common Denominators and Matrices in All QFD Systems for All Manufacturing Industries

The life of an individual is in many respect like a child's dissected map. If I could live a hundred years, keeping my intelligence to the last, I feel as if I could put the pieces together until they made a properly connected whole. As it is, I, like all others, find a certain number of connected fragments, and a large number of disjointed pieces, which I might in time place in their natural connection. Many of these pieces seem fragmentary, but would in time show themselves as essential parts of the whole. What strikes me very forcibly is the arbitrary and as it were accidental way in which the lines of junction appear to run irregularly among the fragments. With every decade I find some new pieces coming into place. Blanks which have been left in former years find their compliment among the undisturbed fragments.

(Oliver Wendell Holmes)

In search of a holistic perspective in product innovation

In today's product innovation, if it is carried out properly, the problem is often not a lack of relevant information but rather an extreme overload of information — an inability to "see the forest for the trees." For this reason, it is generally difficult to see "the whole" and how individual pieces of information link to each other. Moreover, information that yesterday appeared to be irrelevant could be of utmost importance tomorrow. New pieces of information tend to put previous product information in new perspectives; thus, the above quote is a good picture of the day-to-day complex working environment for professional product innovators. It additionally bring forward the argument that there is a need for supporting innovation methodologies, and in particular for the matrix approach in the use of the QFD methodology as an instrument for organizational learning. The QFD matrices are thus outstanding instruments to grasp "the whole" in product innovation, and the gradual upgrade of the matrices over time will capture new patterns emerging from the new amalgamated information. The QFD matrices thus function as containers of synthesized tacit and explicit product and process knowledge.

In other manufacturing industries, the first step in the use of the QFD methodology is usually a "preplanning" phase. Since assembled products normally have a number of hierarchic structural levels of a large number of components and sub-components, it is vitally important to identify early on what part of the product to focus on in the use of the QFD methodology. This is generally carried out in a component *Preplanning Matrix*, in which customer requirements are related to a single component's selected number of sub-components. In the process industries, which generally produce non-assembled products of a more or less homogenous nature, such a preplanning phase is unnecessary, since the use of the QFD methodology will directly address the total product ("the whole"). This is a truly important advantage in the use of the QFD methodology in the process industries that has been grossly neglected in previous presentations of its industrial use.

Moreover, in other manufacturing industries, and in order to keep matrix sizes manageable, not only is it necessary to select a single component, but also it is often recommended to focus only on what is "new, important and difficult." However, it is argued in this book that it is essential to know not only which *Customer Requirements* are of customer importance but also what is "not so important" to the customer. For example, in a "frugal innovation" approach, this information is crucial to consider in the development of simplified product concepts for customers with low purchasing power. However, in the use of the QFD methodology in the development of non-assembled products (in which all Product Attributes are more or less interrelated), one must abandon the "new, important and difficult" mantra and take a holistic perspective on total product attributes — that is, the whole.

An operational QFD definition selected for this book

There is a saying in France that there are as many recipes for the fish soup Bouillabaisse as there are restaurant chefs; the same is true for definitions of methodologies like QFD. The overall criteria for good definitions, however, is that they should assist argumentation and improve communication. This book uses what is called an "operational definition," which acts as a guide for how to use the methodology. The following definition of QFD includes the three areas:

"Voice of the Customer"

Quality Function Deployment (QFD) is a methodology and a work process in which stakeholders' (customers'/consumers') requirements for a product are initially collected, refined and structured. These are afterward evaluated (ranked or rated) by the stakeholders and benchmarked against competitors.

"House of Quality"

Next, the requirements are related in a matrix to measurable properties of the product. The stakeholders' importance ratings are then translated into this new technical dimension, in which a technical benchmarking is performed and new provisional target figures for product development are set.

"Phase progression"

The measurable product properties and their importance ratings and target figures are then further progressed as requirements on the production system and set further backwards into desired raw-material requirements.

One could alternatively describe the QFD methodology in the perspective of expected output as:

A cross-functional approach in product innovation and renovation, and an investment in improved understanding and knowledge creation about customers and competitors.

The overall work process for the total Multiple Progression QFD system

In Fig. P3.1, the overall work process for the first two QFD matrices in the use of the QFD methodology are illustrated. In this part of the book, the work processes for the *Voice of the Customer* and the related *House of Quality* will be detailed. While the *Voice of the Customer* is a market-oriented activity, the construction of a *House of Quality* is a more development-oriented activity. The development of the other matrices will be presented in the following Part IV of the book.

Using the Voice of the Customer for customer interaction and customer understanding

In Chapter 10, you will be introduced to the *Voice of the Customer*, which is the most important building block in the further construction of a solid and reliable *House of Quality*. If you are reading this book, it is probably evident to you that if one wants to develop a new or improved product for a group of customers, one must understand the customers' underlying needs, wishes and dreams. However, this is easier said than done. For that reason, this

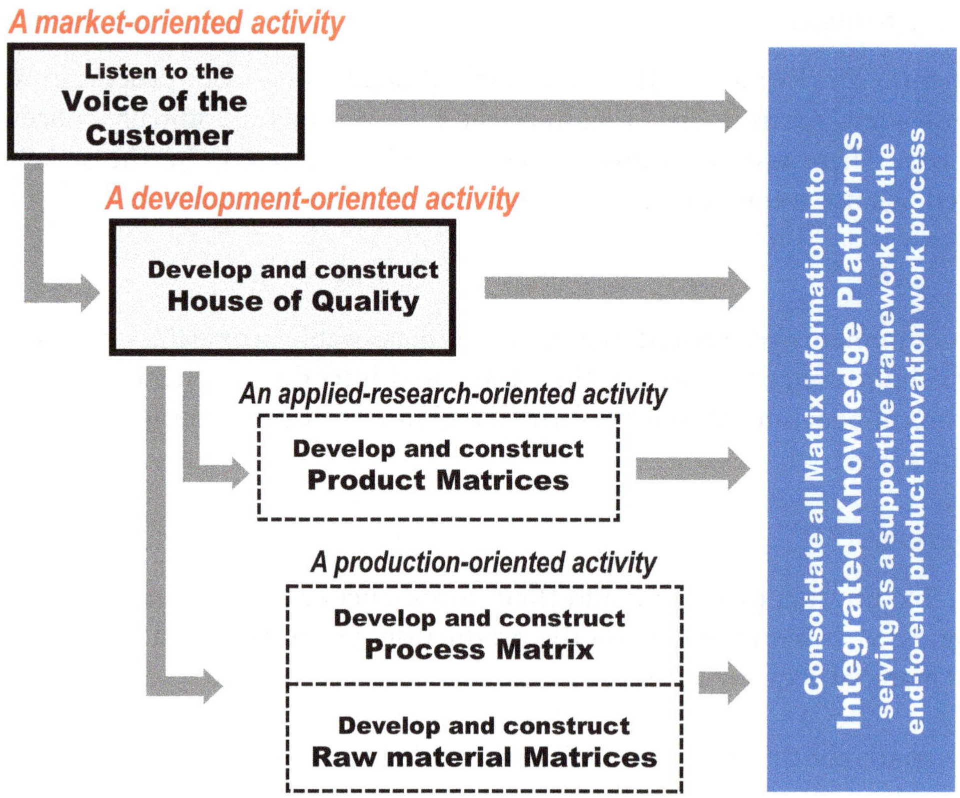

Figure P3.1 An overview of the total mpQFD work process focusing on the *Voice of the Customer and the House of Quality.*

first part of the product development process is often treated by engineers as an unreliable, fuzzy kind of activity that preferably should be left to the company's Sales & Marketing organization. This is wrong!

You will learn that the first step on the road for an improved customer understanding is to identify all different kinds of customers (stakeholders) for the product. You will further learn that customers usually only mention some of their requirements for a product when asked, and you will be briefly introduced to an approach to handle this situation. After being collected, the Customer Requirements must be refined and structured well. In this chapter, you will learn how to do this while ensuring that the *Voice of the Customer* is not misinterpreted or distorted. Since all *Customer Requirements* are not of equal importance, you will learn in a Customer Importance Rating process that one must return to selected customers to collect such information, and you will briefly be introduced to some select tools. Afterwards, the importance of benchmarking all *Customer Requirements* is stressed when all of the information is put together in the *Voice of the Customer.*

Using the House of Quality to move from a "customer dimension" to a "product technical dimension"

In the next Chapter 11, using the *Voice of the Customer* as a point of departure, you will learn in detail how to build a *House of Quality* on a product. In the *House of Quality*, you will move from the customer dimension of the product to a technical dimension and the relationship matrix is the instrument for this transformation. You will discover that the company's previously defined product specifications usually have been resting on a lot of fuzzy ideas or thinking and that the QFD approach will provide "interactive product targets and specifications." In other words, the Target Figures will be linked and related to all pieces of information available in the *House of Quality*. In the *House of Quality*, the customer importance ratings of individual *Customer Requirements* will be translated, using the relationship matrix as an operative tool, into *Product Design Requirement Importance Ratings*. Moreover, you will be acquainted with the systematic Technical Benchmarking room, which is a most supportive instrument for competitive product understanding.

Matrix analysis and provisional target setting for a new or improved product — merging the total product information

Finally, you will learn in Chapter 12 how to do a Matrix Analysis, which is a review of the total information contained in a *House of Quality*. This often neglected part of the methodological use of QFD will take a holistic combinatory view of a huge amount of information in the real spirit of the introductory quotation. There are a number of objectives for the Matrix Analysis, but one of the most important of these is Target Setting. The Target Setting activity involves the use of all of the information in the *House of Quality* to set provisional target figures for all *Product Design Requirements* to be used as a screen in the further development of concepts for new or improved products.

Chapter 10

The Voice of the Customer — The Facilitating Instrument for an Improved "Market & Customer Understanding"

It is not unusual for old-timers (and sometimes rookies) in company R&D and market organizations to feel confident that they already are well aware of customer preferences and the importance of individual customer requirements on a product. However, an expression often used in the process industries is GIGO (garbage in, garbage out), which refers to the influence of inferior raw-materials on the production process and finished products: the quality of output is determined by the quality of the input. The same goes for the use of the QFD methodology in product innovation, underscoring the importance of starting with the *Voice of the Customer* in product innovation work processes of excellence.

Does one really need to listen to customers?

The founder of Ford Motor Company, Henry Ford (1863–1947), once stated that if he had asked people what kind of improved means of transportation they wanted, he was sure that they would have asked for "a faster horse." Expanding on this idea, one might wonder whether there is any use in asking a customer about future product requirement? The answer to this question is certainly yes! Product development includes not only new and innovative products but also the improvement and maintenance of products and product categories that have existed on the market for many years. Additionally, and most importantly, in the search for future customer requirements on a product, one should not inquire about specific product solutions but about customer and consumer needs.

One must therefore try to find the "root customer requirements" on a product. When the customer identifies a "product solution" or desired future "product specifications," one should ask *why* this product feature is important or *why* a certain product specification is desirable. If Henry Ford had been aware of the QFD methodology, he would have tried to find "root requirements" by asking the customer *why* he wanted a "faster horse."

The consumer would most certainly have answered he wanted to have a *faster* form of transportation. This was his "root requirement." The selected "product solution" for Henry Ford was the car, but it could have been an airplane or a motorbike. Similarly, a root requirement for an "electrical start" could be "easy to start," the root requirement for a "screw cork on a bottle" could be "easy to open" or "no cork damage" and the root requirement for "good paperboard foldability" could be "easy to convert in the production process."

Is the QFD methodology unfit for development of radically new products?

In the light of radical product innovation, you will find some product developers of the opinion that traditional market research, including customer surveys, focus groups and clinics, are worthless instruments, since customers usually have difficulties expressing their requirements on a new radical product idea. While it is true that one cannot assume that such an idea (e.g., cell phones) could have been identified in discussions with a customer, a root requirement for more "mobility in communications" could certainly have emerged. In a similar vein, the iPhone and the iPad represent improved customer interfaces that could have been identified if market research had been more observant and listened more carefully to the *Voice of the Customer*. In retrospect, the Apple organization obviously had a better understanding of customer needs than Ericsson or Nokia.

Furthermore, it is also sometimes argued that the QFD methodology is not applicable in the development of radically new products, since there is no *Voice of the Customer* available until a product has been put on the market. However, in reality, new products and product ideas are created in order to satisfy still unmet, albeit possibly unknown, customer needs; thus, in such cases, the product developers must "speak in a provisional internal voice," which is the platform for the new idea. It is likely that the company's internal product development team will benefit a lot from "impersonating external customers" in order to build a provisional *Voice of the Customer* on a radically new product concept. Moreover, such a provisional *Voice of the Customer* can be used in clinics and in demonstrations of product concept in order to strengthen the voice into a more "fact-based" *Voice of the Customer*; a voice which can be further improved and validated during test-market activities. The author of this book has sometimes developed such a "provisional *Voice of the Customer*" together with SMEs (lacking financial means for extensive market research activities), and the experience is generally rewarding. ***Even radically new products will finally meet demanding customers.***

Introducing the guiding work process for the development of the Voice of the Customer

A substantial benefit of the QFD methodology is the development of an internal company common language for a coherent product design process, but it is also, and most

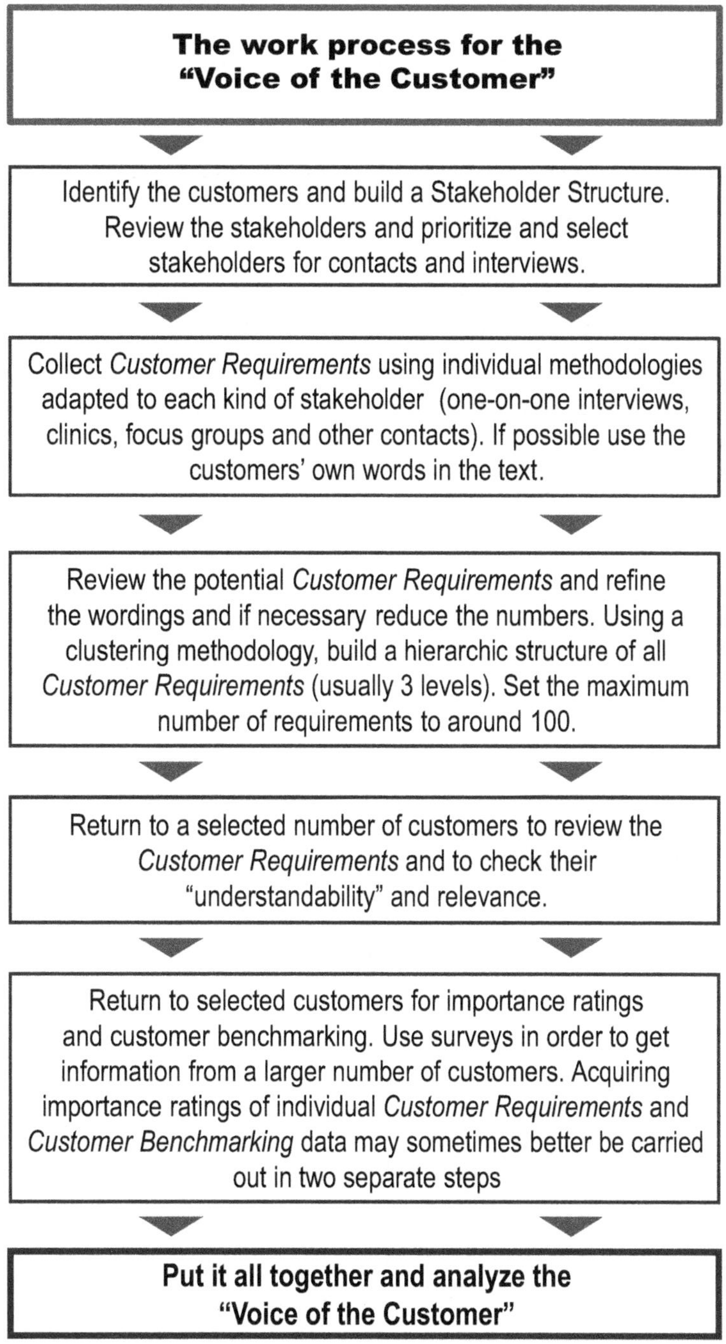

Figure 10.1 The work process for the *Voice of the Customer*.

importantly, fostering the organization's sensitivity to customer demands in general. The overall work process for listening to the *Voice of the Customer* is presented in Fig. 10.1. Reading through this process now will help you to grasp the individual activities of the total work process, while in the following sections you will be familiarized with each of these activities in greater detail. However, it must be observed that the work process and the techniques used to establish *Customer Requirements* naturally must be adapted to the structure of the individual company's clientele and the nature of its products.

10.1 Identify the customers for the product and build a Stakeholder Structure

After waiting a long time in line at the desk at a garage, suddenly one customer blew his fuse and said, "I am a very busy man, and I have now been waiting for half an hour to talk to somebody. I am totally fed up, and now I am going to leave, and I can assure you that I will never set my foot in this place again". The general foreman answered immediately, saying, "This is perfectly OK with me, since in this place we only want to have satisfied customers".

(The fable of the customer-oriented general manager of a garage)

This fable serves as a reminder that each individual in a company generally interprets the construct of being "customer-oriented" or listening to the *Voice of the Customer* differently, which sometimes is also a dilemma within professional product development teams and in Sales & Marketing organizations.

Identify customer categories and market segments

Before the process of collecting *Customer Requirement* and related information begins, it is necessary to ask the following question: WHO are the customers for the product? A traditional segmentation of the market for the product (by geographical location, size, etc.) is in most cases a good point of departure, which usually should be supplemented by additional customer structural dimensions like life-style and other demographic data. Consumers can also be categorized as faithful to the "brand," "function" or "price." From a product innovation strategic perspective, it could also be of interest to categorize customers into past, current and future customers.

Customers come and go. One group of customers that is of particular interest in product innovation is that of "lost customers." It is thus generally most rewarding to identify, locate and listen to "dissatisfied customers" in order to discover why a particular product has failed to attract them or has even disappointed them.

Customers usually have different abilities to articulate present and future product requirements. However, a small group of customers are truly aware of on-going trends. Such customers are often denominated "barometric customers" or "lead users." These customers are of particular interest in product innovation, as will be discussed further in Chapter 25. In addition, a company not only wants "satisfied customers" but also is eager to favour the "loyal customers" who will hopefully continue to return in the future. Additionally, and most importantly, the company must survive on "profitable customers."

Visualize the different product customer categories by building a Stakeholder Structure

Before you embark on the never-ending journey of listening to the *Voice of the Customer* for a company product, it is first advisable to understand whom it is important to listen to. It is thus initially imperative to recognize that a "product buyer" is not necessarily also the "product user." This is often the fact for consumer products, and especially for "fast moving consumer goods," when the shopper is not the single product user. Similarly, for products intended for B2B customers, the purchasing organization and supporting technical personnel are not the real end-users.

In a review of customers for a new or improved product, one usually discovers that there are many different kinds of customers, who may pose different kinds of product requirements. Because both "customers" and "consumers" as well as many other actors may be involved, in this book they are all collectively called "stakeholders." Depending on the complexity of the stakeholder structure, it is sometimes advisable to illustrate this state of affairs in a formal map of stakeholders, also visualizing their interactions and product relations (Gustafsson, 1986). It is the author's experience that such a map usually triggers interesting discussions and opens up new perspectives in customer-centric product innovation. In the creation of such Stakeholder Structures, one can usually recognize three different kinds of "end-users":

- Product consumers and customers.
- An equipment the product will be used in.
- A production process where the product is a raw material.

In the following sections, these three situations will be further discussed.

A Stakeholder Structure for the Perfect Lunchtime Beer — Consumer

A stakeholder structure for the illustrative case is presented in Fig. 10.2. Such a stakeholder structure not only serves as a guide for selecting respondents for interviews but also should be used to prioritize and plan interactions with different stakeholders. In a review of the Stakeholder Structure for a beer, one can notice that it is not only private customers that put demands on the product manufactured by the beer company but also a number of other actors. Naturally, the customer is the most important stakeholder, and consequently it is vitally important to inquire about their preferences. However, it is essential for the beer producer to understand all different kinds of customer requirements within the total supply chain.

Wholesale organizations and resellers are powerful actors, and it is imperative that their requirements are satisfied. For example, in Sweden, the state-owned monopoly and reseller for hard liquor, "Systembolaget," sometimes asks beer manufacturers to deliver styles of

Figure 10.2 A stakeholder structure for the perfect lunchtime beer.

beer products (including their defined product specifications) they would like to offer on their shelves. International and national standards must be adhered to, and the voices of beer professionals and food journalists must always be loud and clear to beer manufacturers.

A Stakeholder Structure for office paper –Equipment as end-user

This specific case has been selected to illustrate that there are many different Stakeholders for a product and that the "end-user" is not always a person but in some cases may be a machine. Referring to Fig. 10.3, the better one understands the Printer's requirements for copy paper, the better one understands the *Voice of the Customer* for copy paper. Unfortunately, Printers do not yet speak, so we have to ask other groups of Stakeholders to assist in the translation of printer demands on a copy paper.

All of the stakeholders in Fig. 10.3 may contribute with important requirements for copy paper, but in search of future customer requirements for copy paper it is important to recognise the printer manufacturers' unique position. In general, it is very difficult to get stakeholders to express product requirements in a long-term perspective. For this reason, the "Voice of the Printer Developers" at the Printer manufacturing company is generally of paramount interest to listen to and interact with.

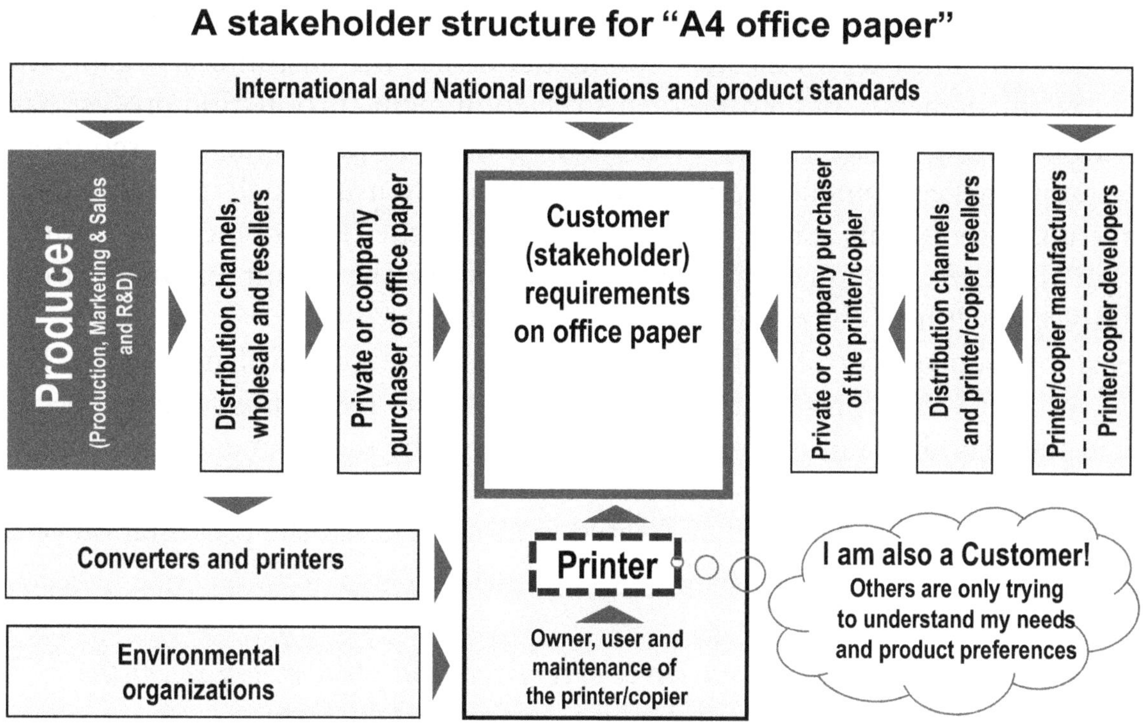

Figure 10.3 The "speaking printer" and a Stakeholder Structure for office paper.

10.1.1 *Malt production for the brewery — Production process as end-user*

Industrial customers are "professionals" that usually prefer to express their requirements on a product as "product specifications." In such cases, it is necessary to try to find out their "root requirement" — that is, to ask customers WHY they want such product specifications. Moreover, if an industrial B2B product is not be used as a finished product or a component in an assembled product, but as a raw material or ingredient in a customer's production process, one must take quite a different stance in customer interactions.

Compared with companies producing consumer products, companies supplying B2B customers generally have a more limited number of customers for co-development, but the process of collecting data for the *Voice of the Customer* is nevertheless crucial for the development of new or improved products. However, in this case it is advisable to select customers that have an interest in both the development of the supplied product and their own production processes, as well as the willingness to spend time in co-development. Ideally, such users are those that are also willing to be involved in later stages of product development and who hopefully will end up as customers for the new product.

If the *Customer Requirements* for a supplied product are the requirements of a "production process," one must try to locate respondents within the customer's organization who can articulate the demands and requirements the equipment/production process has for the product. In inquiring about future "production process requirements" or requirements of a novel equipment, equipment suppliers to the company using such a production process are interesting respondents. Since they are involved in the development of future process technology, they are consequently knowledgeable about the requirements future production technology will have for raw materials or ingredients.

For example, a producer of malt for breweries, as introduced in Part I of this book, not only must interact with representatives from the breweries regarding product specifications but also, in their product innovation, must understand the requirements the "brewing process" places on malt as a prime raw material. In the following Part IV of the book, interaction with raw material suppliers will be treated in more detail in the presentation of the Raw material QFD Matrix.

10.1.2 *Review the Stakeholder Structure, prioritize and plan whom to contact, and select proper company representatives as contact persons*

Stakeholder Structures show that product developers cannot neglect any of the presented Stakeholders and their Stakeholder requirements. However, in the review of multiple stakeholders and in the design of a proper strategy for the collection of *Customer Requirements*, it is recommended to focus on the end-users. In order to make the collection of *Customer Requirements* a manageable exercise, different techniques must be employed for different kinds of stakeholders, which will be discussed in the following sub-section.

10.2 Collect customer requirements from selected stakeholders

Consumers often present "top of mind" requirements; they often express themselves in an unstructured manner and state only part of their requirements, but they usually state true "root requirements" and not "design solutions." On the other hand, the professional industrial customers generally want to express their requirements as "product specifications," preferably measured by internationally accepted test methods. For such specifications, one must further inquire about the "root" requirements.

10.2.1 *Who should listen to the Voice of the Customer?*

The *Voice of the Customer* is, in many respects, a kind of product information that is usually "owned" by a company's Sales & Marketing function. However, unfortunately, Sales & Marketing is often less concerned about the exploration and identification of *Customer Requirements* for new or improved products than about the marketing and sales of the company's existing products. In such cases, Sales & Marketing people neglect to hear and further transmit customer requirements or sometimes even customer complaints to the R&D organization. However, product developers are not really in need of an interpreter of customer WANTs, desires and dreams from marketers. It is thus recommended that they directly, but preferably together with representatives from the marketing organization, listen to what the customers are saying in their own words, and then use this information in the identification and labelling of individual *Customer Requirements* in interaction with Marketing & Sales.

In the use of the QFD methodology and in the development of the *Voice of the Customer*, it must be emphasized that "root" *Customer Requirements* are sought rather than specifications for a new or improved product, since the translation of *Customer Requirements* into *Product Design Requirements* will be addressed later in the *House of Quality*. Representatives from Sales & Marketing usually feel it is liberating not to have to do translation themselves from customer WANTs into product HOWs. However, the development and execution of well-structured marketing surveys to industrial customers, and the related extended dialogues between customer representatives and R&D representatives, will often improve communication and develop into an interesting "relational building instrument" for later use in the marketing and sales of a new or improved product.

In the collection of *Customer Requirements*, one should not be surprised or disappointed to learn that some of them already are well-known to the company. If one listen carefully to the customers, one is likely to discover some really new and exciting requirements; moreover, and more importantly, the product development team will get a holistic perspective on all individual customer requirements, their validity and also their individual importance ratings.

10.2.2 *Notice the difference between "spoken" and "unspoken" Customer Requirements in the spirit of the "Kano model"*

The extended Kano model is used today as one of many market research methodologies, and there are numerous publications supporting its usefulness in product innovation (e.g., (Afshan and Sindhuja, 2013; Matzler and Hinterhuber, 1998; Witell *et al.*, 2013). In customer interviews, the customers usually only articulate the requirements that Professor Kano has termed *Performance Quality Requirements* but not their *Basic Quality Requirements* and certainly not their *Excitement Quality Requirements*, sometimes called "WOWs." The Kano model is presented in Fig. 10.4 using the two dimensions *Customer satisfaction* and *Degree of requirement satisfaction*.

On the *Y*-axis, Customer satisfaction ranges from "Customer very dissatisfied" to "Customer very satisfied," while the *X*-axis ranges from "No customer requirements satisfied" to "All customer requirements satisfied." In in a normal interview situation, customers usually focus and only articulate *Performance quality*, and the relationship can generally be simplified as a straight line. The more Customer Requirements are satisfied in the product, the more satisfied the customer will be.

The customer does not usually even consider the requirements of a *Basic quality* nature, a fact which is important for product developers to remember since such requirements often must be generated internally in the R&D department. If *Basic quality* requirements are not satisfied, the customer will generally be truly dissatisfied! However, if all basic requirements are fulfilled, the customer will only be indifferent. For example, the Mercedes A-Klasse failed to satisfy a Basic requirement in their earliest model because if the car was turned very abruptly, it could flip upside down. Customers do not imagine that this could be possible, and consequently this is an example of an unspoken Basic requirement. Likewise, car windows must not leak, and so on.

Excitement quality Customer Requirements are normally requirements that are totally new and of a kind customers do not (yet) expect to find in products. These requirements are naturally of a kind that product developers of excellence deliver after doing creative work and gaining a deeper understanding of which "root requirements" a customer is interested in and is willing to pay for.

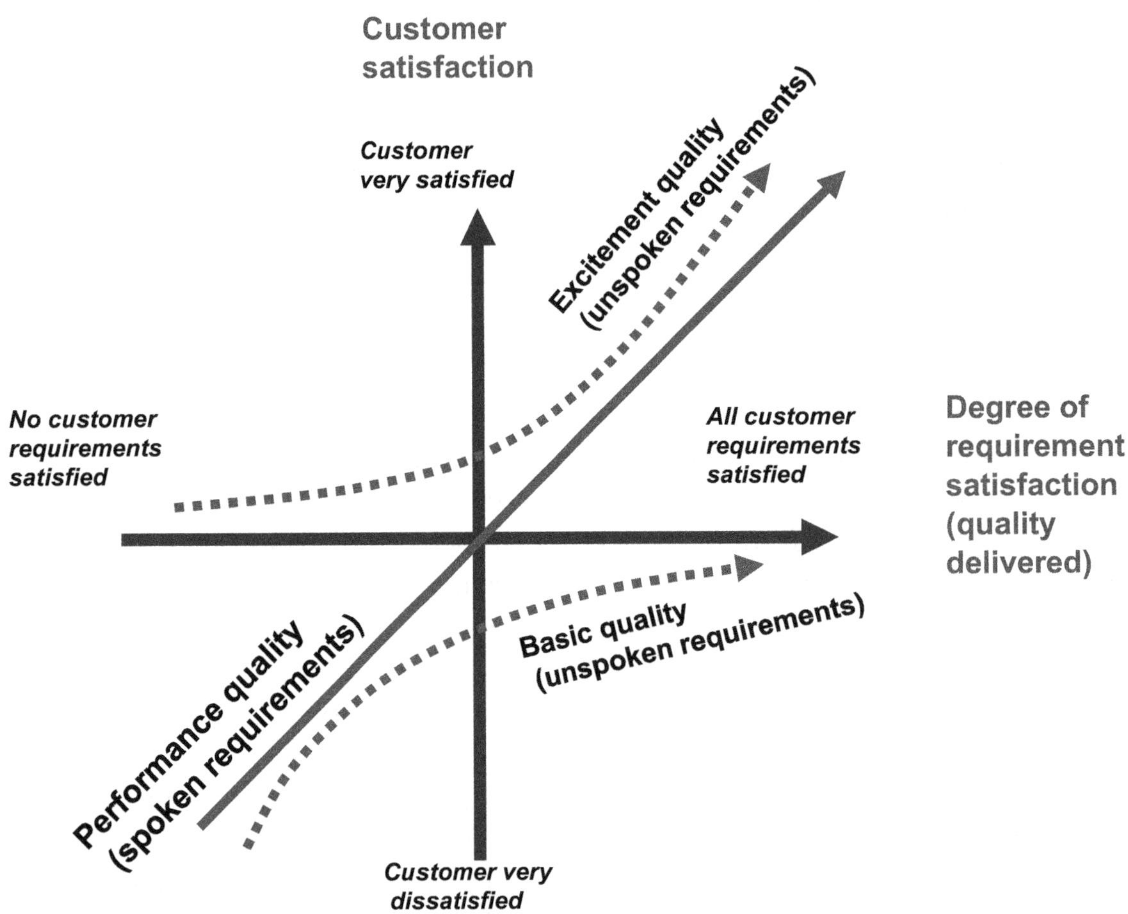

Figure 10.4 A simplified presentation of the Kano model.

10.2.3 *Review and select proper interview settings and methodologies for individual stakeholders*

While classical market research methodologies are generally of a quantitative nature, the collection of the *Voice of the Customer* must focus more on qualitative methodologies. There are a number of different methodologies and techniques to deploy in the development of the *Voice of the Customer*, each with pros and cons. It is strongly advised that the QFD team initially should establish a good relation with the company's Sales & Marketing department and Customer & Technical Support department. Following the previously presented overall work process for the *Voice of the Customer*, the Stakeholder Structure will provide a good platform for the R&D organization, together with Sales & Marketing, to outline and organize how to listen to the *Voice of the Customer*.

How to proceed must naturally be contingent on product characteristics, available company resources and previous customer experiences in this area. There may or may not be ample resources and knowledge to carry out this kind of work within a company. Using external expert marketing research organizations often proves to be a good alternative and a good investment in "out-of-the-box" product innovation. However, the author's general experience is that a well-structured and refined *Voice of the Customer* is an excellent and preferred point of departure for further interactions and use of external market research organizations.

If the company has previously used the QFD methodology on a select product and if the geographical distance from potential customers is very large, this may sometimes incentivize the development team to apply a more "lean" approach in re-drafting an existing B2B *Voice of the Customer*. Using previous knowledge of *Customer Requirements* as a point of departure, such demands can be further reviewed together with in-house company marketers and R&D expertise in an iterative dialogue, gradually refining them into a list of potential *Customer Requirements* in advance of the first meetings with B2B customers.

Afterwards, in interactive discussions with customers, the draft list of *Customer Requirements* can be presented and discussed in an open dialogue, during which each *Customer Requirement* must be reviewed separately. First, each requirement must be discussed from the perspective of how easy they are to understand. Second, they should be challenged; are they still valid *Customer Requirements*? The importance rating of each requirement must be left for a follow-up meeting, when a refined list of *Customer Requirements* can be presented and individual Customer Requirements can be importance rated and benchmarked.

There are many alternative ways of developing and refining a list of *Customer Requirements*, and each company must seek and find its own best way for different kinds of products and customers. There are many alternative approaches for the collection of *Customer Requirements*. In the following a few commonly used techniques are highlighted.

One-on-one interviews

One-on-one interviews (also called face-to-face interviews) represent the preferred technique to use in QFD. The advantage of this approach is that it gives the interviewer full control of the interview situation and personal contact with the respondent during the whole session. An early study at MIT proved the usefulness and cost-effectiveness of this approach compared to focus groups in the identification of *Customer Requirements* (Griffin and Hauser, 1991); around 20–30 such interviews were sufficient to obtain over 90% of all customer needs. A primary review of "post-its"; see Fig. 10.5 used for the *Customer Requirements* together with the respondent will clarify what the *Customer Requirements* are intended to express. However, it is generally necessary afterwards to review all requirements and, in an iterative refinement process, formulate the individual requirements crisply and clearly. One guideline is that the *Customer Requirements* must be understandable for an "outsider" who has not been involved in recording them.

Group sessions in the spirit of one-on-one interviews

One-on-one interviews can be carried out with a single respondent or together with a group of customers (e.g., a few selected representatives from a company). In Fig. 10.5 the procedure is further explained.

Telephone interviews

As face-to-face interviews can be demanding on company resources, a supplementary technique for acquiring information from a larger set of potential product customers is telephone interviews. This kind of market research is best left to professional Market Research Organizations. If used properly, this can be an important supplementary source of information, especially for consumer products.

Focus groups and clinics

Focus groups or "clinics" are interesting instruments that sometimes are used for the collection of *Customer Requirements.* The term "Focus groups" emphasizes that a facilitator should try to focus the discussions on selected areas of interest in a product innovation project (Kreuger, 1998; Kreuger & Casey, 2009; Stewart *et al.*, 2007). If a skilled facilitator is available, the advantage is that some sort of group dynamic creative situation can be achieved which will stimulate "out-of-the-box" ideation. The disadvantage is that the sessions sometimes tend to be dominated by a few extrovert persons; thus, they should be regarded as supplementary to the previously discussed face-to-face technique.

A helpful hint

One-on-one style of group session

- Assemble the group of respondents for the interview and give each of them a block of Post-Its.
- Explain the purpose of finding "root requirements" for the product, and explain that they should list only one requirement on each Post-It.
- Explain that no communication between the respondents is allowed and that they will be given about half an hour to list the requirements.
- Ask them to use the first 20 minutes to list today's requirements and the last 10 minutes to list future requirements.
- Since after 30 minutes many respondents consider themselves to be ready, explain that they should try to brainstorm for 10 minutes more, since many important requirements tend to appear at the end of the exercise.
- Go through the Post-Its individually with each respondent to agree on the meaning and check readability.
- For presented "design requirements", ask the respondent WHY he or she wants a given solution.

Figure 10.5 Interview with a group of customers.

Ethnographic investigations

In order to develop a more intimate understanding of a customer's use of a product, it is advantageous to study customers' behaviour in their use of a product in its natural environment — an ethnographic product innovation approach sometimes described as "walking in the customer's shoes." In Japanese, this approach is called "GEMBA," and it is represented by Kanji characters that integrate the symbols for "a king overlooking the land" and "watching swine-herds working in the sun":

現場

> *If you lived the lives of the customers, perhaps you would behave as they do and understand their needs. ... Some organizations require design team members to work in the customer's organization for six months.*
>
> (Terninko, 1997)

This methodology focuses on a thorough understanding of the customer and the customer's behaviour together with the product in his specific context. However, the collection of sound demographic data on the users is always a necessity. To follow this interesting approach in a product innovation project, product developers need to live close to the customer or sometimes work together with the customer. A product developer can spend several months with the user in order to experience how the customer utilizes the product — a kind of study that is also called "going native." ***Product developers of products for industrial customers may work together with the customers in their production plants in order to experience how the product is used in the customer's production process.***

10.3 Build a hierarchic structure of all *Customer Requirements*

It is sometimes claimed in a Total Quality perspective that a customer's opinion is always "right" and important. Hopefully the customer is expressing his true feelings, but that is not to say that he always is right! Nevertheless, right or wrong, a customer who is not satisfied with a product has the prerogative to reject it.

Building a true hierarchic structure of *Customer Requirements* requires a skilled internal or external facilitator experienced in clustering methodology and in the use of affinity grouping techniques, see Fig. 10.6. Such product requirement structures of excellence are important company documents and an intellectual property of importance to refine, use and protect. It was discovered and argued in an early MIT study (Griffin and Hauser, 1991) that the best architects for building *Customer Requirement* hierarchic structures were actually the customers themselves. Even so, since it is generally experienced that customers usually are not willing to engage in such time-consuming exercises, it is recommended that the clustering activities should be carried out by the product development team in collaboration with representatives from Sales & Marketing. In the development of such hierarchic structures, it is generally advisable to differentiate the *Customer Requirements* into 2-4 levels for a proper structure. Three commonly used levels are:

- *Primary (strategic) needs*: One to five top-level headings that are used by the QFD team to set the strategic intention for the product.
- *Secondary (tactical) needs*: Five to ten sub-headings that explain how the product must perform in order to satisfy each of the primary needs.
- *Tertiary (operational) needs*: Ten to twenty detailed Customer Requirements that satisfy each of the secondary needs.

In the clustering of *Customer Requirements*, it is important to obtain a good balance among the often very large number of individual *Customer Requirements*. Creating a balanced hierarchic structure of all *Customer Requirements* is thus essential, since only the Tertiary Needs will be used for importance ratings. It must be observed that the hierarchic

A helpful hint

Build a hierarchic structure of Customer Requirements following this procedure:

- Put the requirements on Post-Its and use different colors for different kinds of customers that you have identified in the product stakeholder structure.
- Spread out all Post-Its on a very large table in front of a large wall, which you can use later on for the Post-Its.
- Let the team review all the requirements and put duplicates on top of each other, and just scrap the Post-Its that are unreadable or not understandable.
- Let each team member pick up some of his preferred type of requirements on Post-Its and start building clusters of similar kinds of requirements.
- Select a heading for each cluster (sometimes you will find a good name among the existing Post-Its).
- Put clusters together and establish a good structure, but try to end up with the same number of levels in each cluster.
- Edit the text on each Post-It into easily readable and understandable requirements.
- Insert the "Voice of the Customer" into the QFD software, and afterwards review and revise it until it sounds like the true *Voice of the Customer.*

Figure 10.6 The process of building a hierarchic structure of *Customer Requirements.*

structure is only developed to improve the further understanding of the *Voice of the Customer*; since it will not be used in any further matrix calculations. However, a good matrix structure often makes it easier to discover whether any groups or individual *Customer Requirements* have been forgotten at each hierarchical level.

10.3.1 *Starting with the unstructured customer requirements from customer interviews*

Initially, all Post-its with all Customer Requirements should be spread out randomly on large tables. Afterwards, they should be picked up, one by one, during the clustering activity. It is sometimes advisable to use large rolls of paper on the adjoining tables, and preferably also on the walls, to be used to administer and glue the "Post-its" on during the clustering process. When the clustering is finished, the facilitator or team leader can fold them up (remember to use extra tape to secure each Post-It) and bring them back to the office for introduction into the selected QFD software. Such a procedure also facilitates a later "split up" and reorganization of different parts and restructuring of large chunks of Post-Its.

Figure 10.6 shows some helpful hints for how to administer the process of building a hierarchical structure of *Customer Requirements.* As previously noted, one often overlooked issue in QFD textbooks is how to develop a balanced hierarchic structure. In clustering the different *Customer Requirements*, some areas contain a large number of Customer Requirements, while others contain only a few. A simultaneous development of a large number of hierarchical Post-it structures may result in some large clusters with three to four levels and smaller clusters with only two levels. The organization afterwards of the total *Voice of the Customer* with clusters on many levels often tend to be somewhat problematic in the use of QFD software. This is significant, since the relationship matrix will use only individual *Customer Requirements* in the recalculation into *Product Design Requirement* importance ratings. Moreover, when the final *Voice of the Customer* is introduced in the QFD software, only one selected number of hierarchic levels is permitted. An experienced facilitator is well aware of this problem from the outset of this exercise and will supervise the clustering individuals in a step-wise manner in order to finally reach a well-balanced structure with a proper number of levels.

When the clusters emerge in the hands of each individual, the facilitator should encourage them to walk around and review, discuss and assist others. Gradually, all individuals will get a feeling for "the whole" and also adjust to a suitable number of levels in their own clustering exercise.

10.3.2 *Identify and separate different kinds of B2B customer requirements*

In the process industries and in a B2B context, customer requirements can usually be separated into three groups:

- *Good material delivery (logistics).*
- *A good "core" product.*
- *Good service and support.*

Experience tells that if you initially separate these kinds of *Customer Requirements* and treat them in separate matrices, the use of QFD will be vastly facilitated. If you try to build a *House of Quality* on the full spectrum of *Customer Requirements*, you will soon discover that the corresponding number of *Design Requirements* (HOWs) will be huge. This will result in a matrix that will be too large and unmanageable, which is one serious complaint from practitioners using the QFD methodology. Moreover, it is highly advisable to start focusing product development activities on the "core product" and select such requirements for the building of a *House of Quality.* The two other groups of requirements are certainly also important, especially for a B2B customer, but later on the development of only the *Voice of the Customer* for "Good Material Delivery" and "Good Service and Support" could suffice.

Total customer requirements	Customer requirements		How Important? Today	How Important? Tomorrow	How well satisfied by the company?	How important to improve?
Good material in delivery (Logistics)	Important, but to be addressed later on					
A good "core" product	Good material handling properties	Requirement A1				
		Requirement A2				
	Contribute to a good "processability"	Requirement B1				
		Requirement B2				
	Contribute to good end-products	Requirement C1				
		Requirement C2				
Good service and support (Application development)	Important, but to be addressed later on					

Figure 10.7 *Customer Requirements* of different characteristics. The importance ratings of both present and future *Customer Requirements* are selected to illustrate the flexibility in the use of matrix rooms. Furthermore, the benchmarking room "How well satisfied by the company" has been supplemented with the optional room "How important to improve." However, please observe that the figures in all rooms are to be generated in interactions with the stakeholders for the product.

Too large a *House of Quality* must thus be avoided, while on the other hand too small a *House of Quality* usually will become very banal and uninteresting. However, the total format of each matrix will relate to individual product characteristics; some tall matrices have many WHATs and few HOWs, while some matrices are "fat" and contain few WHATs and many HOWs. The separation of the three different kinds of *Customer Requirements* and the further use of the *Voice of the Customer* is illustrated in Fig. 10.7. Especially in interactions with B2B customers and in the review of the *Voice of the Customer*, it is often advisable to explain the initial sole focus on the "core product" to the customer in order to avoid giving customers the impression that other requirements will be ignored in the future. Please notice the initial benchmarking focus on the company's own product and an optional possible direct inquiry to customers regarding which *Customer Requirements* are most desirable for improvements.

10.3.3 *If necessary, reduce the total number of customer requirements*

The number of *Customer Requirements* in the *Voice of the Customer* should not be allowed to be too large, since too many *Customer Requirements* will result in too large a *House of Quality*. All QFD matrices must be kept at a manageable size, since completing very large matrices often tends to wear down the QFD teams. However, and most importantly, this also tends to make it difficult to capture the whole of the matrix information in further reviews and Matrix Analysis. Good chefs in good restaurants often pride themselves on being experts at creating "rich-tasting" stocks for use in preparing sauces. The secret in doing so is to "reduce" the stock to one half or one third of its original volume by cooking, thus greatly increasing the flavour.

If you have too many (i.e., more than ~100) *Customer Requirements*, a similar kind of "reducing" exercise is needed. While such a reduction is often viable, it is important not to lose the original "flavour". For example, in a small team exercise, four similar requirements can most likely be reduced to one strong *Customer Requirement*. However, be careful not to make them "double barrelled" (two requirements inappropriately integrated into one single requirement), and try to stick to the customers' original wordings.

10.3.4 *All requirements must be attended to but not all within the voice of the customer*

Sometimes in the use of the QFD methodology, the question arises of how to handle specific kinds of requirements in the *House of Quality* (e.g., the *Customer Requirement* "product price"). The proper answer to this question is that the product price is set by the marketers and that, when using QFD in product development, one should instead focus on manufacturing costs. However, the manufacturing costs are not an issue in the *House of Quality* but rather in the Process Matrix and the Raw material Matrix. For this reason, using only the *House of Quality* in the development of new products, the manufacturing cost and the design for manufacturability of the products results in grossly neglecting the QFD methodology's potential benefits in product innovation in the use of the upstream & downstream matrices.

10.3.5 *Building a hierarchical structure of customer requirements for the perfect lunchtime beer*

In Fig. 10.8, a number of *Customer Requirements* for the Perfect Lunchtime Beer have been compiled in a totally unstructured fashion. Requirements for the "core product," the beer, have been mixed with requirements for how it is packaged, served and produced. Important requirements are mixed with unimportant requirements. However, please note that this illustrative short-list of requirements is the result of a pre-screening. Non-relevant requirements have already been removed, unreadable or non-understandable requirements (collected on Post-Its) have been deleted, and duplicate or very similar requirements have been combined. One can easily comprehend that going to the brew-master with this information in discussions of a development of a Perfect Lunchtime Beer would not be highly appreciated.

Figure 10.8 A short-list of unstructured *Customer Requirements* for the Perfect Lunchtime Beer.

In Fig. 10.9, the unstructured *Customer Requirements* for the Perfect Lunchtime Beer have been arranged in a clustering exercise in a hierarchic structure. Underneath the single Strategic need "An attractive and well-presented product," the requirements have been separated into five groups of Tactical Needs:

- Good image
- Dependable and safe
- Attractive packaging
- Professionally served
- A product of excellence

However, only the sub-group "A Product of Excellence" will be further used in the construction of a *House of Quality* in this book. The other sub-groups are naturally also important, but they can be developed into four different voices — not necessarily a *House of Quality*, but in this case only four individual *Voice of the Customer*. You can note that the sub-group "Dependable and Safe" does not belong to this *House of Quality*. These are important requirements for the production process and thus belong to the upstream & downstream matrices of the Process Matrix and the Raw material Matrix. Below the Tactical Need "A product of excellence," 13 Operational Needs have been arranged. Those marked

Voice of the Customer (The perfect lunchtime beer)		
An attractive and well-presented product	Good Image	A reliable and well-known brand
		Exciting and new
		Often recommended
		International cachet
		Available in shops
		A "trendy" beer
		A traditional style
	Dependable and Safe	Environmentally friendly (an eco beer)
		Brewed by experts
		Local production
		Brewed on high quality ingredients
		No harmful additives
	Attractive Packaging	Easy to open
		Recyclable pack
		Nice bottle/can
		A package that spells beer
		Reliable product information
		Interesting labelling (decor)
	Professionally Served	Served at a proper temperature
		Served well at the table
		Served in the right type of glass
	A Product of Excellence	Goes well with food
		Refreshing (thirst quenching)
		A beer with some character
		Good nose (bouquet)
		Good finish
		Nice colour
		Not cloudy (clarity)
		Easy to drink
		A low calorie beer (light)
		Non intoxicating (drink and
		Good head
		Light-bodied flavour (not watery)
		Well balanced
		No defects

Figure 10.9 A hierarchic structure of *Customer Requirements* on the Perfect Lunchtime Beer. In the subsequent construction of a *House of Quality*, only the group "A product of Excellence" is to be used. Moreover, in the simplified versions of the matrix, only the *Customer Requirements* marked in red will be used. You can further note that the *Customer Requirements* in the group "Dependable and Safe" can only be met using the Process- and Raw material matrices.

in red are the ones used in the "mini version" of the *House of Quality*, but full *House of Quality* on the Perfect Lunchtime Beer is presented in Appendix C.

All *Customer Requirements* within the sub-group "A Product of Excellence" have been used in the construction of this full *House of Quality*. In the full *House of Quality*, a second level in the structure has been added including the three groups "Attractive look," "A mealtime drink" and "Good taste." The hierarchic structure is not important operationally in the use of the house, but it serves the important purpose of illustrating the different kinds of requirements and supporting an improved "understanding" of consumer preferences.

10.4 Return to selected customers for further importance ratings and customer benchmarking

The importance ratings of *Customer Requirements* can be carried out using many alternative techniques and the selection of samples of customers in many different ways. The focus of this book is not market research, statistics or scale selection, so if you want to acquire a deeper understanding in those areas, please refer to this rich body of literature. However, since there are so many alternative techniques, a basic knowledge and overview is provided here in order to use the QFD methodology well.

Selecting customers (respondents) for importance ratings and benchmarking

Selecting proper respondents for importance ratings of *Customer Requirements* is the first step on the road to get valid information. The selected group of customers for delineating individual *Customer Requirements* in the *Voice of the Customer* is not necessarily the same group as that used for finding out their importance ratings. Depending on company contingencies, the selection of respondents will naturally be a cost issue; however, to identify customer preferences in different market segments, it is important to obtain samples that are representative for each segment.

Later on, when the results are presented in the *Voice of the Customer* and in the *House of Quality*, one can calculate the average figure for all respondents, but it can certainly also be of interest to present and separately analyse the average figures from different market segments. QFD programs usually have an ability to accommodate multiple importance rating rooms, as well as to recalculate the figures from a number of different Customer Importance Rating rooms into a number of Technical Importance Rating rooms in the *House of Quality*. Moreover, the results from collecting Customer Importance Rating figures for individual *Customer Requirements* could additionally be analysed as bar charts showing Customer rating distributions. In other words, if, for example, an average importance rating for a *Customer Requirement* is 5 using an ordinal nine-point scale, this information has a different meaning if all customers have given a rating of 5 than if half of the group gave a rating of 1 and the other half a rating of 9. This kind of information can be

A helpful hint

A facilitating software feature – flag notes

Different QFD software programs may vary in user-friendliness and have different facilitating and operating features. The **QFD Designer™** is one software program that has an important tool that is applicable to all rooms in all kinds of matrices. It is thus possible to insert specific information into all rooms and for all symbols, and the information is advertised with a small red flag in the matrix. When pointing the cursor at a flag, a pop-up window tells what kind of explanatory information it contains.

This is a truly important aspect in the use of the QFD methodology. When the team in the future asks "What did we mean by that", it is easy to recapitulate the previous discussions that have taken place. Newcomers can thus easily be updated, and the explanatory capability of the matrices is vastly improved. In the further chapters, we will revert to this tool, since it is very useful for knowledge building.

Figure 10.10 On the use of "flag notes" as a QFD software feature.

supplied together with comments on individual importance ratings as "flag notes" (see Fig. 10.10).

The preferred instrument to obtain Customer Importance Ratings is a survey. You develop, arrange and present the list of all individual *Customer Requirements* and ask the respondents to estimate their importance. A survey is an excellent tool for this task, and it is also a good opportunity to add "open questions" where the respondent is asked to add comments or any kind of information (e.g., new requirements). The technique for doing surveys and to formulate a good inquiry is a bit of a "science" in itself, and there is not space here for an extensive treatment. In Part VI of the book, you will be acquainted with the concepts of "lead users" and "customer co-development," which can assist in one's selection of respondents. However, there are some important things to remember in doing surveys:

- It is one thing to send out inquiries and another to receive answers. The use of outside expertise is recommended if one is not trained in "social sciences"
- Use the hierarchy of *Customer Requirements* for the inquiry, but check that the questions are understandable and do not contain two questions in one. It is a good idea to pilot-test the inquiry on outside persons to get feedback for improvements.

- If a company has a limited number of customers (especially in B2B contexts), it will greatly improve the validity of the answers if a company representative personally can introduce the inquiry on the customers' premises and assist if questions arise.
- It is possible to combine this activity with a benchmarking of competitive products, but beware of overly complicated inquiries.

Selecting proper scales for importance ratings and benchmarking

Generally, a nine-point ordinal scale should be used for both importance ratings and benchmarking. Scale selection is a difficult area, but the author's practical experiences favour such a scale (an alternative scale could be a five-point scale). The previous text has referred to this kind of scale including only explanatory text to end figures (1 = low, 9 = high; 1 = poor, 9 = excellent). An alternative nine-point scale offering more guiding supplementary text for importance ratings:

Crucial (decisive for buying) 9
Very important 7
Important 5
Less important 3
Unimportant 1

In a similar vein, an alternative nine-point scale with more guiding supplementary text for benchmarking is as follows:

Superior product (world class) 9
Very good product 7
Good product 5
Not so good product 3
Poor product 1

There are alternative techniques for finding out importance ratings like the Conjoint Analysis (Green *et al.*, 2001; Katz, 2004; Silayoi and Speece, 2007), which in itself is a superior market research instrument, but unfortunately it only allows for working with a maximum of around 10 *Customer Requirements*. The weakness of rating *Customer Requirements* on an ordinal nine-point scale is that customers usually consider all requirements important. Since we do not get interesting information from a respondent rating all answers at a 9, it must be explained and stressed in the inquiry, or by the person introducing the inquiry, that the respondent must try to use the full scale and must focus on finding correct importance ratings, since this will be an important guide in future product development.

When there are few respondents and interaction is possible, it can be interesting to ask the respondents to provide a supplementary ranking of only those importance ratings with

high rating figures (e.g., 8–9). This is particularly advisable if customers have selected very high ratings for a large number of Customer Requirements. Such a complementary exercise usually is not difficult for the respondents and offers a greatly improved understanding of the most important product requirements.

10.4.1 *Customer Benchmarking — in search of an improved competitor understanding*

The QFD methodology is unfortunately often presented as a tool only for improved customer interaction and understanding, neglecting its vitally important capability for competitor understanding. Consumers often find benchmarking exercises interesting and stimulating, while industrial customers tend to be more suspicious. Benchmarking competitive products in B2B relations is sometimes not possible, because the customer purchasing organization may suspect that it could disturb future price negotiations with the supplier.

In the selection of competitive products for *Customer Benchmarking*, it is recommended to select the products that are the strongest competitors on the market. It is also advisable to discuss a future *Technical Benchmarking* and acquisition of competitive product samples, since it is most advantageous to do both kinds of benchmarking exercises on the same competitive products. If customers do not want to benchmark the company product against competitors, the simplified procedure illustrated in Fig. 10.7 can be used. In this procedure, for the top-rated and afterwards top-ranked *Customer Requirements*, the customer is asked to give advice on their importance for improvements.

10.5 Develop and design supplementary rooms of interest

Following the sound advice of the late Professor Akao to adapt the use of the QFD methodology and its matrices to project needs, each company should develop and design supplementary rooms of particular interest for its kinds of products, product portfolio character and general product innovation needs. Available QFD software makes such "tailormade" design of matrices very easy even for inexperienced QFD users. The following tentative list of supplementary rooms has sometimes been used by the author in different QFD projects; please use the list as a "Swedish buffet" and choose those you prefer. However, note that many of the listed rooms are adapted to the voice of B2B customers and benefit from interactive discussions with professional industry respondents.

Room for ranking of top-ranked Customer Importance Ratings (ratings above 8 on a nine-point ordinal scale)

It has previously been pointed out that Customer Importance Ratings sometimes "mis-fire" when respondents rate all requirements very highly. A further ranking of all such highly rated Customer Requirements is generally found to provide important guidance in product innovation.

Room for Customer Importance Ratings in a long-term perspective

This room is of particular interest for process-industrial development, since the lead-time for more radical product innovation projects is often very long (5–10 years). The B2B customer perception of how *Customer Requirement* importance ratings will change in the time perspective of 3–5 years is therefore truly important. In other words, this room will provide "an early warning" of future changes in customer needs and expectations.

Room for "how important to improve" — customer's own rating or ranking

The *Voice of the Customer* articulates *Customer Requirements*, their importance ratings and how well the customers perceive that the product performs compared with competitive products. Company product developers afterwards analyse this information and decide on development directions and strategy for a new or improved product. In a more "open innovation" approach, it is also often rewarding to inquire about customers' direct opinion regarding which Customer Requirements they would like the company to improve on; this is the room for such complementary information.

Room for explanatory customer comments

In interactive interviews and in open questions in surveys, comments from the respondents should always be encouraged and sought. For example, referring to the previous supplementary room, customers may not only express their recommendations for product improvements but may also have ideas and suggestions for how such improvements could be made.

10.6 Putting it all together in the "Voice of the Customer"

Drawing conclusions from an analysis of the *Voice of the Customer* ahead of the construction of the *House of Quality* can be an interesting activity, but it is generally recommended to wait until the full *House of Quality* has been erected. The total *Voice of the Customer* includes the combined information from well-articulated Customer Requirements, their individual Importance Ratings, and the results from benchmarking the company product against competitors. ***It is thus important to recognize that the Voice of the Customer is an instrument for not only improved "customer understanding" but also improved "competitor understanding."***

Now is the time to position the company product against competitors and to articulate customer "value added" properties of an improved or new product. This is the time to identify the competitive position of an improved company product and to differentiate the product offerings against competitive products. Which *Customer Requirements* should the company's Sales & Marketing articulate well in the promotion of the new product to be developed?

10.7 The illustrative case of development and manufacturing of beer — The "Voice of the Customer"

The "full version" of the *House of Quality*, including the "full version" of the *Voice of the Customer* is presented in Appendix C of this book. In the "mini version", only six *Customer Requirements* have been selected from the "full version" to illustrate important issues in the use of the *Voice of the Customer*. The selected *Customer Requirements* for the "mini version" have been marked red in the "full version". As competitive products for the development of the Perfect Lunchtime Beer, three different products, representing three different beer styles, have been selected for the Customer Benchmarking. The figures for Customer Importance Ratings were developed in a mini-survey, and the Benchmarking figures were developed in a real Customer Benchmarking exercise. Some figures are, however, slightly modified in order to illustrate certain methodological issues of importance.

The "mini version" of the Voice of the Customer

The "mini version" of the *Voice of the Customer* is presented in Fig. 10.11. The following beer styles were selected for the survey and benchmarking exercise:

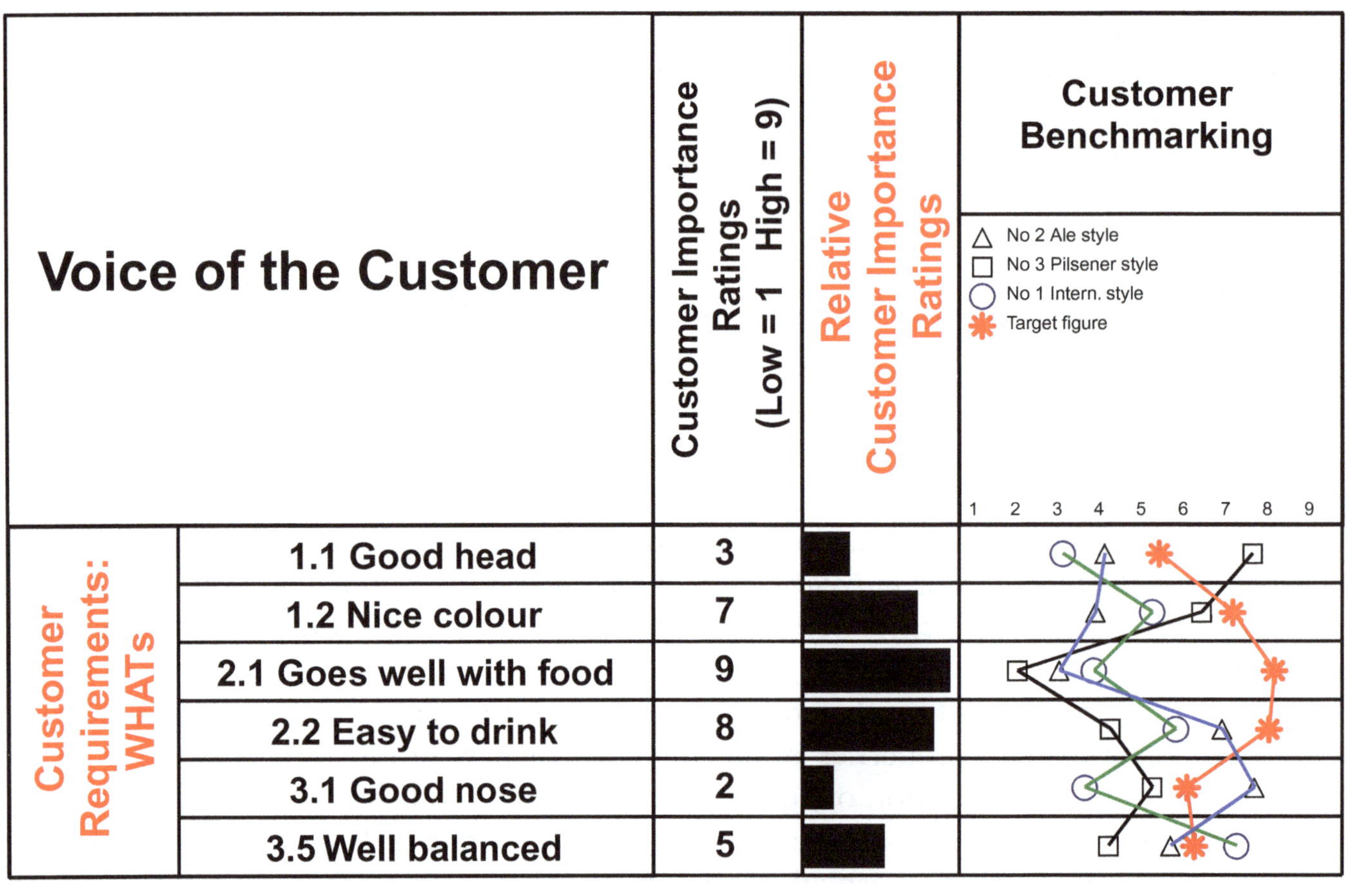

Figure 10.11 The simplified mini version of the *Voice of the Customer* for the Perfect Lunchtime Beer.

Beer No 1 (symbol ○)
A light-bodied lager-style beer, very light in colour, slightly sweet and fruity, with rather a low alcohol content and no bitterness; an "International" style of beer.

Beer No 2 (symbol △)
A beer with medium-high bitterness but a malty character, mahogany in colour with full-bodied flavour and fruitiness and a high alcohol content; a "Northern England Ale" style of beer.

Beer No 3 (symbol □)
A lager beer with a bright clear darkish golden colour and medium alcohol content. A beer with a hoppy, somewhat "tight bitter" taste; a "Pilsner Czech" style of beer.

Target figures for the Perfect Lunchtime Beer" symbol ★

Customer Requirements

In the following text, the individual *Customer Requirements* will be discussed in the perspective of a Perfect Lunchtime Beer; however, remembering that different kinds of food sometimes require specific styles of beer. The Target Figures for the Perfect Lunchtime Beer are thus set in search of a "good product compromise".

Good head

It is generally experienced that a good head on a glass of beer will improve the tasting experience. The Pilsner beer was rated highest in this aspect in the benchmarking, while the Ale did not perform as well. The International beer initially had a very good head, but it did not last very long.

Target Figure: The low Customer Importance Rating figure (3) indicates that it is not necessary for a beer to be a top performer in this dimension.

Nice colour

The Ale was considered to be too dark for a Perfect Lunchtime Beer, and since it was also a bit cloudy the rating figure in benchmarking was low. On the other hand, the International style of beer was deemed to be too light in colour. The Pilsner beer was rated the highest, but none of the beers performed really well.

Target Figure: The high Customer Importance Rating figure indicates that the Target Figure for the Perfect Lunchtime Beer should be "best in class".

Goes well with food

A fruity taste generally goes well with food, but a beer with too strong a malty or hoppy taste is generally not considered to go well with food.

Target Figure: None of the beers performed well in the benchmarking, and since the Customer Importance Rating figure (9) is very high, this requirement should be targeted as a "sales point" for a Perfect Lunchtime Beer. That is to say, a beer that is positioned very well in this very important dimension will give a strong marketing advantage.

Easy to drink

A dark beer is generally perceived as "not so easy to drink", as is a cloudy beer. A fruity, less hoppy beer is often thought as easy to drink. The International style of beer had a "carbonic bite" (too much CO_2), and was consequently not rated easy to drink in the benchmarking. The Pilsner beer was rated too bitter, while the Ale was generally viewed to be a most "drinkable" beer in spite of its dark colour; however, it had too high an alcohol content for a proper "lunchtime beer".

Target Figure: The very high Customer Importance Rating (8) makes this dimension very important to explore in detail in the further development of the *House of Quality*.

Good nose

The Ale was rated high in this category (even though the head was poor), while the International style of beer was judged "flat" and was rated low in the benchmarking. The Pilsner had too hoppy an aroma for a Perfect Lunchtime Beer.

Target Figure: The very low Customer Importance Rating figure (2) suggests that this requirement is not a very important characteristic for a well performing "lunchtime beer".

Well balanced

A well balanced beer should have a proper ratio between its "bitterness" and its "sweetness". The International style of beer was rated high in this dimension while the Pilsner beer was rated low.

Target Figure: The Customer Importance Rating figure (5) was medium high, indicating a *Customer Requirement* that must be taken into account in product innovation.

Chapter 11

The House of Quality — The Facilitating Instrument for an Improved Product & Competitor Understanding

The Burgundy Wine Castle in Fig. 11.1 is a proper metaphor for a *House of Quality* in the QFD methodology. This castle was built many centuries ago but is still a usable and functional building. Similarly, a *House of Quality* is an important company asset that should be developed, refined, updated and used for many years to come in product innovation. No two castles look exactly the same; likewise, each *House of Quality*, if properly constructed, will contain a number of standard rooms as well as rooms particularly developed and designed in order to capture specific knowledge related to individual products.

Moving from a Customer Dimension into a product Technical Dimension

In the previous chapter, it was argued that, in order to acquire a profound understanding of customers' present and future needs and desires, one must initially focus on customers' "root" requirements. In order to succeed in such an endeavour, one must utilize sophisticated techniques for capturing such qualitative information, preferably in close interaction with the customers and potential users of a product. However, in product innovation of excellence, this knowledge base, totally embedded in the "customer dimension," must afterwards be further translated into a "product technical dimension." The *House of Quality* is the prime instrument and QFD matrix for this translation, in which the previously developed *Voice of the Customer* is accommodated and well-integrated.

Some scholars advocate that it is sufficient to develop a *Voice of the Customer* to be successful in product innovation, and it is unfortunately also officially often assumed that a company can state that they deploy the QFD methodology in product innovation if they use only that part of the methodology. However, in today's product landscape and complex product innovation life, this is untrue. If successful product innovation only needed a

Figure 11.1 A real house of quality: a burgundy wine castle.

thorough understanding of customer needs and preferences, the work of the R&D department would be very easy.

Ignoring the necessity to convert the qualitative information into well-defined and measurable *Product Design Requirements* is not only abandoning the "core philosophy and soul" of the QFD methodology but is a most dangerous simplification of the complex innovation process. The ever increasing cost-innovation challenges in product innovation not only necessitate well-structured product design Target Figures in the use of a *House of Quality,* but also necessitate addressing product manufacturability in the use of the upstream & downstream QFD matrices.

This matrix (component), which is deployed in all QFD systems, has been designated Product & Competitor Understanding in this book, and the primary object of this matrix is to translate the qualitative *Customer Requirements* WHATs for the product into a company's technical dimension as measurable Product Design Requirements HOWs. All the rooms that are used in a *House of Quality* and that will be presented in this chapter can be used in all other QFD systems, which have been briefly introduced in Part 1.

11.1 Introducing the House of Quality and the guiding work process for matrix development

In Fig. 11.3, the *House of Quality* is depicted, including the standard rooms in such a matrix. You will find the "full version" of the *House of Quality* for the Perfect Lunchtime Beer in Appendix C, and in Section 11.9 you will also find the "mini version." During the course of

A helpful hint

Different kinds of rooms in the matrices

The software QFD Designer uses the following types of rooms:

- *Text rooms:* Rooms where any kind of text can be inserted in a similar way as in a Word document.
- *Numerical rooms:* Rooms where numbers can be entered and, if needed, used in numerical calculations, also related to other numerical rooms.
- *Symbol rooms:* Rooms where different symbols can be entered from the software library.
- *Benchmarking rooms:* Rooms where data from the company and from competitors can be entered and represented by "snake plots" in an ordinal value scale.

Usually all rooms can be expanded or shrunk according to the QFD user's wishes. Most programs have the ability to change assigned values and additionally include other kind of symbols.

Figure 11.2 Different types of rooms deployed in the QFD matrices

the discussions of the individual rooms in this matrix, you will sometimes be referred to the "full version" and sometimes to the "mini version."

However, please observe at this point that this matrix can be designed in different ways in order to be adapted to individual products and company needs. Using a versatile software like the **QFD Designer**™ will greatly facilitate such exercises and promote a creative and customized design of the whole *House of Quality* and its individual rooms.

A brief review of different kinds of rooms and functions usually available in QFD software

In Japan, standard Excel spreadsheets are sometimes used for the design of QFD matrices. In Fig. 11.2 you will find a helpful hint about different kinds of rooms usually deployed in QFD software. However, it is highly advisable to invest in a flexible and easy-to-use software. The importance of using different kind of fonts and easy-to-read symbols in the use of the matrices, and during the Matrix Analysis will be discussed later in this book.

There are only three mandatory rooms: *Customer Requirements*, *Product Design Requirements* and the Relationship Matrix. The hierarchical arrangement and denomination of the three separate areas, in Fig. 11.3 of *Product Design Requirements* is optional, since in each project the structure and denominations should be adapted to product specificities.

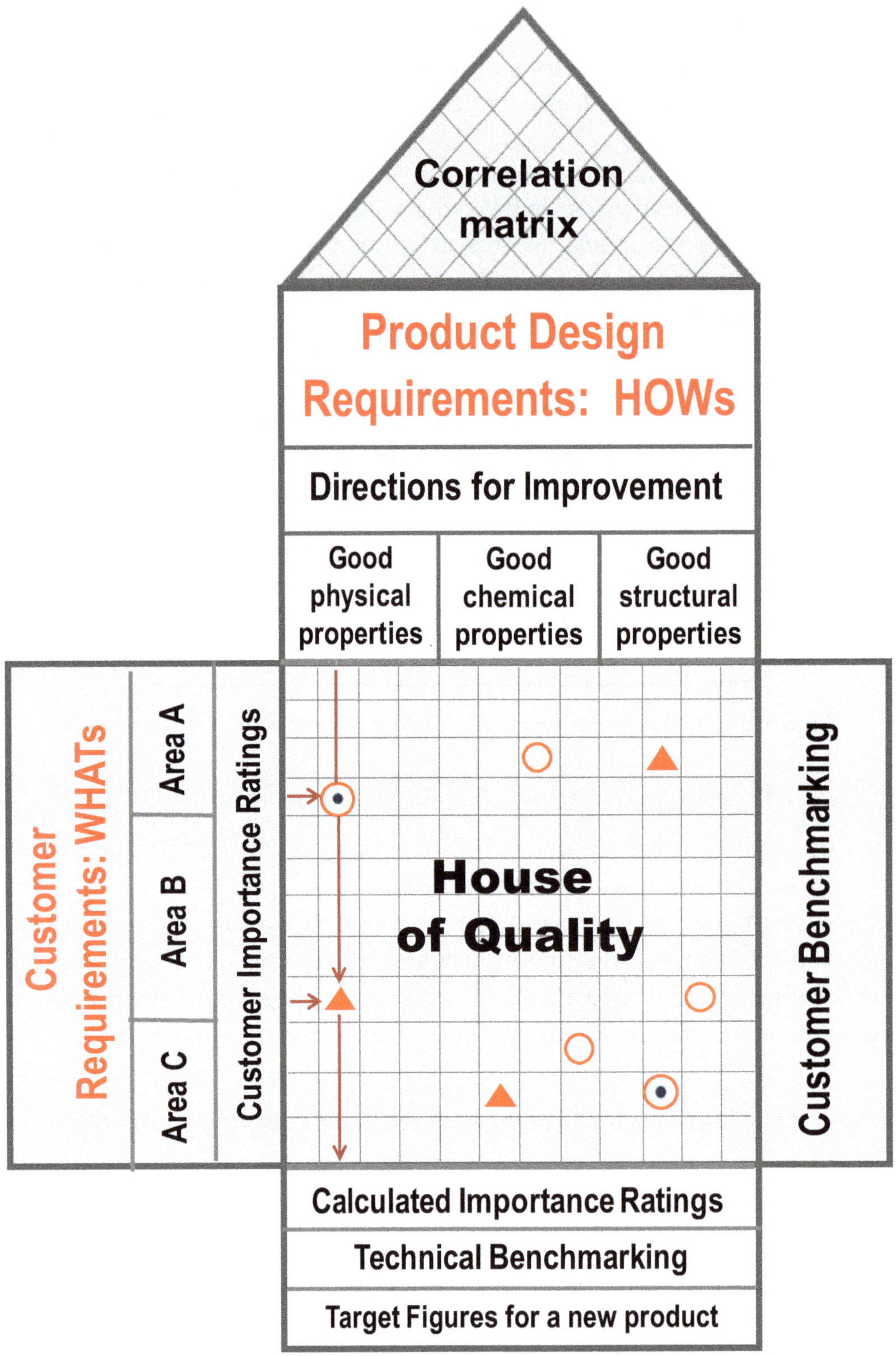

Figure 11.3 The *House of Quality* for a product, and its standard rooms.

Consequently, some users prefer to cluster and categorize the matrix HOWs as "measured online," "measured frequently" and "measured on daily or weekly basis." An alternative categorising option for the HOWs in the matrices could be "easy to measure," "difficult to measure" and "do not know how to measure." Chapter 1 briefly introduced the standard rooms, but in this chapter you will learn more thoroughly how to develop and use all individual standard rooms and optional supplementary rooms.

11.1.1 *Introducing the guiding work process for the construction of a* House of Quality

In Fig. 11.4, the overall work process for the development of a *House of Quality* is presented using the previously introduced *Voice of the Customer* information. Please read through the full work process, and you will get a first general understanding of how such a matrix is developed and the individual steps involved.

The design and erection of a full *House of Quality* is a somewhat time-consuming exercise but it must always be remembered that the most time-consuming activity is the collection of novel information about the product. The collection of customer information, including product customer benchmarking, will be supplemented in the *House of Quality* with the truly important *Technical Benchmarking*. If the company has no previous experience in this area, the collection of samples of competitive products and laboratory testing usually takes a considerable amount of time. However, please observe that this is NOT a time-consuming QFD disadvantage, but only a consequence of company neglect in Business Intelligence.

When the erection of the *House of Quality* starts, it is important that the project manager/facilitator initially explain all rooms presented in Figure 11.3 and afterwards introduce and review the total work process. These activities should be carried out in a separate kick-off meeting. The review of all individual rooms in the perspective of the total work process will greatly facilitate and promote a simultaneous and integrated development approach of all rooms, which will speed up the design of the *House of Quality* substantially. However, the company implementation and use of the Quality Function Deployment methodology will be extensively treated in Part VI and the chapters therein.

11.2 Identify *Product Design Requirements* and define related test methods and metrics

"Applying Quality Function Deployment (QFD) to existing product lines, cross-functional teams developed innovative design requirements which are now used to measure product and service performance. By faithfully recording the voice of the customer, teams discovered that traditional industry standard methods for measuring product performance had little relevance to the end use, and even less to satisfying the customer. Using the discipline of QFD, we were able to identify the weaknesses in current test methods and create new tests that directly measure performance characteristics of most interest to customers."

(Hanson, 1993)

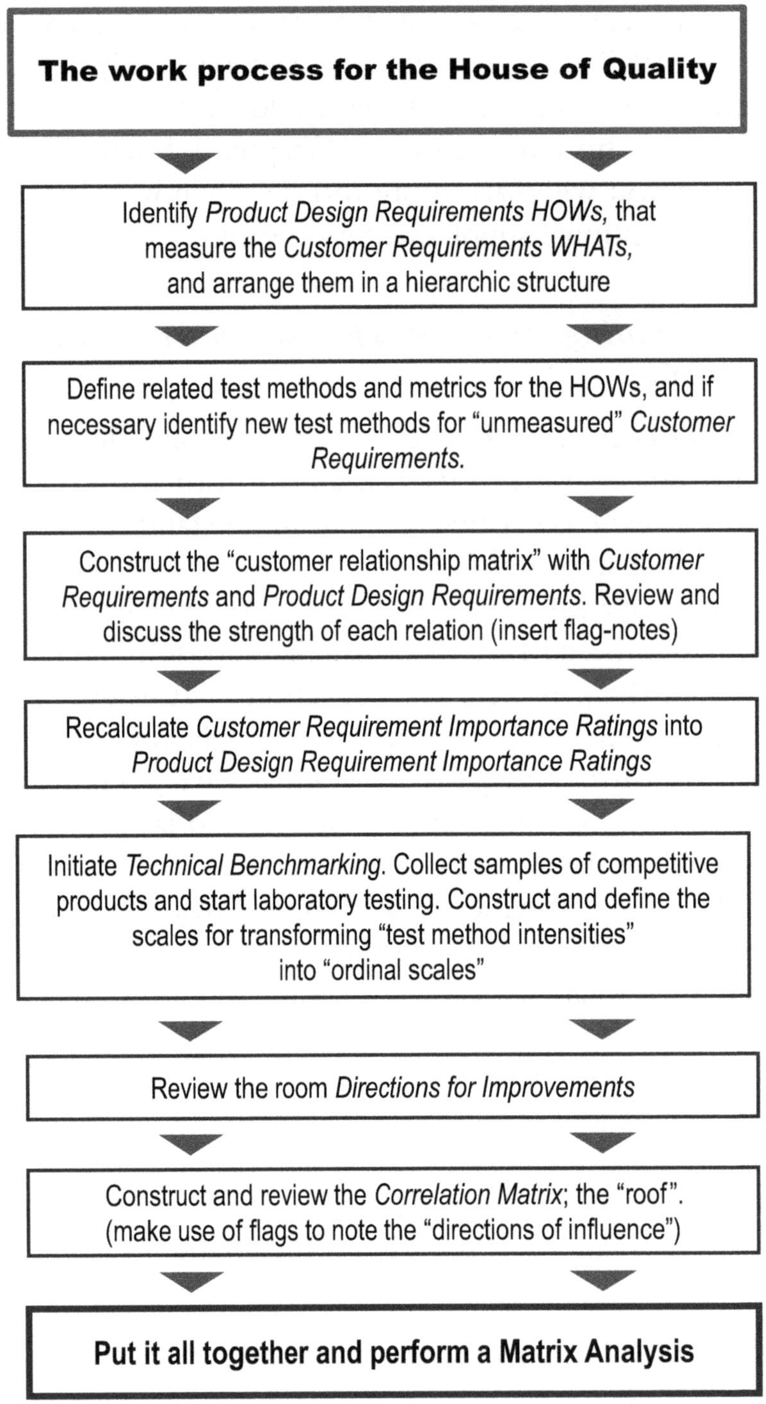

Figure 11.4 The work process for building a *House of Quality*.

This early quote regarding the use of QFD in the Forest Industries highlights the often experienced fact that existing test methods do not adequately measure *Customer Requirements*, and in properly conducted QFD projects it is thus very commonly discovered that necessary test methods are lacking. Nonetheless, this discovery is one most valuable outcomes of the use of the QFD methodology, since:

> Developing new or improved products without well-defined properties that cannot be measured with well-defined, reliable and valid test methods and related metrics is a futile undertaking and is doomed to end up in confusion.
>
> (Anon.)

The translation of *Customer Requirements* into reliable *Product Design Requirements* is thus one significant advantages in the use of the QFD methodology.

11.2.1 *Identify Product Design Requirements*

If one wants to be true to the original QFD methodological ideas, one should develop *Product Design Requirements* in a thorough review of each *Customer Requirement* and select a corresponding number of test methods and proper metrics. This generally works very well, especially if a radically new product is to be developed. However, in the refinement and renovation of existing products, one can alternatively use the often large number of already existing test methods and assaying techniques as a point of departure, structuring them as potential HOWs in the *House of Quality*. Experience tells that the latter approach is advisable; if there are too many pre-existing tests, it is easy to reduce and select the most important ones later on. On the other hand, if proper test methods for some *Customer Requirements* are lacking, they can afterwards be developed and supplemented.

11.2.2 *Define related analysis, test methods and associated test units (metrics)*

The old saying that "what gets measured can be improved" is very true here; in the development of *Product Design Requirements*, unmeasurable HOWs should generally be avoided or even disregarded. However, for example, in the Food & Beverage Industries, *Customer Requirements* like "good smell" from product benchmarking exercises in clinics are today generally easy to measure and convert into measurable *Product Design Requirements* using expert sensory laboratories and expert panels. In Fig. 11.5 this is further illustrated.

One could initially inquire what characterizes a good test method and its related metric. The following quote could initially be a guiding principle:

> A variable is a valid measure of a property if it is relevant and appropriate as a representation of the property. Does the process measure what you want it to do? To discuss the issue sensibly we must ask validity for what purpose and validity for what population.
>
> (Moore, 1991)

An explanatory example

Measuring product attributes of a beer

In the "full version" of the *House of Quality*, one can identify 12 metrics where the measuring instrument is a test panel. For consumer products, this is not an unusual situation. Today, professionally selected and trained test panels must be regarded as just as reliable as any other test method.

Yesterday's visual inspection of the colour of the beer and the brewmaster's tasting of its bitterness and sweetness has been replaced by international standard test methods illustrated in the "full version"; e.g., the European Breweries Convention (EBC) Colour Unit Measurement (Absorbance at 430 nm).

Figure 11.5 Measuring product attributes on a beer.

The concept of "construct validity" is used to describe how well the operational measure corresponds to the property one wants to measure (the definition of that property). The concept of "reliability" is about repeatability and lack of measurement errors in the measuring instrument and measuring process; a result cannot be valid if it is not reliable. The ability to measure/characterize product properties is fundamental, especially in the process industries, in order to:

- Secure the quality control of products from the production process
- Define product properties for marketing and sales
- *Guide product development*

The author's experience, and especially his evidence from running a large number of QFD projects in the process industries, tells that selected test methods and metrics for the above activities not necessarily are the same.

Product attributes that are measured in order to manufacture products at set product specifications and within specified tolerance levels are often of a different kind to test methods and metrics for product innovation; the metrics for *Product Design Requirements*. On the other hand, those test methods and metrics are sometimes different from metrics used in marketing and sales, which necessarily must adhere to internationally accepted standards.

Company internal test methods and assaying practices that are specifically designed and used in product innovation activities are sometimes lacking, which may necessitate new test methods for "unmeasured" *Customer Requirements* to be developed. This situation will be accentuated when the Customer Relationship Matrix has been developed and if one discovers that some *Customer Requirements* do not have any relation to any of the *Product Design Requirements.* However, even if it is recognized that a *Product Design Requirement* needs to be measured, it is usually not directly evident whether it can be measured, how easily it can be developed, and how important it is to develop. This is something that must be assessed in separate reviews and assessments. The kind of information that must be supplied in supplementary rooms to the *House of Quality* which will be commented upon later on in this Chapter.

In the process industries, there is a kind of test method of vital importance for product innovation; such test methods are designed to measure the product's "functional properties." These methods try to simulate how the product behaves when subjected to different process conditions, simulating different customer production process conditions and use. In the use of the QFD methodology, one must always be on the outlook for improved measurements of product functional properties, since such properties, if measured, can often be improved and develop into a formidable "sales-point." Analysing/assaying methods, by contrast, measure a product property as-is (chemical analysis, physical analysis, etc.). Such an analysis is thus not related to any kind of specific operating "environment" for the product.

Build a simplified hierarchic structure of the Product Design Requirements if this will improve understanding

It is usually experienced in clustering exercises that it is beneficial to arrange *Product Design Requirements* in a hierarchic structure. In Fig. 11.3, they have only been arranged in two levels, the second level including "Good chemical properties," "Good physical properties" and "Good structural properties."

11.3 Construct the Customer Relationship Matrix and calculate *Product Design Requirement Importance Ratings*

The Customer Relationship Matrix connects the horizontal *Customer Requirements* and the vertical *Product Design Requirements* like a well-woven piece of fabric.

If any kind of fabric is to be robust and durable, both the horizontal and vertical threads must be strong and well interlinked; the relationship matrix must be similarly designed. It is advisable to keep relationship matrices not too large, since too large a matrices will involve too many discussions and decisions during matrix development. In the previous presentation of the QFD methodology in Part I, you were introduced to the construction of this matrix and its usually deployed symbols.

It is generally perceived that the best way to complete the relationship room is initially to review the relations "column wise"; see Fig. 11.6. The recommended procedure is to review each *Product Design Requirement* and examine whether and how any of the *Customer Requirements* are affected if the intensity of the *Product Design Requirement* is changed

A helpful hint

The relationship rooms in the different matrices need to be revised many times to get in good shape.

- In a "lean" QFD methodology approach, a relationship room should be completed as a draft by the "matrix owner" only.
- In a first review of the draft relationship room together with all QFD team members, each *Product Design Requirement* should initially be reviewed "column-wise".
- In a second review of the relationship matrix together with all QFD team members, each *Customer Requirement* should be reviewed "row-wise".
- In a final revision of the relationship matrix with all team members, the focus should be on all strong relations, and during the review these should be given "flag notes".

Figure 11.6 Matrix reviews and revisions.

from a lower figure to a higher figure. The first object is to find out if there is any relation, and the second is to determine the strength of such a relation.

You have already been acquainted with the commonly used symbols in relationship matrices (double ring = 9, strong relation; single ring = 3, medium strong relation; triangle = 1, weak relation; empty = 0, no relation at all). These symbols are generally experienced by QFD teams to facilitate the identification of "relationship patterns." Important *Customer Requirements* ought to have at least one strong relation or several medium relations to a *Product Design Requirement.* The gaps in symbol strength between the different symbols follow a "Pareto Principle" in an effort to single out "the vital few" among the "trivial many" relations between *Customer Requirements* and *Product Design Requirements.*

Initially, it is important to remember that the symbols only show how strongly an individual *Product Design Requirement* relates to an individual *Customer Requirement*; no difference is made if a change in the HOW intensity relates positive or negative to a WHAT. Some practitioners find this use of the methodology and of the relationship matrix difficult to comprehend. The reasoning behind this practice is that one is interested in calculating the total influence of a selected HOW on Customer Requirements, either negative or positive; see further Fig. 11.7. However, in the review of the matrix, it can be of interest to distinguish between positive and negative influences, using differently coloured symbols. Since QFD software programs usually do not incorporate such facilities, one can alternatively comment on this for each strong relation using "flag notes."

An explanatory example

Use of Relationship Matrix symbols

Referring to the "full version" of the *House of Quality*, there is a strong relation between "Bitterness" and "Goes well with food". A high bitterness intensity has a negative relation to the Customer Requirement. However the "Bitterness" relation to "A beer with some character" is of a more positive nature.

"Carbon dioxide" has a strong negative relation to the Customer Requirement "Easy to drink", but on the other hand it has a strong positive relation to the Customer Requirement "Refreshing (thirst quenching".

Figure 11.7 Relationship symbols can have either a positive or a negative relation to Customer Requirements.

It can sometimes be advisable to split up a very large QFD matrix into a number of more manageable sub-matrices that can be developed by separate QFD teams, which also can further facilitate discussions together with different customers using selected areas of a matrix. Matrices can be split up horizontally or vertically dependent on matrix content and structure into sub-matrices in the use of the **QFD Designer™** Software. Sometimes, when certain parts of the matrix get too "thick" during matrix development (too many HOWs) or too "tall" (too many WHATs) it is recommended to "compress" the matrix in order to get a better balance among the HOWs or the WHATs. On the other hand, sometimes an individual part of a matrix could benefit from a "matrix expansion" in order to facilitate development of Target Figures and product design activities.

In larger QFD teams, it is advisable not to get stuck for too long on a single relationship in the review of the relationship matrix. Instead, it is better to move on and insert question marks in the matrix to indicate a questionable relationship to be returned to later on (a question mark is preferred over a triangle in this case). Sometimes the software provides an option to show which parts of the matrix relations have already been reviewed; if not, it is possible to select a symbol for this purpose to be inserted in the individual rooms. Relations that are uncertain but most likely of considerable interest to discuss and review further could also be assigned a special symbol.

The software usually has a function that can calculate the total "fill" in the relationship matrix, which in a well-structured matrix is often between 30 and 50 %. A very low fill indicates that the QFD methodology has not been used properly (sometimes with only one relation for each requirement). In other words, the possible multiple relations between *Customer Requirements* and *Product Design Requirements* have been overlooked. If there are no such multiple relations, one does not really need to use a relationship matrix at all!

Furthermore, low "fill" in different areas of the matrix and in a specific row (especially if a row is empty) indicates that *Customer Requirements* are not measured at all. If such empty rows have high Customer Importance Ratings, one must inquire whether new or improved Product Design Requirements and associated test methods should be developed and introduced. On the other hand, a very high "fill" could indicate that the requirements have been developed on a too high a hierarchic level and that both the *Customer Requirements* and *Product Design Requirements* should be developed into more differentiated, specific requirements (i.e., the operational hierarchic level should be expanded).

In general, the development of a good and valid Customer Relationship Matrix should be a continuous on-going activity within a company's R&D organization, which should be organized and governed by a "matrix owner" as a part of an Integrated Knowledge Platform; this will be further discussed in Part V and VI of this book. Thus, it is usually necessary to review the matrix in a number of iterative sessions; the matrix must be "well kneaded" in interaction with specialists from different company functions representing different knowledge bases. The individual matrix areas and individual relations will then be gradually improved and refined.

Recalculate Customer Requirement Importance Ratings into Product Design Requirement Importance Ratings

Do not fall into the trap of believing that this recalculation is the most important result from the use of the QFD methodology! In reality, it is only one of many important outcomes. There are some weaknesses in this kind of calculation; the most significant of these is that if several *Customer Requirements* are very similar in character or overlap, these relations will give too-high figures in the *Product Design Requirement* dimension. In a similar vein, if several of the HOWs are of a similar nature (i.e., if they measure more or less the same characteristics of the product), they will provide somewhat redundant and confusing information. However, while the calculated figures should be carefully reviewed and discussed among QFD team members, they generally provide important and reliable input for use in company product development activities.

11.4 Review the room Directions for Improvement

The symbols in the room Directions for Improvement can be viewed as overall representations of the underlying, often complex and sometimes contradictory relations for each Product Design Requirement and all Customer Requirements in the Customer Relationship Matrix.

In the room Directions for Improvement, the direction for the improvement of each *Product Design Requirement* is presented; notably, this information is only from the

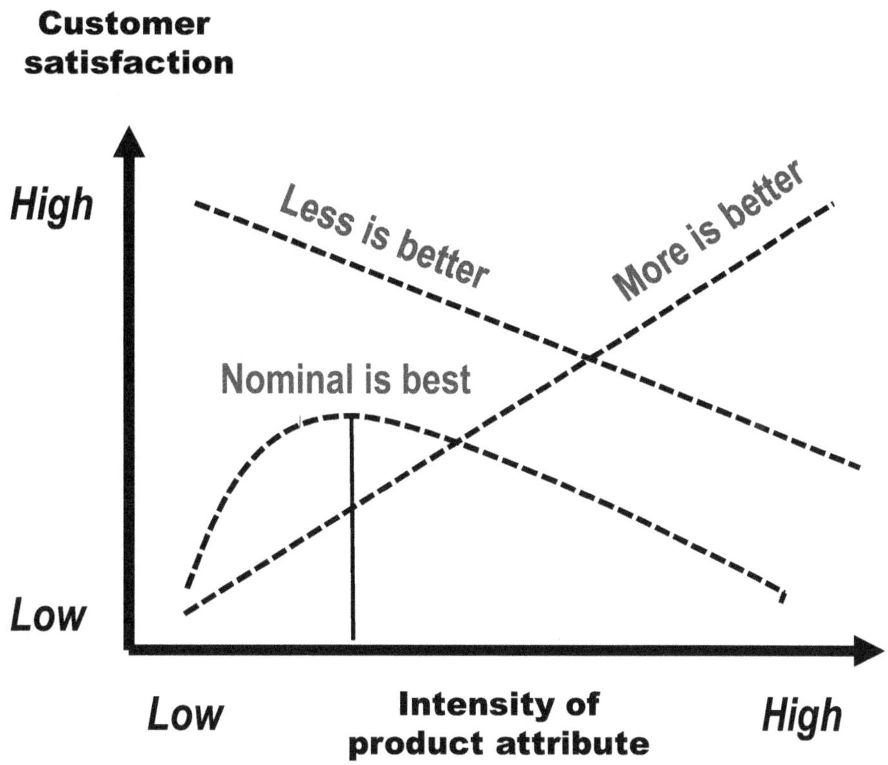

Figure 11.8 Directions for product improvements.

perspective of the customers. In this room, an upward arrow symbolizes "more is better for the customer," a downward arrow symbolizes "less is better for the customer," and a ring symbolizes "target figure is best for the customer." The objectives of this room are, first, to indicate customer preferences for the directions for improvement for each individual *Product Design Requirement*, and second to assist in the further development of the *Correlation Matrix* (i.e., the "roof" of the house).

It must initially be observed that the directions for change in this room certainly do not indicate any sort of "decided change" in intensity for product attributes (*Product Design Requirements*); they only indicate if a change could be of interest in a new or improved product, and what direction such a change should take in order to please the customer. Moreover, when the direction for improvement is to be decided on for an individual product attribute (*Product Design Requirement*), you must consider this individual requirement in isolation (a sterile environment for the selected attribute) while disregarding possible influences on other *Product Design Requirements.*

Such influences will afterwards be in focus and taken care of in the development of the *Correlation Matrix*. Further on, you must not take a stand regarding whether or not improvements are technically possible and whether or not a change is desirable. Fig. 11.8 illustrates that the underlying relation between customer satisfaction and the

Customer Requirements	Customer Importance Rating	Relation strength	Calculated relational importance	Type of relation
Requirement No 1	4	⦿	36	↑
Requirement No 2	9	○	27	○
Requirement No 3	7	△	7	↓
Requirement No 4	3	○	9	↑
Selected summary Direction for Improvement				↑

Figure 11.9 Selected summary of Directions for Improvement of an individual *Product Design Requirement.*

intensity of a select product attribute often is a curve which can take many directions and forms.

Completion of this room is generally experienced by a QFD team to be a fairly simple exercise based primarily on company representatives' "gut feeling" of customer preferences. In other cases, if the relationship room indicates many *Customer Requirement* relations of different kinds for a single *Product Design Requirement* (both positive and negative influences), this work can be complex. The recommended work process for this room is to review each *Product Design Requirement* and assign a symbol for the obvious ones, while leaving question marks for the difficult ones. Afterwards, you can go back separately to the difficult ones and review the total picture in the associated relationship room as illustrated in Fig. 11.9.

In Fig. 11.9, the underlying complexity of the development of each Direction for Improvement is illustrated. For each *Product Design Requirement*, there are a number of individual *Customer Requirement* relations that may behave differently if the intensity of the *Product Design Requirement* changes. The selected summary Direction for Improvement is thus composed of a number of individual Directions for Improvement that the QFD team must decide upon.

11.5 Initiate the Technical Benchmarking

Technical Benchmarking of competitive products is an important area in a company's general "Business Intelligence." Selecting the strongest competitive products from previous Customer Benchmarking and acquiring representative samples for internal laboratory testing along all

Product Design Requirements dimensions should be company high-priority activities. These activities should be a part of continuous product innovation and the creation of the company's *Integrated Knowledge Platforms* for enhanced product innovation.

Technical Benchmarking starts with the selection of competitive companies and products and is followed by the acquisition of representative product samples. This area requires intimate collaborations with the company's Sales & Marketing department, which sometimes can be difficult to accomplish because of internal barriers. In Part VI of the book, Chapter 25 is solely dedicated to this important issue in product innovation. If laboratory testing of competitive products is not carried out as a continuous company activity, this area of QFD work must be given top priority, since it will be time-consuming and sometimes difficult to collect these kind of competitive data.

When laboratory testing have been carried out on company and competitive products, the results must be reviewed, compiled and prepared for inclusion in the *House of Quality*. It is recommended that the results from the technical benchmarking be initially inserted in separate rooms, with each one representing a competitive product (see Fig. 11.10).

However, in the *House of Quality*, one room is usually designed for the presentation of all results from the *Technical Benchmarking* using snake plots. Since the laboratory data are in the form of intensities of different individual metrics, the data must be transformed from such "intensity scales" into a common "value scale." It is recommended to employ a nine-point ordinal value scale similar to the scale used in Customer Benchmarking in the Technical Benchmarking room. The recommended process for this transformation is presented in Fig. 11.11.

An explanatory example

Technical Benchmarking room using product "intensity scales"

In the "full version" of the *House of Quality*, you can find, in the lowest part of the matrix, the three Technical Benchmarking figures for the three selected competitive products. Additionally, marked in red, you also find the Target Figures for the Perfect Lunchtime Beer.

For example, for the Product Design Requirement "Colour", and using the absorbance figures on the EBC Colour unit scale (0 – 80), the darkest beer (Ale) measured 30, Pilsner 15 and Intern. Style 3.5.

Figure 11.10 Technical Benchmarking figures on "Colour" for the Perfect Lunchtime Beer using intensity figures.

A helpful hint

- Select one *Product Design Requirement* and review the laboratory data. If this *Product Design Requirement* has a upward-pointing arrow, select the intensities for "A product of excellence = 9" and "A poor product = 1). Afterwards, it should not be too difficult to select the intensities for the intermediate figures in the value scale.
- If the Product Design Requirement has a downward-pointing arrow, you can proceed in a similar fashion.
- If the Product Design Requirement has a target figure (a ring in the room Direction for Improvement), this indicates that a target figure for the intensity of a "world-class product = 9" must be set. Starting from this intensity, one can set intensity figures (higher and lower) for the consecutive figures on the value scale until you reach the two different intensities for "A poor product = 1".
- Afterwards, repeat this process for all Product Design Requirements" and you can add these figures in supplementary rooms into the *House of Quality.*
- You have now designed a transformation from "intensities" to a "value scale", and you can now review each Product Design Requirement and position your own product relative to all competitive products in the Technical Benchmarking room. It is generally experienced that connecting "snake plots" improve the visibility a lot in this room.

Figure 11.11 The process for transformation of "intensity scales" to "value scales."

In Fig. 11.12 the procedure for the translation from "intensity scales" to "value scales" is further illustrated. The transformation process greatly improves the visibility of benchmarking data, and the review of these results often provides food for thought in process innovation. In Fig. 11.12, for the product attribute A, the company product performs better than the competitive product, and the transformation scale is as follows: 10 intensity figure = poor product (1); 90 intensity figure = superior product (9). For product attribute B, the company product is inferior to the competitive product, and the transformation scale is as follows: 300 intensity figure = poor product (1); 100 intensity figure = superior product (9). Finally, for product attribute C, the company product has a similar performance to the competitive product, but they obtain this rating from different positions on the intensity scale.

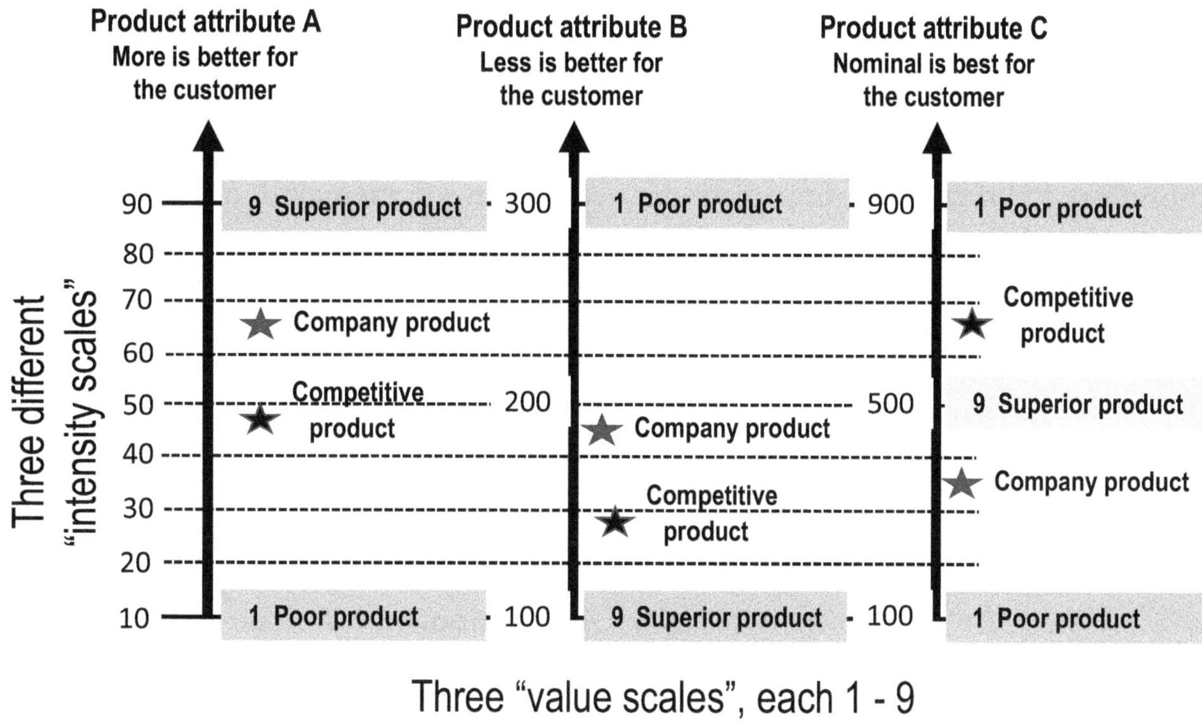

Figure 11.12 The process for transformation of "intensity scales" to "value scales."

The "full version" of the *House of Quality* shows, for example, the Technical Benchmarking figures for "Colour" in the value scale room; the situation here is to that of Product C in Fig. 11.12.

11.6 Construct and review the Correlation Matrix

Since relatively homogenous products are produced in the process industries, changes in the intensity of one product attribute (Product Design Requirement) will, in many instances, influence other Product Design Requirements (positively or negatively). For that reason, the Correlation Matrix room is far more important in the process industries than in other manufacturing industries.

In the *Correlation Matrix* room, it is delineated whether and to what extent individual *Product Design Requirements* interact, including whether the product attributes are "friendly" or "hostile" to each other. If an increased intensity of one *Product Design Requirement* (if that is the preferred direction of improvement) results in an increased intensity of another *Product Design Requirement* (if that is the preferred direction of improvement), it will be noted as a positive correlation with a ring symbol in this matrix room. If it is a strong correlation, it will be marked with a double ring.

On the other hand, in an increased intensity of one *Product Design Requirement* (if that is the preferred direction of improvement) results in an increased intensity of another *Product Design Requirement* (if that is NOT the preferred direction of improvement), this will be noted as a negative correlation with a cross symbol in this matrix room. If it is a strong negative correlation, it will be marked out with a double cross.

The relations between *Product Design Requirements* that are of a "nominal is best" kind and that are characterized with a ring are more intricate to sort out. The same goes for mixtures of arrows and rings. Furthermore, QFD software programs usually employ those symbols but do not generally differentiate whether or not the interaction is bidirectional. In other words, an intensity change in *Product Design Requirement* A can influence another *Product Design Requirement* B, but an intensity change in B does not necessarily result in an intensity change in A (i.e., the interaction may not be bidirectional).This important difference can be further explained using related "flag notes" for the individual symbols.

The recommended work process is to select one *Product Design Requirement* at a time and find out if there are any relations to other *Product Design Requirements.* Sometimes it is better only to mark if there is a correlation and later on review what kind of relation it entails, its strength, and the direction of changes.

Later, in the analysis of this room, it is of interest to focus on strong positive correlations, since an improvement of one *Product Design Requirement* automatically will benefit another. On the other hand, strong negative correlations indicate a development problem; however, if such negative correlations can be by-passed, this can offer an important and significant opportunity in product improvements. The relationships in the Correlation Matrix are further illustrated in Fig. 11.13.

11.7 Develop and design supplementary rooms of interest

For the *House of Quality*, there are a large number of optional rooms to develop and design. These are briefly introduced below. Nevertheless, please do not get stuck on only the listed supplementary rooms; instead, use your creativity to develop and design rooms that are important for your specific product innovation project.

Room for customer complaints

Quality can be defined as: When the Customer comes back to the supplier and not the product.

An explanatory example

On the use of Correlation Matrix relational symbols

Referring to the "full version" of the House of Quality, the relation between "Sweetness" and "Bitterness" is clearly a strong negative and bidirectional correlation. A change in "Bitterness" will result in a change in "Sweetness", and vice versa.

An increased "Sweetness" will generally result in an increase of "Calories", and a decreased "Sweetness" will generally result in a decrease of "Calories". The chosen symbol is a strong cross. However, in the other direction, a decrease in "Calories" does not necessarily influence the "Sweetness" (not bidirectional).

An increase of "Carbon dioxide" will certainly result in an increase of "Frothability", and a decrease in "Carbon dioxide" will also result in a decreased "Frothability". The relation is marked as a strong positive correlation (double ring). However, an increased or decreased "Frothability" will certainly not influence the "Carbon dioxide" content (this is not a bidirectional correlation).

The "Turbidity" will often influence the "Froth stability", and the relation is marked as a negative cross; however, the correlation is not bidirectional, since changes in "Froth stability" do not generally influence the "Turbidity".

Figure 11.13 On the use of correlation matrix symbols.

This room can be very rewarding to include; however, please be aware that product development in the past tended to put too much emphasis on customer complaints, neglecting customers' future wishes, desires and dreams. One can develop this room in the *Voice of the Customer* WHAT dimension or alternatively in the Technical Dimension related to the *Product Design Requirements* (the HOW dimension). The figures to include for each requirement may be related to "number of customer complaints over the past year" or alternatively to "company cost for complaints." Both alternatives are naturally of interest. The number of complaints is more related to the customer perspective, while the cost for complaints is of no interest for the customer but certainly is of interest for the company. Available information and advice from the Sales & Marketing department will determine how to proceed and how to set up this room; in some cases, both kinds of rooms can be of interest to use, and presentation with bar charts is generally advisable.

Room for present Target Figures and evenness of Target Figures

In the process industries, a radically improved product may be a product with the same product specifications but with radically improved and lowered tolerances.

Sustainability room

The sustainability room is a kind of room that is of growing importance because of companies' focus on sustainable production and products and on environmental issues in general.

Room indicating how difficult it is to contain present product target figures

This is an assessment of how difficult it is today to contain present product specifications and tolerance levels. This is often a most interesting room to develop in interactions with representatives from the company's production organization.

Room indicating how difficult it is to improve

This is an assessment of how difficult each *Product Design Requirement* is to improve, from either a technical or an economic perspective. Sometimes it is advisable to put this information into two separate kinds of rooms.

Room indicating how important it is to improve in a new or improved product

This is an assessment of how important each *Product Design Requirement* is to improve as a conclusion from the first Matrix Analysis exercise and after setting provisional Target Figures for a new or improved product.

11.8 Managing packaging integration for consumer product development

In other manufacturing industries and in modular design, there is a strong need to design products with components that interface well. In the process industries, which lack a component structure, there are no such interfaces to handle. However, in the case of, for example, fast-moving consumer products, the packaging is and preferably should be developed interactively during the development of the "core product." This is because the packaging serves multiple functions and sometimes interact with the content. Figure 11.14 illustrates how such a "system requirement" could be handled using a separate *House of Quality* and using a Correlation Roof to interface Product *Design Requirements* for the "core product" and the packaging.

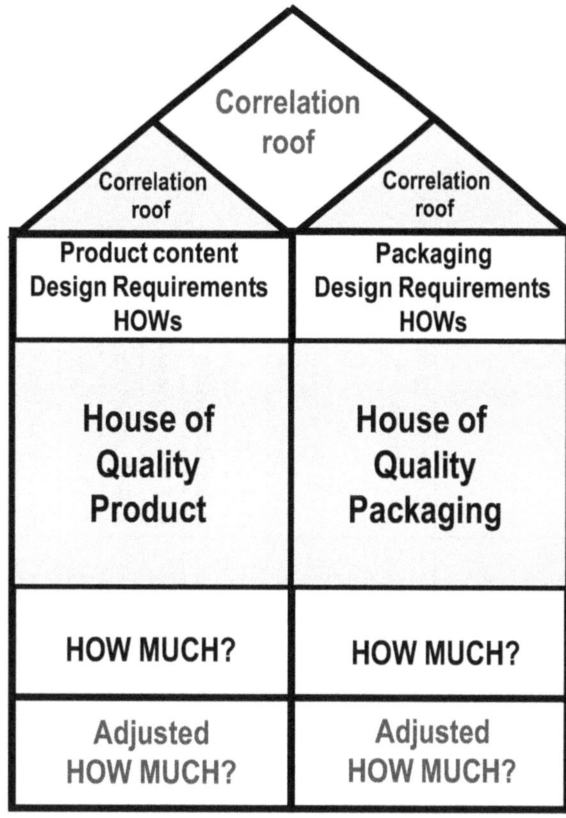

Figure 11.14 Integration of *Product Design Requirements* and *Packaging Design Requirements*, using a connecting Correlation Matrix.

11.9 The illustrative case of development and manufacturing of beer — The House of Quality

You have previously been introduced to the "full version" of the *House of Quality*. Now, the "mini version" of the *House of Quality* is presented in Fig. 11.15. Please observe that recalculated Technical Importance Ratings in the "mini version" will not correspond to the figures in the "full version", since there are additional Customer Requirement relations included in the calculation of the figures in the "full version".

In the *House of Quality*, we move from a Customer dimension into a Technical dimension. Only eight of the total *Product Design Requirements* have been selected from the "full version", even if there are additional such requirements that also relate to the *Voice of the Customer*. The *Voice of the Customer* has previously been presented in Fig 10.11; thus, in this section only the new information in the technical dimension will be discussed.

Product Design Requirements and related metrics

The selected Product Design Requirements and their associated metrics are presented in Fig. 11.15; however, you will find complementary information about test methods and metrics in the "full version" presented in Appendix C.

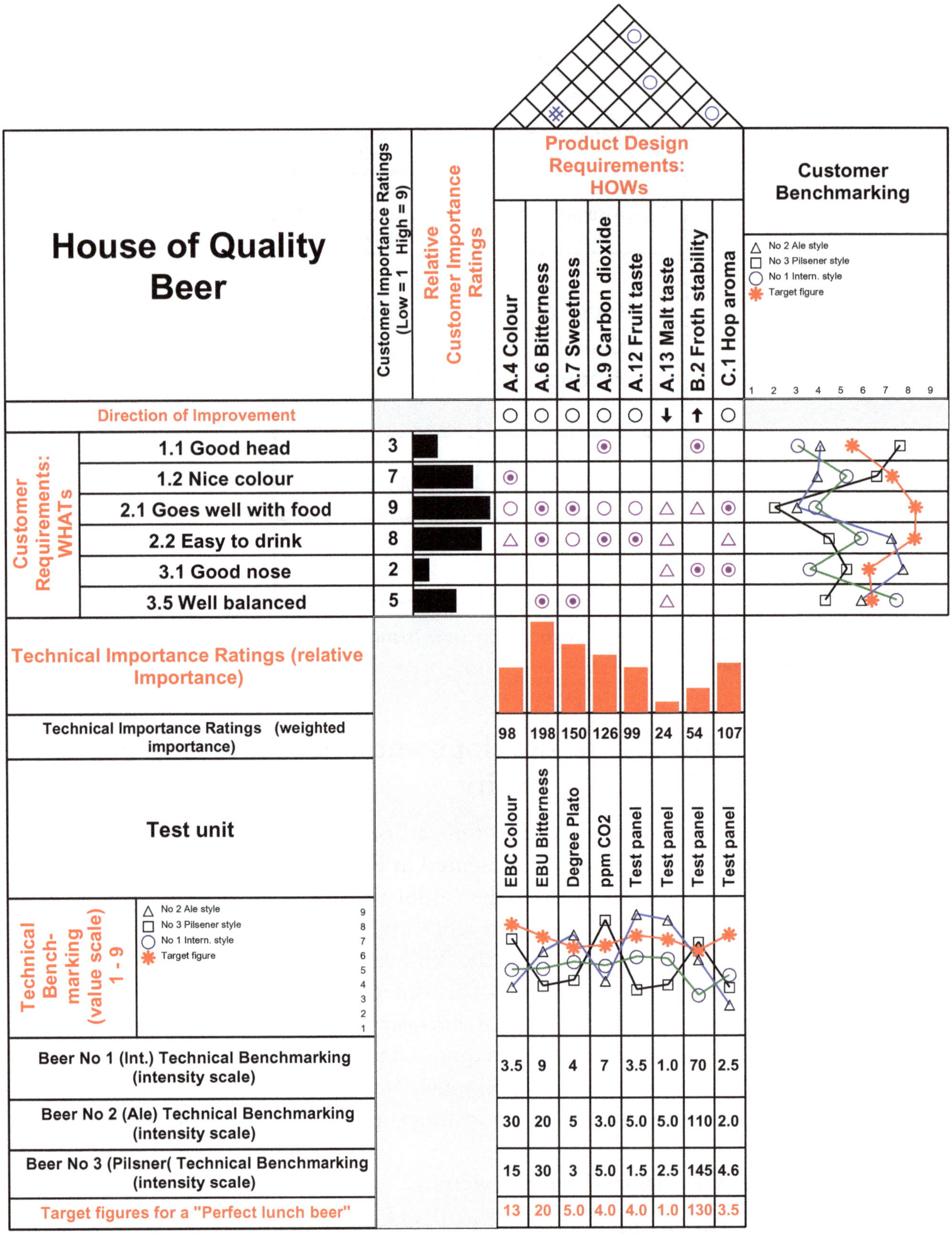

Figure 11.15 The simplified "mini version" of the *House of Quality* for the Perfect Lunchtime Beer.

The Relationship Matrix

In the Relationship Matrix, you will find the previously presented symbols indicating the strength of each relation. For example, the Customer Requirement "Good head" is thus strongly related both to the *Product Design Requirement* "Carbon dioxide" and to "Froth Stability". In a similar vein a fruity, not so hoppy beer is often regarded as a beer that is "Easy to drink", a fact which also applies to a beer with a low alcoholic content (see the "full version"). Please remember that relationships can be either positive or negative.

Technical Importance Ratings

Using the relationship matrix, Customer Importance Rating figures are recalculated into Technical Importance Ratings by multiplying each importance rating figure with each relation in the relationship matrix and adding up all figures. For example, in the calculations for Technical Importance Ratings for the *Product Design Requirement*, "Colour": $7*9 + 9*3 + 8*1 = 98$. In Fig. 11.15, all numeric figures for absolute Importance Ratings are presented, as are their Relative Importance Ratings with bar charts. The red bar charts representing Technical Importance Ratings show that the *Bitterness*, *Sweetness*, *Carbon dioxide* and *Hop aroma* of a beer are important *Product Design Requirements* in ranking order.

Direction for Improvement

The *Direction for Improvement* shows that most of the directions are of a Target Value kind, while only *Froth stability* has an upward arrow (the more the better) and *Malt taste* a downward arrow (the less the better).

The Correlation Matrix

In the Correlation Matrix (the roof), the *Bitterness* of the beer and the *Sweetness* of the beer are strongly "hostile" to each other. In contrast, a good *Froth stability* is rather "friendly" to *Hop aroma* (i.e., a good head is favourably to a good aroma of a beer). In the opposite direction, however, a good *Hop aroma* will not influence the *Froth stability.*

Technical Benchmarking

The figures for the Technical Benchmarking are presented in the lower part of the *House of Quality*, both as original laboratory data on intensity scales and as snake plots on a value scale. For example, the low Customer Benchmarking figure on *Good head* for the International style of beer was revealed in the Technical Benchmarking to relate to poor "Froth stability" (even if the "Frothability" figure as such was rather high; see the "full version"). This beer also did not perform well in the customer dimension "Easy to drink". The Technical Benchmarking showed that the "Carbon dioxide" content was very high.

The Product Design Requirement "Colour" confirmed that the Ale type of beer was the overall darkest beer, whereas the International type of beer measured "very light" on the EBC Colour Unit scale. The value scale presented as a snake plot illustrates well that the desirable Target value for the "lunchtime beer" is a bit darker than the International style of beer but not as dark as the Ale style of beer. The rather high figure for the "Bitterness" of the Pilsner style of beer (EBU Bitterness unit) indicates that the lower bitterness figure for the Ale type of beer would be more appropriate for the Perfect Lunchtime Beer.

Chapter 12

Matrix Analysis and Target Setting for New or Improved Products

In the course of our work, we have become convinced that better displays are a major avenue to valid qualitative analysis. The displays discussed in our book include many matrices, graphs, charts and networks. All are designed to assemble organized information into an immediately accessible compact form so that the analyst can see what is happening and either draw justified conclusions or move on to the next step of analysis the display suggests. As with data reduction, the creation and use of displays is not separate from the analysis, it is a part of the analysis.

(An Expanded Sourcebook: Qualitative Data Analysis, Miles & Huberman)

The above quote from top scholars in social sciences underscores the unique capability of the QFD matrices as instruments for collection and analysis of a huge amounts of information in product innovation. In one's development of the QFD matrices, one must thus observe the importance of designing the rooms using different fonts, different colours and selected symbols facilitating the display and analysis of important information.

12.1 Putting all information together in a Matrix Analysis

When you have duly arranged your facts in a logical order, it is like an oil lamp you have made, filled and trimmed, but which sheds no light unless first you light it.

(de Saint-Exupéry, 1984)

The Matrix Analysis is a grossly neglected part of the QFD methodology; and an oversight that may be partially attributed to the lack of proper presentation in textbooks. As poetically explained in the above quote, you have collected all the facts during the development and design of the *House of Quality*, but now is the time to put it all together and reflect and conclude on the "whole." However, when a QFD team has finished the erection of a *House of*

Quality, they are often mentally exhausted while also being satisfied with their achievement, and thus they sometimes cannot mobilize the necessary strength to undertake a Matrix Analysis.

However, paraphrasing Winston Churchill, this state of affairs "is not the beginning of the end but the end of the beginning"; that is, the Matrix Analysis of a *House of Quality* is only the important beginning of the proper use of the QFD matrices and information. Now is the time to reap the fruit of all previous efforts. However, it is recommended not to start such an analysis too early, specifically during the erection of the individual rooms, since one can easily get trapped in premature conclusions before the "whole" is available.

12.1.1 *Organizing for a stimulating event — who should attend?*

The Matrix Analysis is a QFD activity of utmost importance, and thus the preparation and planning for the event must be thorough. The most important aspect is to decide whom to invite for this stimulating but demanding exercise. One should not mix the activity of "disseminating information" within the company organization with the quite different activity of a Matrix Analysis. Both activities are naturally important, but to bring in new people with no previous knowledge about the QFD methodology and the actual project usually tends to "torpedo" such an event. Further on, if during the execution of the project you come across a team member totally lacking interest or even showing mistrust in the methodology, it is recommended not to invite this person to such an event. Rather, explain that the outcomes of the matrix analysis will be presented to him or her afterwards, hopefully providing an opportunity for him or her to reconsider this mistrust of the methodology.

12.1.2 *Arranging a creative environment — where to do it?*

The Matrix Analysis should be a really creative event, and thus ample time must be provided. It is generally recommended to arrange this event outside company premises during a two-day workshop. After the first full day of such a Matrix Analysis, a get-together should be organized in the evening in order to stimulate further discussions, as well as to hinder individual members from leaving for late-night office work distractions.

The meeting room should be open and large, and separate rooms for group-work must be provided for. The workshop is recommended to be chaired by the responsible project manager, but if she/he wishes to have more "private time" for reflection during such an event, a separate facilitator (either "internal" or "external") can be assigned to this task. Since this is an "extreme learning event" including the collection of both tacit and explicit information, a separate secretary (this could be a skilled team member) should be assigned to make notes on all discussions during the meetings.

If the Matrix Analysis is managed well, it is usually a truly stimulating and creative event in which new technical solutions and new product ideas also will emerge. Even if the Matrix Analysis is not a forum for designing new product solutions, but rather is used to set Target Figures for future concept development, such product improvement ideas that pop up during discussions must be carefully collected and preserved by the secretary and listed on a separate flip-board during this event. Please notice that the suggestions listed in the following text generally only relate to the mandatory matrix rooms. If several supplementary rooms have been developed, the analysis of such rooms must be carefully integrated into the agenda.

12.1.3 *Day 1 Review the total House of Quality and its individual parts*

Part I (Morning session) Take a "guided tour of the House" and briefly review the information together

The first activity after "cooling out" from other kinds of work is the introduction of all members and a review of the program for the Matrix Analysis. The first step is to review the total *House of Quality* together. Since the project has been going on for a considerable period of time, many team members have already forgotten why and how some activities were performed; thus, now is the time to visit the House and recapitulate and review all information together. This update should allow time for questions and open discussion, but the analysis should wait until group work has been carried out properly.

It is recommended first to start with a thorough review and recapitulation of how the previous collection of *Customer Requirements* and interaction with consumers/customers were arranged and completed, as well as how Customer Importance Ratings were collected (e.g., survey) and, finally, how Customer Benchmarking was accomplished.

Part II (Afternoon session) Analyse different parts and the content of the rooms in separate groups

The information contained in a fairly large *House of Quality* is generally so extensive that it is recommended to split up the analysis into three separate areas and to designate a small group of suitable persons for each area:

- *The Voice of the Customer*
- *The relationship and correlation matrices*
- *The technical dimension*

The structure for each area and the questions to be answered are presented in the three following sub-sections. A group leader and presenter of the results should initially be appointed within each group, and the results from the analysis are to be presented in slide-shows to facilitate further discussions and later documentation. The *House of Quality* should be available in the form of a very large print-out (in colour and at least double A1 size) on the wall in each group room. During the analysis, it is sometimes preferable to put the enlarged matrix on a table.

The use of such large matrix versions, and when people congregate around the matrix is experienced to greatly improve group-dynamics. A computer screen on a wall is a necessary complement for reviewing flag notes and other kinds of supplementary matrix information. The group leaders must encourage the group not to get stuck on overlong discussions of individual issues but rather to take notes, both directly on the provided large paper versions of the matrix and also in separate documents to be introduced into the software later on.

Group A — Analysing the *Voice of the Customer*

- *How good is the total hierarchical structure of* Customer Requirements? *Have we forgotten any requirements, or should any be removed?*
- *Are the* Customer Requirements *well-explained and understandable for customers? Are there any "double barreled" questions containing two or more questions in one?*
- *Mark the 20% most important* Customer Requirements *in red (following a Pareto Principle, these should they get most attention during the analysis), and mark the 20% least important in yellow (can they be given less attention in product design?). Discuss and reflect!*
- *If there are different groups of customers (reported in separate columns for Customer Importance Ratings), discuss how well Customer Importance Ratings agree or disagree.*
- *Are any horizontal areas in the matrix lacking "fill," indicating that the company has not measured a certain* Customer Requirement?
- *Put individual* Customer Requirements, *their importance ratings and benchmarking results together and reflect on the following questions: How well are the company and its competitors doing on the most important WHATs? Should the company position be improved for any* Customer Requirements? *How well are the company and its competitors doing on the least important WHATs? Should the company position be reduced for some* Customer Requirements? *Are there any important Customer Requirements where nobody is doing well (in terms of sales points)?*
- *Make some general comments on the total* Voice of the Customer. *Does it make sense?*
- *Are there suggestions for further activities?*
- *Should the* Voice of the Customer *be cut up into smaller complementary parts?*

Group B — Analysing the "product relationship matrix" and the "correlation matrix"

Product Relationship Matrix

- *How much "fill" is there in the matrix? Is there too much or too little?*
- *Review the "flag notes" for strong relations, and discuss the available information.*
- *Are there any areas of significant "weak knowledge" or important areas represented by "white spots" on the map, indicating potentially interesting areas for further R&D?*
- *Should the Product Relationship Matrix be cut up into several "sub-matrices"? Which areas should be further discussed, reviewed and validated in "internal" company forums, and which together with selected customers?*

Correlation Matrix

- *Review the room for Directions for Improvement. How valid are the selected symbols?*
- *How much "fill" is there in the correlation matrix?*
- *Review the "flag notes" and discuss the available information for strong positive or negative correlations.*
- *Are there any areas of significant "weak knowledge" or even important "white spots" on the map? Focus on strong positive and strong negative areas.*
- *Should the matrix be cut up into several "sub-matrices"? Which areas should be further discussed, reviewed and validated in "internal" company forums?*
- *Which matrix areas should be further discussed, reviewed and validated together with selected customers?*

Group C — Analysing the technical dimension of the *House of Quality*

- *How good is the total structure of* Product Design Requirements? *Have we forgotten any requirements, or should any of them be removed?*
- *Assess the validity of available test methods. Should the company improve some, develop new ones or make some existing ones obsolete?*
- *Mark the 20% most important* Product Design Requirements *and the 20% least important ones with different colours. Discuss and reflect!*
- *Review the individual* Product Design Requirements *and their test methods and metrics. For the most important* Product Design Requirements, *are the test methods good, valid and reliable? Does the company lack good test methods for any Requirements? Should the company stop doing some tests? Should the company develop supplementary new tests (internally or internationally)? What about multivariate relations?*
- *Do any vertical parts of the matrix lack "fill" or contain only weak relations, indicating that such measured product attributes do not really relate to the articulated* Customer Requirements?
- *Review the transformation from intensities to ordinal scale in the "basement" of the matrix, and review the scales in the perspective of the matrix content.*
- *How good is the company at Technical Benchmarking? Does the company have access to competitive products for Benchmarking (and how are they acquired)? How often is Benchmarking carried out? How well are the Technical Benchmarking results related to Customer Benchmarking results?*
- *Review the total information on individual* Product Design Requirement, *their calculated importance ratings and benchmarking results on an ordinal scale. How well are the company and its competitors doing on the most important and the least important HOWs? Are there any important* Customer Requirements *with which no one is doing well (in terms of sales points)?*
- *Should the total* House of Quality *be cut up into complementary matrices?*

12.1.4 *Day 2 Presentation of group work and a final synthesis of the results*

Part I (Morning session) Presentation of the results from the group work

The presentation of the group work should be carried out by the selected work group-leader using prepared slides and complementary comments on the large paper version. Free discussion should be encouraged, and the workshop secretary should make notes. These notes can be revised later on and presented in a comprehensive and digestible form as a report from the Matrix Analysis.

Part II (Afternoon session) Discussion and synthesizing of the results in a holistic fashion and deciding on further actions

During the previous group work, one has been moving within and between the individual rooms of the *House of Quality*. Now is the time to move between its three different larger parts and to take a holistic view of the total information; the "whole." The following list of questions to be answered and actions to be taken are only tentative, since each project may require separate activities:

- *Are existing company product specifications (internal or external) usually contained for the delivered products? If not which ones are difficult to achieve?*
- *For an existing product, review present* Product Design Requirements *and existing product specifications and make a preliminary review and assessment of provisional Target Figures.*
- *For a completely new kind of project, set preliminary target figures for* Product Design Requirements.
- *What kind of internal theoretical or experimental work should be carried out? Is it possible to synchronize such work with already on-going activities or programs.*
- *What kind of external theoretical or experimental work should be carried out? Are there any suggestions for collaborative partners?*
- *What kind of input from the Matrix Analysis is especially of interest for Sales and Marketing? Are there any actions that should be taken?*
- *What kind of input from the Matrix Analysis is especially of interest for Technical Support? Are there any actions that should be taken?*
- *What kind of input from the Matrix Analysis is especially of interest for Production? Are there any actions that should be taken?*
- *Should the* House of Quality *be enlarged or possibly reduced?*
- *Review the* Product Design Requirements *and make a preliminary assessment of which should be progressed into a Process Matrix, a Raw material Matrix and a Product Matrix.*

12.2 Consolidate matrix information for provisional and final target setting

Experience tells that "target setting" for product innovation is one of the most difficult issue in R&D, and that further product development is more of a craftsmanship

nature. Target setting in innovation is thus an issue of utmost importance and the "heart of the matter" for corporate innovation of excellence.

(Curt Nicolin, former ABB CEO and Chairman of the board)

Target setting has been briefly mentioned in the previous chapters, and the Matrix Analysis is the point of departure for this activity. During the Matrix Analysis, a large number of comments, ideas and conclusions will emerge that afterwards must be compiled and structured into a Matrix Analysis document. During the Matrix Analysis, it is recommended as a finishing activity to set "provisional targets" as a concluding activity for the workshop. After distribution of this material among the product development team, it is recommended that they should get together and during a half-day session discuss and agree on final Target Figures for a new or improved product.

The selected figures should be introduced in the *House of Quality* together with the selected Target Figures for upper and lower tolerance levels. The motives should also be presented in an internal report, but the overall motives should also be inserted as "flag notes" with the associated target figures into the matrix. If associated Target Figures for associated process capabilities (Process Matrix) and raw materials (Raw material Matrix) have been developed, it is strongly recommended that a separate workshop should be arranged in which the total information is integrated and possible target figures are adjusted into final Target Figures. Finally, all Target Figures will then guide the further development of new Product Concepts and Process Concepts.

As usual, it must always be observed and noted whether different Customer Segments may require different product varieties. One must thus avoid falling into the "Majority Trap," which involves designing a product for an imaginary "average customer" without actually satisfying any real customer. This phenomenon is related to the issue of product variety and platform-based design, which will not be discussed further in this book.

12.2.1 *The* House of Quality *as one part of an* Integrated Knowledge Platform

The *House of Quality* and the results from the Matrix Analysis are afterwards to be included in the *Integrated Knowledge Platform* for the product or product family to be deployed in further company product innovation. This will be presented and further discussed in the following Part V of this book.

Chapter 13

Summing Up and Some Issues to Reflect Upon

In this part of the book, you have learned the rationale for the often-used mantra of "listen to the voice of the customers", and you have been presented with a methodology that can assist you in this endeavour. Some complementary ways to interact with customers have been presented, and it is argued that sometimes several of them can be successfully deployed simultaneously in one project. The most important message, however, is not how to do it but simply the importance of doing it.

Product development of excellence cannot do without an intimate understanding of customer needs, wishes, dreams and unmet desires. Securing a company system that provides product innovation projects with such information on a continuous basis is thus of unparalleled importance, and it is argued in this book that no other methodology has yet been able to perform this task better than the QFD methodology. In summary, listening to customers' "murmurs" and clearly expressed needs includes:

- *Developing of a Stakeholder Structure*
- *Collecting* Customer Requirements
- *Refining and structuring* Customer Requirements
- *Estimating the importance of individual requirements*
- *Assessing how well the company meets requirements compared with competitors*
- *Reviewing the combined information of each requirement*
- *Taking a holistic perspective of the total* Voice of the Customer

QFD, however, is not only about customer understanding but also, and most importantly, about competitors and competitive product understanding. For this reason, you have been presented with a system for Customer Benchmarking and a related system for Technical Benchmarking. The collection of information on competitive products is naturally very important, and information on future competitive products must also be addressed by company business intelligence. The importance of reliable and customer-oriented test methods for product functionalities has been stressed, and experience tells that the QFD methodology generally is a superior instrument to delineate and outline the company needs, as well as a reliable strategy for test method improvements and development.

However, it is not sufficient for future product innovation only to listen to the *Customers' Requirements* and transform them into *Product Design Requirements* and finally into target figures. Future products must also be produced in cost-efficient production processes using functional raw materials. Thus, the *Customer Requirements* must be progressed through the *House of Quality* into the company production system. In the following Part IV, of this book you will be shown how this can be accomplished using the QFD phase progression methodology and the deployment of the upstream & downstream matrices (mpQFD).

After reading this part of the book about QFD theoretical and methodology use, it is now important that you try to relate this information to the perspective of your company's product innovation behaviour and philosophy. To facilitate this, please reflect on the following questions:

- Do you believe that your company's Sales & Marketing department is fully aware of future customer requirements and desires for company products?
- In some cases, the Sales & Marketing department focuses more on marketing and sales of current offerings than on investigating customers' present and future product preferences. What is the situation in your company?
- Do you believe that Sales & Marketing systematically progress new customer information into the R&D organization as an on-going activity, or is it more of an ad hoc process?
- Select one important product in your company and assume that company product developers have a long list of Customer Requirements to fulfil. Can you identify the three most important and the three least important requirements for this product, or do you believe that they all are of equal importance?
- Regarding the selected product, can you name the three strongest competitive products and their manufacturing companies? Identify each competitive product's most important "sales point" from a customer perspective.
- Using the same product, identify and list in ranking order the three most important common customer complaints on the product.
- How much of the total company R&D budget is allocated to the improvement of available product test methods or the development of new test methods? How "customer-oriented" and how reliable are your present test methods?
- In the perspective of the usually strong importance of "evenness" of each Product Design property in the process industries, how much attention is this kind of product property getting in the overall product innovation budget?
- The room Directions for Improvement presents only a simplified picture, since the relation between customer satisfaction and intensity of product attributes is generally better represented by a curve. Using the selected product, draw up such a curve using Customer satisfaction and the intensity of three important *Product Design Requirements.*

- Interaction between *Product Design Requirements* is intricate and often difficult to fully understand for many sectors of the process industries. Did the review of the correlation roof in the *House of Quality* trigger some ideas for further company-related activities?
- Using your previously selected product, in the perspective of the correlation roof, please name three pairs of *Product Design Requirements* that are strongly "hostile" to each other.
- Reviewing the suggested questions and actions presented for the Matrix Analysis, do you agree that such an analysis is important and really requires a two-day working session? Alternatively do you believe that such reflections could be left to individual product developers?
- If you do not currently use the QFD methodology, do you believe that such a holistic review of products is really necessary, or do you have a different best practice?

Part IV

The Advanced Multiple Progression Quality Function Deployment System (mpQFD) — Introducing the Upstream & Downstream Matrices Adapted to Process-Industrial Conditions for Innovation

The new mpQFD methodology really glued the different sub-projects together, and the use of classical QFD methodology to create a strong foundation for market research was unquestionable. The results showed that new test methods for product properties needed to be developed in order to measure unspoken consumer demands for good finished dish and cooking functionality. The new matrices in the system pointed out the necessity to improve fundamental knowledge in order to give a stronger foundation for further competitive new product development.

(Lager and Kjell, 2007)

In the previous review of the process-industrial use of the QFD methodology in Part II of the book, and in the further acknowledgement of the QFD methodology as a knowledge-building instrument in the following Part V, it will be noted that, unfortunately, the industrial use of the QFD upstream or downstream matrices is scarce. One likely cause for this state of affairs is that after the development of a *House of Quality*, a QFD team will sometimes experience fatigue and thus prefer to take a pause before proceeding with further methodology implementation and use. Surveying phase progression in the literature, including other manufacturing industries, the development of the customer-oriented *House of Quality* is often in focus, possibly related to QFD users' lesser knowledge of industrial production systems and process technologies. However, it is argued in this book that in neglecting the opportunity of phase progression in product innovation, an important advantage in the industrial use of the QFD methodology is overlooked.

The multiple progression QFD system (mpQFD), referred to in the above quotation, was used in an industrial R&D project aiming at an improved fundamental knowledge of "cooking cream usability and functionality" at the ArlaFoods company, a major European dairy producer. The experience from the use of the QFD methodology as such, and the new mpQFD system in particular, was generally considered by the QFD team to be positive and fruitful (Lager and Kjell, 2007). In this project, four advantages of the system were identified: cross-functional organizational benefits, creation of a product innovation focus in market research, provision of a fundamental understanding of new consumer requirements, and, ultimately, creation of a platform of knowledge for subsequent product innovation projects. However, the mpQFD system has so far been primarily used by companies in the European process industries, but some recent case studies present encouraging perspectives and incentivize its further dissemination and process-industrial use (Tottie *et al.*, 2016).

From customer understanding to design for "processability" — an overview of the mpQFD system architecture

The original mpQFD system that was presented 10 years ago (Lager, 2005b) has been, in this book, further improved and refined into an advanced instrument for product innovation in the process industries. The development of the Advanced Multiple Progression System (mpQFD) has been guided by the analysis of necessary QFD system capabilities for enhanced process-industrial product innovation, as previously presented and discussed in Chapter 8.

The ambition has been to design a flexible QFD system that can be deployed by experienced practitioners as well as by new users of the methodology. Nonetheless, in the spirit of late Professor Yoji Akao (Akao, 2002, 2003), the system is only intended to give structural guidance and should not to be followed strictly as a "cookbook". Since an early analysis of company usability of the QFD methodology (Lager, 2005a) ascertained that the aspect of "knowledge creation" is one of its most important outcomes, the matrices in the mpQFD system have consequently been given related complementary names to emphasize this methodological capability.

The *Voice of the Customer* and the *House of Quality* have already been introduced and have been further explained as marketing and development-oriented work process activities. Since production process parameters, unlike requirements on raw material properties, obviously cannot be progressed into deeper levels of the production system, but will pose demands on raw materials, the upstream mpQFD system has been split into two separate matrices, as illustrated in the following Fig. P4.1. In the production-oriented Process Matrix, *Product Design Requirements* from the *House of Quality* are related to *Product Influencing Process Parameters* in the production process, and this matrix is thus also designated as ***Process Understanding***.

In the Raw material Matrices, *Product and Process Requirements on Raw materials* are related to *Raw material Design Requirements,* and these matrices are thus designated as ***Raw material Understanding***. The split of process parameters and raw material properties

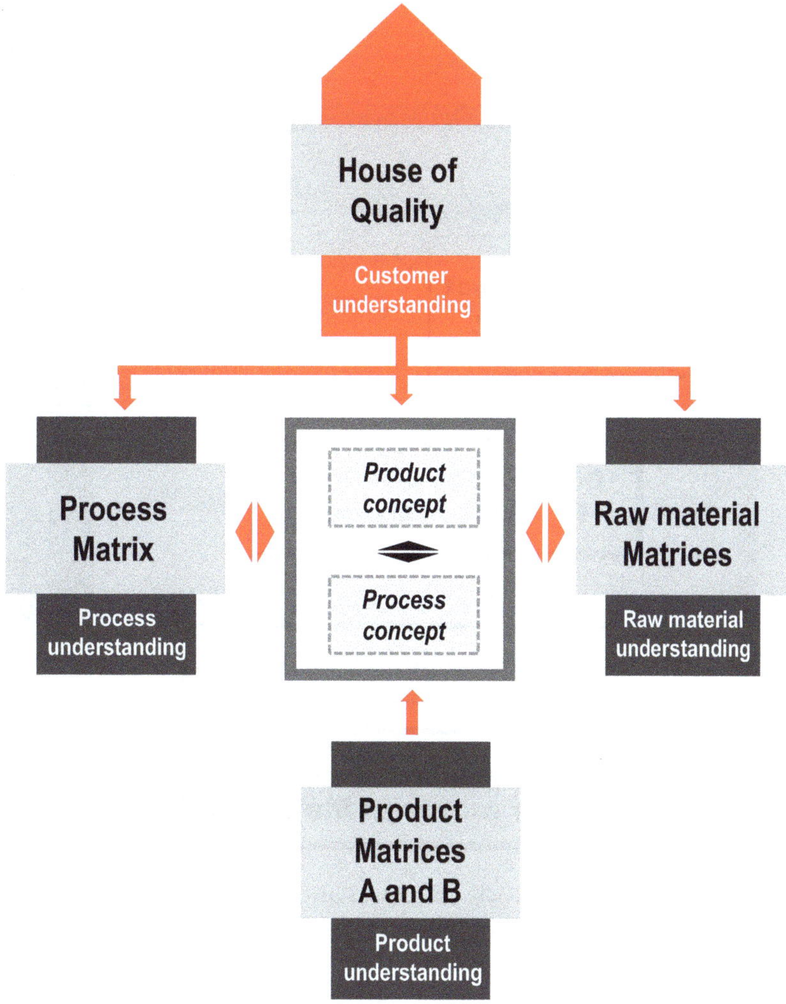

Figure P4.1 The principle QFD matrices included in the Advanced Multiple Progression Quality Function Deployment System (mpQFD). The *House of Quality*, the *Process Matrix*, the *Raw material Matrices* and the *Product Matrices A and B* are standard components of the system.

facilitates the construction and use of the matrices but also enables a multiple progression of *Raw material Design Requirements* into the upstream raw material supply chain. The complementary Product Matrix A provides a more profound understanding of the relationship between *Explanatory Product Characteristics* and related *Product Design Requirements* from the *House of Quality*, while in Product Matrix B *Explanatory Product Characteristics* are related to *Product Influencing Process Parameters* from the Process Matrix. These two matrices are thus designated as ***Product Understanding***.

The overall work processes for the upstream matrices

In this context, it is essential to note that "Global Product Innovation" is not really synonymous with "Global Manufacturing". Thus, understanding different customer preferences and the use of product platforms for the delivery of product varieties on various

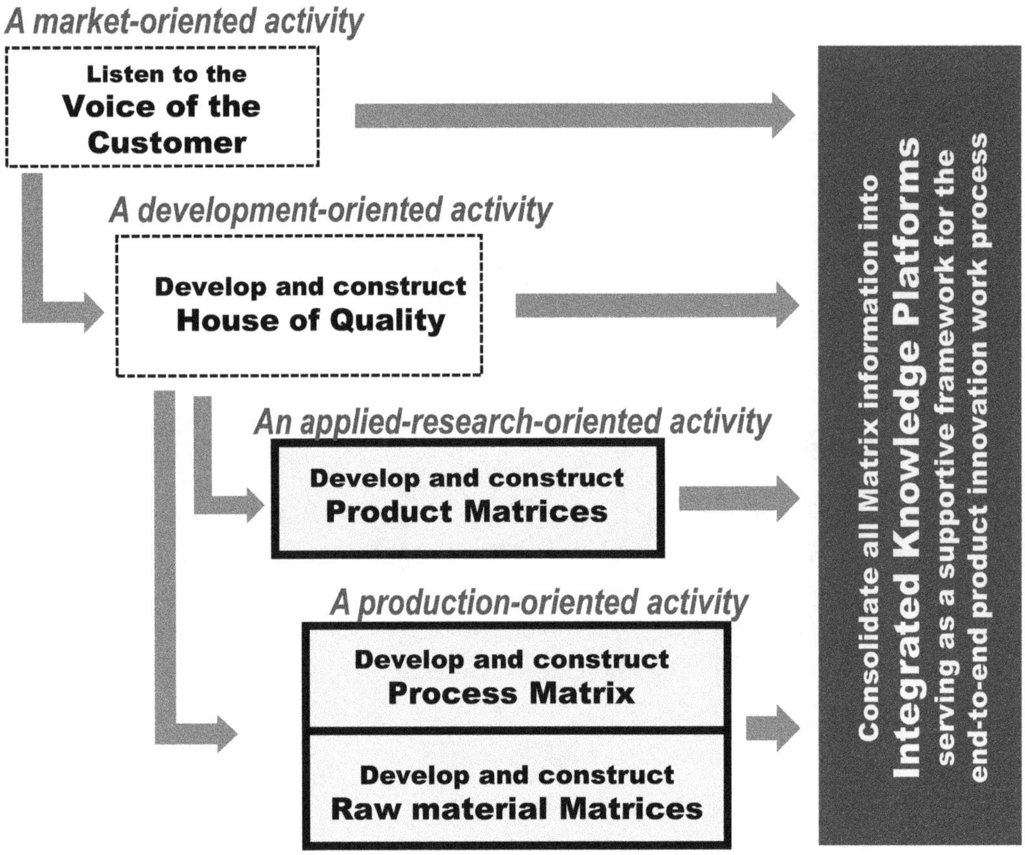

Figure P4.2 An outline of the mpQFD work processes focusing on the upstream matrices. The *Voice of the Customer* is the "foundation" on which the *House of Quality* is built. Using the HOWs from the *House of Quality*, the other matrices can afterwards be erected and the total matrix information can be consolidated in an *Integrated Knowledge Platform* (a concept that will be further explained in Part V of the book).

geographical markets is not the same as being able to globally manufacture products with identical product specifications. However, both objectives require the use of the upstream QFD matrices in order to secure such production system capabilities in the introduction of new or improved products.

In the previous Part III, the matrices and related work processes for the *Voice of the Customer* and the *House of Quality* have already been introduced, and in Fig. P4.2 the work processes for the remaining upstream matrices are briefly illustrated. After *Product Design Requirements* have been selected and delineated, the construction of both the Process Matrix and the Product Matrix A can be initiated. However, the development of the Process Matrix must precede the development of Product Matrix B and the Raw material Matrices, which will be further explained in Chapter 15.

In the next Chapter 14, you will initially learn how to construct the Process Matrix and its individual rooms; an activity that can be characterized as a "production-oriented" work process. The Process Matrix is usually developed following the material flow in the related production process. Depending on the complexity, size and structure of the company's internal supply chain and production system, one must initially decide which part of the production process to select for the construction of the Process Matrix. However, for a fairly uncomplicated process configuration, one can select the total production process, from raw material introduction to delivery of the final product.

In the following Chapter 15, the development of the Raw material Matrix is described. In the development of this matrix, you will learn that one must progress selected *Product Design Requirements* into this matrix as WHATs, as well as some requirements from the production process on the raw materials. The development of the Product Matrices are presented in Chapter 16 as activities characterized as "application-research-oriented". The use of these matrices in process-industrial applications is often rewarding and sometimes even experienced by the QFD team as revelatory from a product innovation perspective. Finally, in Chapter 17, multiple progression within and outside company demarcations is demonstrated. That chapter will further explain how the use of the upstream multiple progression can cascade *Raw material Design Requirements* from the Raw material Matrix into deeper levels of the supply chain. Moreover, it will clarify how downstream multiple progression also can facilitate an improved end-user understanding in B2B relations.

Chapter 14

The Process Matrix — The Facilitating Instrument for an Improved "Process Understanding"

The further use of integrating tools into the formal work process, like the QFD methodology, will be another avenue to follow. But because this tool and its House of Quality is primary a tool for bridging the R&D–Marketing interface, its further use in bridging the R&D–Manufacturing interface through phase progression into the production process will be another area of interest to explore.

(Gamal, 2015)

In a recent study of an improved packaging innovation work process at Nestlé, and in reference to the above quotation, the use of the QFD methodology as an integrating tool between company products and its production system emerged as an interesting instrument to further explore. While seeking technical solutions is the major concern in product design, it is at the production stage that product costs are actually committed, product quality is finally determined, and lead times for product launch are set. This implies that the rationale of cost-efficient and successful product innovation can only be fulfilled at the production stage (Jianxin *et al.*, 2007).

Because of the previously noted strong relation between product functionalities and the production system in the process industries, it is thus essential for a QFD system to be able to translate and progress *Product Design Requirements* from the *House of Quality* into the production process. Phase progression with the mpQFD system is the proper instrument to achieve such an objective, which is accomplished in the development and use of the Process Matrix and the associated Raw material Matrices.

In Chapter 25, the use of the QFD methodology as an instrument in bridging company R&D–manufacturing interfaces will be discussed more in detail, but in the present chapter it will be explained how the Process Matrix can be developed and used for the support of the company product innovation work process. Furthermore, a detailed knowledge of how the often numerous process parameters in the production process influence the many *Product Design Characteristics* is still usually not well understood, even for production processes that have been operational for hundreds of years. For that reason, the Process Matrix is an important component in the mpQFD system for the establishment of solid *Integrated Knowledge Platforms* in search of an improved process understanding.

In this book, the Process Matrix is primarily presented as a tool for product innovation, but the matrix is certainly also usable in other company innovation-related activities, which is articulated in early QFD methodology use at the Kimberly-Clark company (Scheurell, 1994):

> *QFD is a dynamic process and matrices must be modified to meet customer needs.... The methodology can be used in many different projects such as new products and processes, new products with existing processes, and cost reduction programs to name a few.*

Moreover, in the production of new or improved products, the usability of the Process Matrix during production plant start-ups has also been proven (Mongeon, 1996). This underscores the fact that the QFD methodology is an important tool in the pre-product development phase, the product development phase, and the post-product development phase, as well as in the subsequent marketing and sales of new products. This is a significant but generally overlooked aspect of the use of the QFD methodology, since product innovation projects are often delayed because of production start-up problems. However, it must also be noted that the Process Matrix can be used not only at the production planning stage but also as a facilitating tool in training plant operators in advance of plant start-ups (Scheurell, 1993):

> *QFD has made us realize that "process redesign", after start-up due to poor planning can be a big part of the product development cycle time line.... It points out areas, where the process was not totally thought out yet, usually transition points between processes.... It was during the development of the "performance matrix" that we discovered a major problem in the design of the equipment in its ability to meet key customer requirements.*

Fortunately, the development of a Process Matrix is generally less demanding on company resources compared to the construction of a *House of Quality*, mainly because it does not require the collection of new or updated customer and competitor information. However, the production-oriented Process Matrix in the mpQFD system needs to be developed in a truly cross-functional approach, and in collaboration with representatives from both the production organization and R&D. Since the production organization generally is difficult to interact with during time-consuming product development activities, it is recommended that a QFD representative from the R&D organization should prepare an outline matrix structure before it is introduced, discussed and further developed together with production.

14.1 Introducing the Process Matrix and the guiding work process for matrix development

The Process Matrix is the facilitating instrument for improved Process Understanding and will answer the following question: "How do *Product Design Requirements* from the *House of Quality* relate to *Product Influencing Process Parameters* in the Process Matrix?"

From a production perspective, the above question can be rephrased as follows:

- *Taking into account the* Product Design Requirements *for the development of a new or improved product, what are the most important* Product Influencing Process Parameters *to consider and possibly improve in the development of a new or improved product in order to contain the Target Figures from the* House of Quality?
- *How can the company production organization use the Process Matrix to develop into an organization with a clear focus on customer benefits — bringing customer preferences online?*
- *Considering the often long and complex process-industrial production process and its large number of* Product Influencing Process Parameters *in the perspective of the large number of* Product Design Requirements, *how can the Process Matrix guide the development of improved conceptual models of this complex multivariate system, and thus contribute to an improved process understanding?*

In Fig. 14.1, the general structure of the Process Matrix is depicted. In the design of this matrix, *Product Design Requirements* from the *House of Quality* are selected as WHATs, and the *Product Influencing Process Parameters* are used as HOWs in the overall guiding structure. Not only selected *Product Design Requirements*, but also their related Importance Ratings and associated Target Figures, must be progressed into the Process Matrix from the *House of Quality.*

14.1.1 *The overall work process for the development of the Process Matrix*

The overall work process for the development of the Process Matrix is outlined in Fig. 14.2 It is initially recommended that you briefly review this total work process, which in the

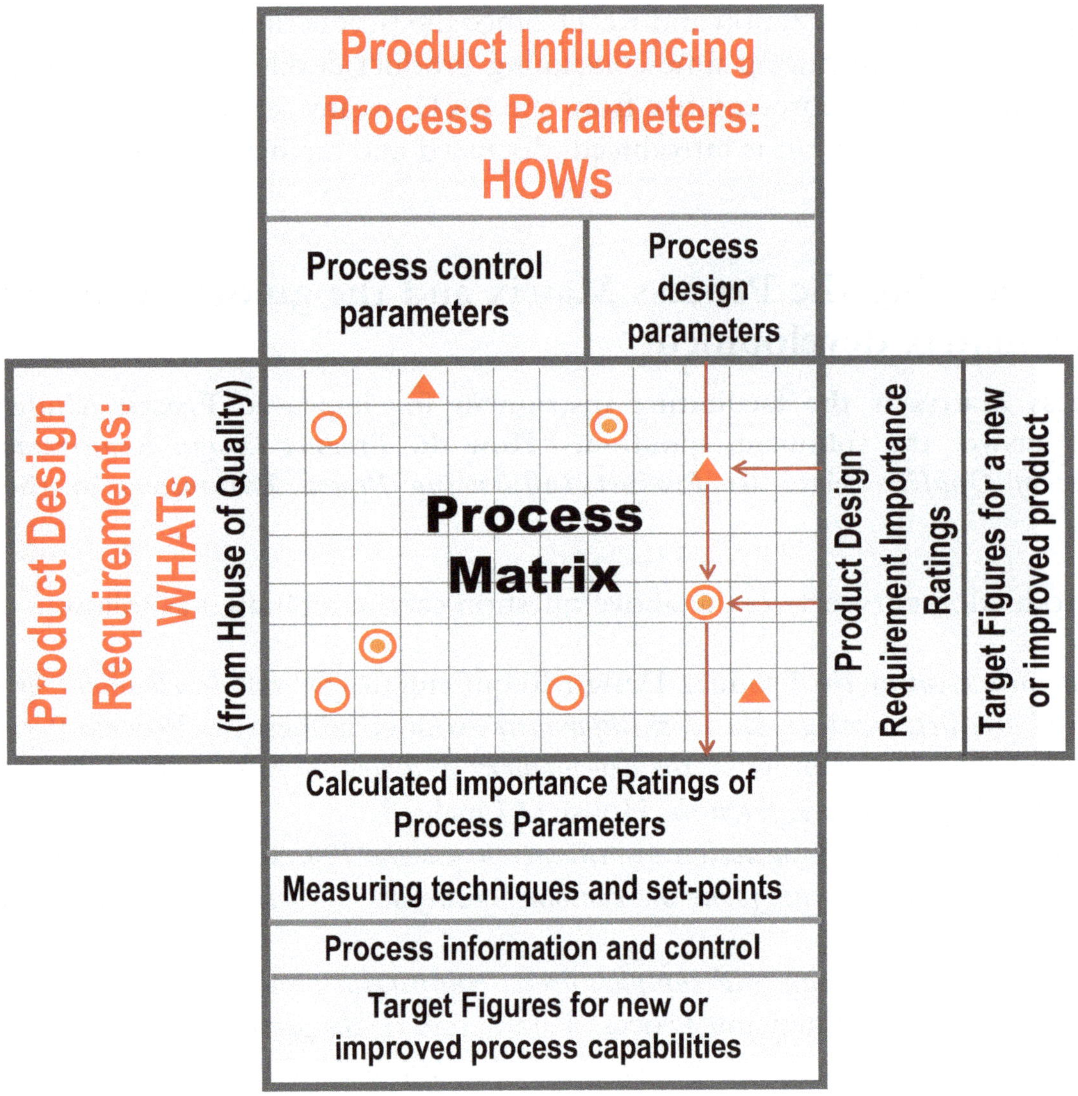

Figure 14.1 A general structure of the Process Matrix and associated default matrix rooms.

following sections will be discussed in more detail. Some QFD software programs have automatic phase progression mechanisms available from the *House of Quality* to the upstream matrices, but it is advised that such functions should be ignored, since the selection of *Product Design Requirements* from the *House of Quality* is a complex activity, which will be discussed further in Section 14.2.

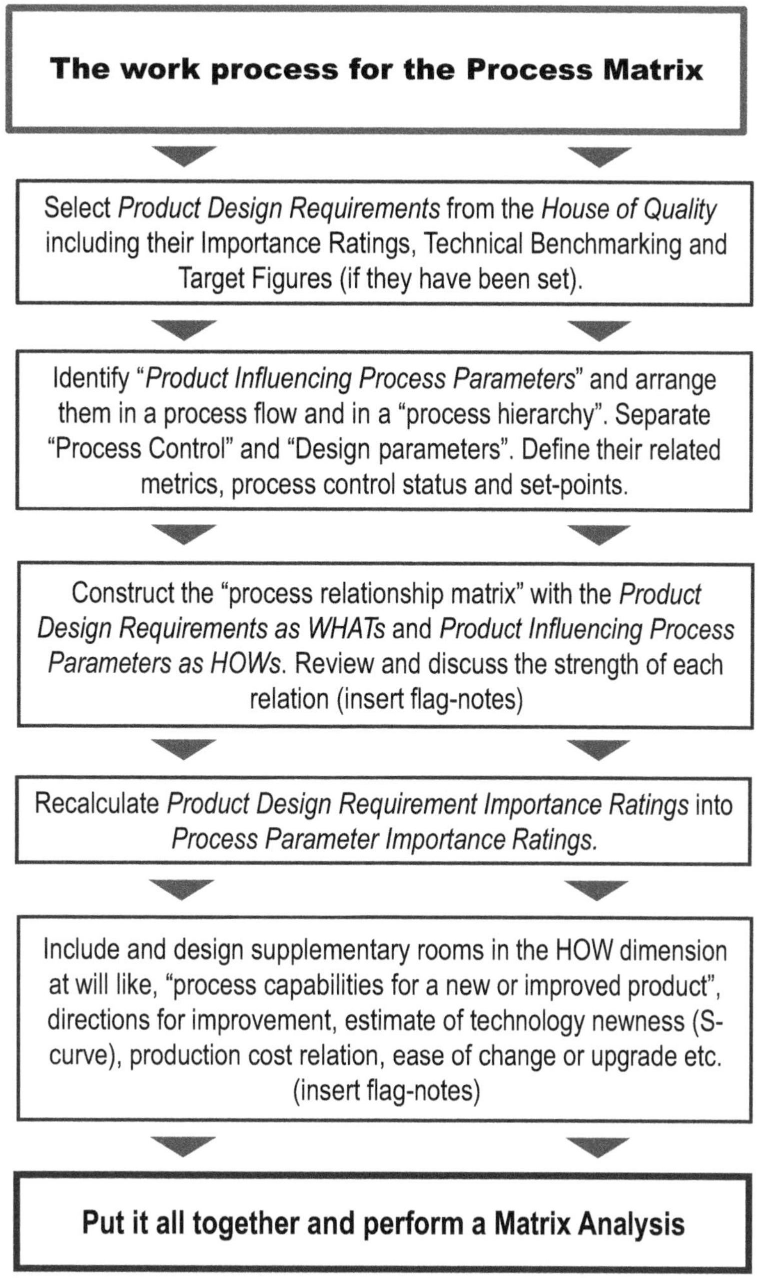

Figure 14.2 The overall work process for the development of a Process Matrix.

14.2 Select Product Design Requirements from the House of Quality — the WHATs in the Process Matrix

In Fig. 14.3, the linkage between the previously presented *House of Quality* and the Process Matrix is illustrated. In other manufacturing industries, it is sometimes advised to select and progress only those *Product Design Requirements* that are new, important, and difficult for phase progression. However, in the process industries, manufacturing more or

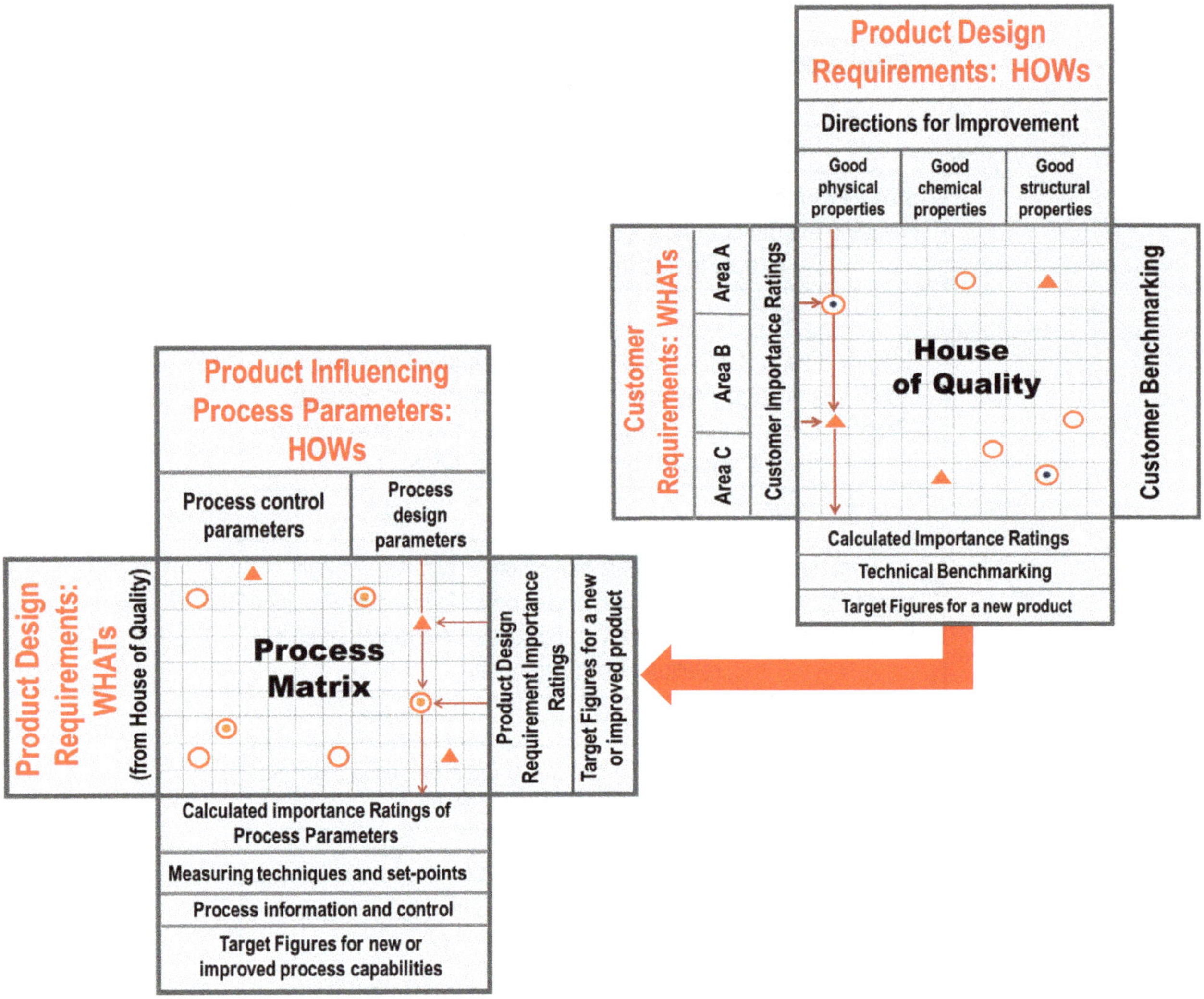

Figure 14.3 Using selected *Product Design Requirements,* or HOWs, from the *House of Quality* as WHATs in the development of the Process Matrix.

less homogenous products, such an approach is not feasible because of the strong interrelationships among all *Product Design Requirements.* Consequently, if the number of HOWs in the *House of Quality* is below 20, all of them should generally be progressed into the Process Matrix. However, try to avoid multiple *Product Design Requirements* of a very similar character, in order to keep the Process Matrix limited to a manageable size and format.

If the number of HOWs in the *House of Quality* is larger than 20, it is recommended to progress a selected numbers of HOWs into the Process Matrix as WHATs, and the following criteria can be used:

- *Focus on "functional product attributes" with high Technical Importance Ratings in the* House of Quality. *In this context, such product attributes are characteristics that cannot be "measured directly on the product" but are measurable product attributes in the "customer use of the product"; mark those WHATs in a normal black font in the Process Matrix.*

- *Select all* Product Design Requirements *that are used by the production organization for online control of the operation of the production process; mark those WHATs with a red font in the Process Matrix.*
- *For consumer products, product specifications communicated with customers (tables of product content) must be considered; mark those WHATs with a green font in the Process Matrix.*
- *For products intended for B2B customers, the focus should be on product attributes that are communicated with the customers in relatively well-specified contractual terms, written product specifications or supplied metrics on-line; mark those WHATs with a green font in the Process Matrix.*
- Product Design Requirements *that during the Matrix Analysis of the* House of Quality *have been identified for improvements in the development of a new or improved product are naturally important to progress into the Process Matrix; mark those WHATs with a bold font in the Process Matrix.*
- *Another guiding principle is to review Technical Benchmarking results in the* House of Quality, *which, if properly assessed for individual HOWs, should reflect some important product attributes that could be of interest to further explore in production in a product competitive perspective; mark those WHATs with an underline in the Process Matrix.*
- *If, in the development of the HOWs in the* House of Quality, *new* Product Design Requirements *have been identified including new test methods, such HOWs are truly important to progress into the Process Matrix in order to explore production relational aspects; mark those WHATs with an italic font in the Process Matrix.*

The QFD team must consider all of the above criteria and decide which *Product Design Requirements* should be selected for phase progression into the Process Matrix. However, the overall guiding principle for selection of *Product Design Requirements* and the use of the Process Matrix is to make the company production system more customer-oriented.

14.2.1 *An alternative short-cut in the development of the upstream matrices*

If a company, in pure B2B relations with its customers, is located within a chain of external industrial customers, and suppliers, a simplified and unorthodox approach in the use of the QFD methodology and in the development of the upstream matrices is to disregard the development of the WHATs in the *House of Quality*.

The point of departure for this alternative is to directly use the *Product Design Requirements*, the HOWs, and their associated importance ratings and product specifications (sometimes jointly set by the company and its customers) for the development of all upstream matrices. In this unconventional approach, the *Voice of the Customer* is left out, and the customer product specifications are directly deployed as the WHATs in the development of the related Process Matrix and Raw material Matrices.

Each QFD user within such a long production chain can either, in a true QFD spirit, develop a full *House of Quality* or, alternatively, decide to start directly with the measurable *Product Design Requirements* (product specifications). The selected route to follow is dependent on whether the company is planning to develop the product further or whether it simply wants to progress product specifications as *Product Design Requirements* into the production process.

14.3 Identify Product Influencing Process Parameters — the HOWs in the Process Matrix

After *Product Design Parameters* have been selected from the *House of Quality* as the WHATs for the development of the Process Matrix, the HOWS for this matrix must be clarified and selected as *Product Influencing Process Parameters* for the production process. It is generally advised to structure the HOWs in two separate matrix areas: Process Control Parameters and Process Design Parameters.

The Process Control Parameters are the production process parameters that are used to control individual process equipment, unit processes or total chunks of process structures, or new process parameters that could be of interest to develop in a future improved production system for new or improved products. In the review and identification of such process control parameters, the QFD team should follow the process logic using existing "process flow-sheets" and lay-outs, and should follow the material flow from initial raw material introduction into the production system, until the finished product is manufactured. Raw materials (ingredients) and their specifications should also be included here following the process flow; see Fig.14.4.

Process Design Parameters are part of all *Product Influencing Process Parameters* related to equipment characteristics which are of a relatively fixed nature and often cannot be changed during continuous operations. Such Process Design Parameters can be totally fixed or manually adjustable, but if manually adjustable often in need of a process shut-down. Example of fixed equipment characteristics are the diameter of a rotating mill or the temperature range of a furnace. Examples of manually adjustable Process Design Parameters are a screen that can be operated with decks of different apertures or the adjustable speed of rotating equipment. In summary, Process Design Parameters are related to equipment that, in the development of new or improved products, is likely to require investments for upgrades or minor modifications or even replacements.

It is recommended to arrange *Product Influencing Process Parameters* in a well-structured hierarchical order, which later on will facilitate the use and understanding of the matrix. Such hierarchical structures of WHATs and HOWs improve the visibility and process understanding but nevertheless will not be used for matrix calculations. If only minor product improvements are under consideration, it is recommended that the Process Design Parameters part of total *Product Influencing Process Parameters* in the Process Matrix should

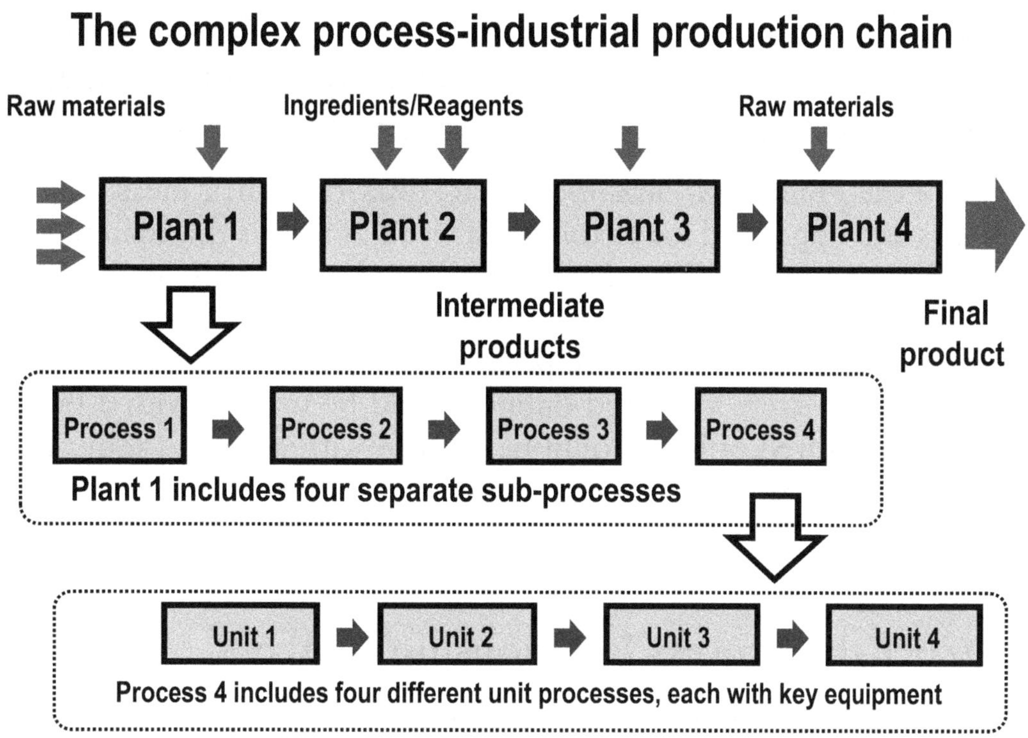

Figure 14.4 A schematic (Russian doll) model and an example of the complex production chain in the process industries (Tottie and Lager, 1995).

be omitted in order to make the matrix smaller and easier to develop and use. If, however, a completely new product is being considered, design parameters must be included, but with a focus on those parameters that have a control span and a pronounced impact on future improved product properties.

If the process system is a very long extended chain of several interrelated sub-processes, as illustrated in Fig. 14.4, it is generally advisable to select only the part of the process chain where most of the product functionalities are created. In such a case, upstream processes could better be considered as internal raw material suppliers.

However, if the decision has been made to treat the total process as one integrated chain of processes, try to identify target figures for "intermediate product deliveries", or in-between production process transition points. Neglecting such important interfacial points in the Process Matrix may seriously hamper the use of the matrix. Such sub-process interfaces could sometimes benefit from the development of a "mini Raw material Matrix" as a facilitating instrument between internal product suppliers and internal customers.

In a review of existing *Product Influencing Process Parameters*, new potential parameters are sometimes discovered that are difficult to control using today's available process technology or sensors. Because such parameters can be of vital future importance if they could be developed into supplementary controllable new *Product Influencing Process Parameters,* they should also be included in the Process Matrix, preferably designated with a special font or colour.

14.4 Identify Product Influencing Process Parameter's measuring instruments, metrics, set-points and upper and lower tolerance levels and raw material specifications

After a selection of the *Product Influencing Process Parameters*, their measuring instruments, metrics, set-points and upper and lower tolerance levels as well as raw material specifications should be identified and inserted in separate supplementary rooms or as "flag notes"; see Fig. 14.5.

It is advisable to mark the text for the parameters in the Process Matrix with different font styles. Process parameters measured online should be marked with a bold font, while process parameters measured offline should be marked with a normal font. The identification of Tolerance Levels for set-points is important in the perspective of future improved evenness of product specifications for new or improved products.

14.5 Construct the Process Relationship Matrix

The Process Relationship Matrix can be developed in a similar fashion to the Product Relationship Matrix in the *House of Quality*. It is thus recommended to start filling in the relations for the Process Matrix "column-wise". The general course of action is to select one process parameter and discuss, assess and determine whether and how changes in the

An explanatory example

Process control parameters and metrics

In the "full version" of the Process Matrix, you will find some Process Control Parameters of different kinds in the mashing process. The "Mash thickness", "Mash pH" and "Filtering temperature" (often called sparging) are process parameters that are easy to measure and control on-line. Meanwhile, "Particle size" and "Water hardness" are usually measured off-line.

The parameter "Particle size after crushing" is in reality a complete screen curve, and the top size and fines are only two points on the curve, which is produced in a laboratory. Small-scale brewing processes need large husks for good filtering properties (high top size), while large scale brewing processes with high tech filtering equipment are not dependent on the husk particle size; thus, they favour a fine grind in order to improve "extract recovery".

Figure 14.5 An Explanatory Example of Process control parameters and related metrics and set-points.

An explanatory example

The Process Relationship Matrix

In the "full version" of the Process Matrix, you will find some Process Control Parameters of different importance for the mashing process. The "Mash thickness" is a process parameter that is easy to measure and control on-line, and the metric is Liter of water per Kilogram of wort; a proper figure for "Mash thickness" is 2.7. However, the influence of this process parameter on *Product Design Requirements* is small (few relations in the relationship matrix).

On the other hand, the "Filtering temperature" (also called "Sparging temperature") is a parameter that may substantially affect the *Product Design Requirements*. After mashing, the grains are washed with hot water in order to recover residual wort. Unfortunately, some undesirable compounds (astringents) may be collected at the same time, which can contribute to beer staling and haze. The higher the water temperature, the higher the risk.

The "Mashing pH" is another process parameter that is easily measured and controlled on-line. A favourable range for the "Mash pH" is 5.2-5.5. A lower pH will decrease enzymatic activities, while a higher pH may result in dull malt flavours (a medium strong relation in the matrix).

Figure 14.6 The process relationship matrix.

parameter intensity (lower or higher) will have an influence on any *Product Design Requirement* (see Fig. 14.6). The use of the HOWs in the Process Matrix as "knobs for the intensities" is generally a good explanatory image for this kind of exercise. If there is an influence (for better or for worse), it must be estimated how strong the relation is, with explanatory "flag notes" if needed.

Previously presented relationship symbols for the *House of Quality* are to be used, and, after a "column-wise" review, it is strongly recommended to do a similar check-up in a "row-wise" revision. The development of the Relationship Matrix, which is the most important part of the Process Matrix, is generally felt to be a true learning experience. However, it usually requires a number of consecutive reviews and iterative improvements, preferably in sessions with different company expert teams from both production and R&D. Separate reviews in collaboration with external experts from different knowledge areas are generally rewarding complementary activities in the development of the Process Matrix into an instrument for an improved process understanding.

14.6 Recalculate product design requirement importance ratings into process parameter importance ratings

Using the calculated Technical Importance Ratings from the *House of Quality*, the relationship matrix can now recalculate those figures into Process Parameter Importance Ratings. They can be presented as figures, but is generally advised to illustrate the results as bar charts in the bottom of the matrix.

14.7 Review production process control status

What is measured and controlled in the production process today, and what should be measured and controlled in the future?

In the mpQFD system, related process control rooms are integrated into the Process Matrix and directly inserted below the relationship matrix, and these rooms are primarily intended to indicate how each process parameter is controlled today. In addition, it is of interest to review all *Product Influencing Process Parameters* in the Process Matrix and indicate which ones are presently controlled online (mark the text with a bold font) and which ones that are not controlled today but are possible to control in the future (italic font). Complementary rooms related to the process control status could include "how difficult to control", "how important to control" and "direction for process control improvements".

It must be observed that the Process Matrix and associated rooms for process control do not constitute a process control model, despite the fact that an analysis of the Process Matrix certainly can give important input for such developments. The large number of *Product Influencing Process Parameters* and their relations to the large number of *Product Design Requirements* create a multivariate picture that can stimulate and challenge existing (or non-existing) company models for process control. The development of a Process Matrix can thus give suggestions and input for the further development of "conceptual explanatory models" for multivariate process control and an improved process understanding.

14.8 Develop and design supplementary rooms of interest

Depending on the character of the production process and its related structural complexity, there are a number of optional supplementary rooms in the HOW dimension that could be of interest to develop and use depending on the intended purpose and future use of the Process Matrix:

A room for Technology Newness — technology position on the S-curve

The S-curve concept developed by Foster (1986) can serve as a useful tool in discussions of the maturity and competitive strength of existing company production equipment, unit process technology and overall main production technologies. The S-curve analyses the technology in the dimensions of "effort to improve" and "performance outcome of improvements".

In such estimates, it can be advantageous to insert supplementary rooms categorizing process technologies (Dussauge *et al.*, 1987, Granger) as:

- *Mature technology that is difficult to improve with further process innovation.*
- *New technology that can be developed with a high improvement potential using internal process innovation (core process technology).*
- *New generation technology that is available on the market but still not in use by the company (basic process technology).*
- *Technology that is in progress but still not available for purchase on the market (emerging process technology).*

Positions for selected *Product Influencing Process Parameters* should be commented on with "flag notes."

A Correlation Matrix as a top roof

The different *Product Influencing Process Parameters* naturally relate to each other, often in a very complex way. Adding a roof on top of the Process Matrix can serve the purpose of exploring the interplay of chunks of the process system, and the roof can function as a complementary room to the Relationship Matrix.

A room for the assessment of process parameter influence on the recovery of valuable materials in the production process — product yield

In some production processes in the process industries, the two most important and measurable outcomes from a production process are "product quality" and "product yield".

A room for the appraisal of the present use of equipment production capability

This room is intended to give an estimate of a selected equipment's available capability for increased production volumes, as well as a guiding reference for removal of "production bottlenecks".

A room for the relative frequency of equipment outages

This is an important room in the Process Matrix, and it can sometimes be supplemented with other kinds of information related to equipment stoppages using flag-notes.

A room for the display of intermediate product interfaces

The purpose of this room is to depict important production system interfaces and to present product specifications for intermediate products using “flag notes”. For very long internal company supply-chains, such interfaces are also organizational, and it can be of interest to develop a small Raw material Matrix or even a *House of Quality* in order to straighten out the relation between internal suppliers and internal customers in selected process interfaces.

A room for estimated relative yearly operating cost

This room could be supplemented with absolute operating costs for individual equipment or unit processes.

A room for estimated relative yearly capital cost

This room could be supplemented with absolute capital costs for individual equipment or unit processes. This is in line with the concept of “fully burdened part cost” in other manufacturing industries. For the production of homogenous products, this is clearly more difficult to calculate compared to assembled products, but the issue is nevertheless not of any less interest to address.

A room for necessary improved production capabilities for a new or improved product

This room is used to define and pinpoint the need for new operating conditions and/or necessary improved process capabilities and design in the production of new or improved products.

A room estimating how difficult the Product Influencing Process Parameters will be to change or improve

In this room, the QFD team will give a rough estimate how difficult it would be to change this *Product Influencing Process Parameters* in terms of engineering or, alternatively, from a cost perspective. It can be interesting to separate the assessments into process parameters for “transitional change” and process parameters for “discontinuous change”.

A room for directions for improvement of Product Influencing Process Parameters for the production of a new or improved product

After a matrix analysis, the overall conclusions and decisions can be presented in this room, which will be of the character of Directions for Improvements.

14.9 The illustrative case of beer development and manufacturing — The Process Matrix

You have been introduced to the "full version" of the Process Matrix in the previous sections of this chapter, presented in detail in Appendix D. Now, you will find the "mini version" of the Process Matrix presented in Fig. 14.7. Please observe that the recalculated Importance Ratings from the *House of Quality* presented in the bottom of the matrix in the "mini version" will not correspond to the figures in the "full version", since there are additional Product Design Requirement relations included in the calculation of the figures in the "full version".

In the Process Matrix, we move from a product dimension into a Process dimension. Only eight of the *Product Design Requirements* have been selected from the "full version". In the process dimension, only eight Process Influencing Parameters have been selected for the construction of the "mini version" of the Process Matrix.

The *Product Design Requirement* importance ratings presented in the "mini version" are recalculated with the Process Relationship Matrix into Product Influencing Process Parameter Importance Ratings. For example, the importance rating of the parameter "Fermentation scheme" is as follows: 150*9 + 126*3 + 99*9 + 54 = 2673.

Product Influencing Process Parameters

In a review of the "full version" of the Process Matrix, the importance of the Fermentation part of the total brewing process stands out. The Product Influencing Process Parameters with the highest importance ratings are thus "Type of Yeast", and "Fermentation scheme"; a result that is well in line with the introductory presentation of the total brewing process, when it was acknowledged that the Fermentation Process is one of the most important but difficult and complex part of the brewing process.

The raw material "Caramel malt dosage" parameter naturally relates strongly to the product attributes "Colour" but also the "Malt taste". The low importance rating of the adjunct dosage can vary, and is naturally dependent on which kinds of adjuncts that are added. During the mashing process, the parameter "Mashing scheme" relates strongly to both the "Sweetness" and the "Malt taste" product attributes. However, there are many different mashing schemes, and this simplified version does not include the process parameter "Mashing temperature" and complex temperature/time profiles.

During the wort-boiling process, the importance of the "Bitter hop dosage" process parameter is naturally recognized and strongly related to the product attribute "Bitterness". For all that, it must be recognized that the hopping of beer may take many forms and relate to many parts of the total brewing process. During fermentation, the importance of "Type of yeast" and the overall "Fermentation scheme" stands out, and both process parameters relate strongly to the "Sweetness" and "Fruit taste" of the beer.

Process Matrix Brewing		Product Design Importance Ratings	Relative Product Design Importance Ratings	Product Influencing Process Parameters: HOWs							
				Type of pilsner malt	Caramel malt dosage	Adjunct dosage	Mashing scheme (temp. & time)	Bitter hop dosage	Type of yeast	Fermentation scheme	Bunging pressure
Selected Product Design Requirements: WHATs	A.4 Colour	98		◉	◉						
	A.6 Bitterness	198						◉			
	A.7 Sweetness	150		△	○	△	◉		◉	◉	
	A.9 Carbon dioxide	126							△	○	◉
	A.12 Fruit taste	99							◉	◉	△
	A.13 Malt taste	24		◉	◉	○	◉				
	B.2 Froth stability	54		○	○		◉		△	△	○
	C.1 Hop aroma	107						◉			
Weighted Importance				1410	1710	222	2052	2745	2421	2673	1395
Relative Importance											

Figure 14.7 The simplified "mini version" of the Process Matrix for beer manufacturing.

Supplementary rooms in the Process Matrix

There are a number of supplementary rooms inserted in the "full version" of the Process Matrix, but they have not been filled in, since the focus of this book is not on brewing science but on the use of the methodology in the construction of the downstream QFD matrices.

14.10 Put it all together and perform a Matrix Analysis of the Process Matrix and target setting

14.10.1 *Matrix analysis procedures and set-up*

A Matrix Analysis of the Process Matrix can, in many respects, follow and be guided by the previously presented recommendations for Matrix Analysis of the *House of Quality.*

Day 1 Review the total Process Matrix and its individual parts

Part I (Morning session) Take a "guided tour of the Process Matrix" and briefly review the information together

The first activity after "cooling out" from other kinds of work is the introduction of all members and a review of the program for the Matrix Analysis; afterwards, the first activity is to review the total Process Matrix together. Since the project usually has been going on for a considerable frame of time, many team members have already forgotten how some activities were performed; thus, now is the time to visit the matrix and recapitulate and review all information together. During this update, questions and open discussion should be allowed, but the analysis should wait until group work has been completed.

Part II (Afternoon session) Analyze the different parts and content of the rooms in separate group work

Group A — Analyzing the Product Design Requirements (product attributes) and the Relationship matrix

- *Review the total structure of* Product Design Requirements, *WHATs, according to the previously presented list in Section 14.2. Discuss the different kinds of* Product Design Requirements *in the perspective of their selection criteria (fonts and colours).*
- *Do any horizontal parts of the matrix lack "fill" or contain only weak relations, indicating that such Product Design Parameters, WHATs, do not really relate to any Product Influencing Process Parameters, HOWs. In other words, are there no process strategies for controlling those WHATs?*
- *How much "fill" is there in the total matrix? Too much or too little? In consequence, should some WHATs or HOWs be expanded or condensed for clarification or possibly removed?*

- *Review the total process structure, the HOWs, and discuss at what stages of the process important product functionalities are created. This information is partly on display in each matrix row as strong relational symbol patterns in the relationship matrix for functional product attributes (use a marker pen).*
- *Are there any areas of significant "weak knowledge" or even important areas of "white spots on the map" indicating potentially interesting areas for further experimental work?*
- *Review the "flag notes" (focus on the strong relations) and discuss the overall content of available information.*
- *Should the Process Relationship Matrix be split up into several sub-matrices? Which areas should be further discussed, reviewed and validated in selected internal company forums or with external experts?*

Group B — Analyzing the Process Parameter dimension of the Process Matrix

- *How good is the total structure of* Product Influencing Process Parameters*? Have any been forgotten, or should any be removed?*
- *Mark the 20% most important and the 20% least important Product Influencing Process Parameters with different colors. Discuss and reflect!*
- *What are the most important "knobs" in the production system? Which ones are available, and which "knobs" could one wish to have?*
- *Do any vertical parts of the matrix lack "fill" or contain only weak relations, indicating that such Process Parameters do not really relate to any* Product Design Requirements *(product attributes)? In other words, should any of them not be included among the Product Influencing Process Parameters?*
- *In the review of the rooms for process control, the following aspects should be discussed for Product Influencing Process Parameters with high importance ratings:*
    - *Are there any difficulties in controlling this parameter?*
    - *What is the frequency of necessary changes?*
    - *What is the severity of an "out-of-control" condition?*
    - *What is the organization's ability to detect an "out-of-control" condition?*
- *Review the information in all supplementary rooms. If there is a large number of supplementary rooms inserted in the Process Matrix, it is recommended that a third group C should focus only on that part of the matrix and that the second day for Matrix Analysis should be extended into a full day.*
- *Should the total Process Matrix be split up in complementary matrices in the vertical dimension?*

Day 2 Presentation of group work and final synthesis of the results

Part I (Morning session, 2 hours) Presentations of the results from group work

The presentation of the group work should be carried out by the selected work group leader using prepared slides and in a complementary review of the paper version. Free discussion should be encouraged, and the workshop secretary should take notes. These notes should later be revised and presented in a more digestible form as a report from the Process Matrix Analysis.

Part II (Morning session, 2 hours) Discussion and synthesis of the results in a holistic fashion and decisions on further actions

Now is the time to move between the matrix parts and to take a holistic view of the information. The following list of questions to be answered and actions to be taken is only tentative, since each project may require separate activities:

- *Are existing company product specifications (internal or external) usually contained in delivered products? If not, which ones are difficult to contain?*
- *What kind of theoretical or experimental work should be carried out? Is it possible to synchronize with already on-going activities or programs?*
- *What kind of external, theoretical or experimental work should be carried out? Are there any suggestions for collaborative partners?*
- *What kind of input is of interest for Sales and Marketing? Are there any actions to be taken?*
- *What kind of input is of interest for Technical Support? Are there any actions to be taken?*
- *What kind of input is especially of interest for Production? Are there any actions to be taken?*
- *Should the Process Matrix be enlarged or possibly reduced?*
- *Review the* Product Influencing Design Parameters *and make a preliminary assessment of what new kinds of Production related data should be defined and progressed into the Raw material Matrix.*
- *Set provisional target figures for* Product Influencing Process Parameters *in the perspective of "Necessary improved production capabilities for a new or improved product".*
- ***Discuss and try to answer the three important questions initially presented in the previous Section 14.2.***

Provisional Target Setting and inclusion of the information in an Integrated Knowledge Platform (IKP)

The Process Matrix and the results from the Matrix Analysis should next be included in an *Integrated Knowledge Platform (IKP)* for the product or product family to be deployed in further company product innovation. IKPs will be presented and further discussed in the following Part V.

Chapter 15

The Raw Material Matrices — The Facilitating Instrument for an Improved "Raw Material Understanding"

> **Although it is possible to achieve substantial benefit from the implementation of only the first out of several phases of QFD, the greatest benefit can be realized only when all phases of the concept are implemented in an integrated manner ... without the active involvement of suppliers, all the effort invested in previous phases may not produce optimal results.**
>
> **(Ansari and Modarress, 1994)**

In other manufacturing industries, the integration of component suppliers in the development of new or improved products is crucial; consequently, it is currently common product innovation best practice. However, in process-industrial product innovation, the integration of raw material and ingredient suppliers in product innovation is unfortunately still often in its infancy. This chapter will discuss how Raw material Matrices can be developed, designed and used to lend overall support to company product innovation in the process industries.

It has previously been recognized that incoming raw materials in the process industries not only influence the choice of unit operations and the design of the total production system but also may influence the properties of finished products (i.e., the process embodies the product). However, even if R&D representatives from many sectors of the process industries often are fully aware of this situation, the details of how multiple raw material characteristics influence production process conditions and product attributes are often still not well-understood and sometimes lack fundamental fact-based knowledge.

Furthermore, in a sustainability perspective on product innovation and renovation, selection and use of raw materials produced under stringent environmentally acceptable conditions is of paramount importance today. For that reason, the Raw material Matrices are important components in the total mpQFD system for the establishment of a solid

Integrated Knowledge Platform supporting the overall product innovation work process. Neglecting the development of the upstream Raw material Matrices is thus a serious oversight in QFD methodology use.

Similar to the development of the Process Matrix, the production-oriented *Raw material Matrices* must be developed together with the production organization. If the raw materials are delivered from external suppliers, this is likewise an interesting opportunity to develop such matrices together in a truly open innovation approach. Moreover, representatives from the company's raw material purchasing organization should certainly also be involved, but beware of this kind of product development activity being turned into a forum for price negotiations.

The "full version" of the Raw material Matrix for Pilsner Malt is presented in Appendix A, while the "mini version" is introduced in Section 15.10.

15.1 Introducing the Raw material Matrices and the guiding work process for matrix development

A Raw material Matrix is the facilitating instrument for an improved Raw material Understanding and will help to answer the following question: "How do *Product Design Requirements* from the *House of Quality* and Process Requirements from the Process Matrix relate to *Raw material Design Requirements* in the Raw material Matrices?"

From the perspective of internal or external raw material suppliers, the above question can be rephrased as follows:

- *How good are the individual characteristics (specifications) of raw materials used in the production process today, and how do they compare with raw materials from competitive suppliers or with possible substitute materials?*
- *Setting aside today's specifications of supplied raw materials, what would the overall specifications be for a number of "perfect future raw materials" for a new or improved product (also setting aside whether and how it is possible to obtain such specifications)?*
- *Considering the* Product Design Requirements *for the development of a new or improved product, what are the three most important* Raw material Design Requirements *for each raw material to improve in order to achieve the provisional product Target Figures for a new or improved product?*

In Fig. 15.1, the general structure of a Raw material Matrix is depicted, when *Product and Process Requirements* on *Raw materials* are used as the WHATs, and *Raw material Design Requirements* are used as the HOWs. The *Product Design Requirements* are selected and progressed from the *House of Quality*, and the complementary process requirements are the internal requirements that the production process makes on raw material properties. These

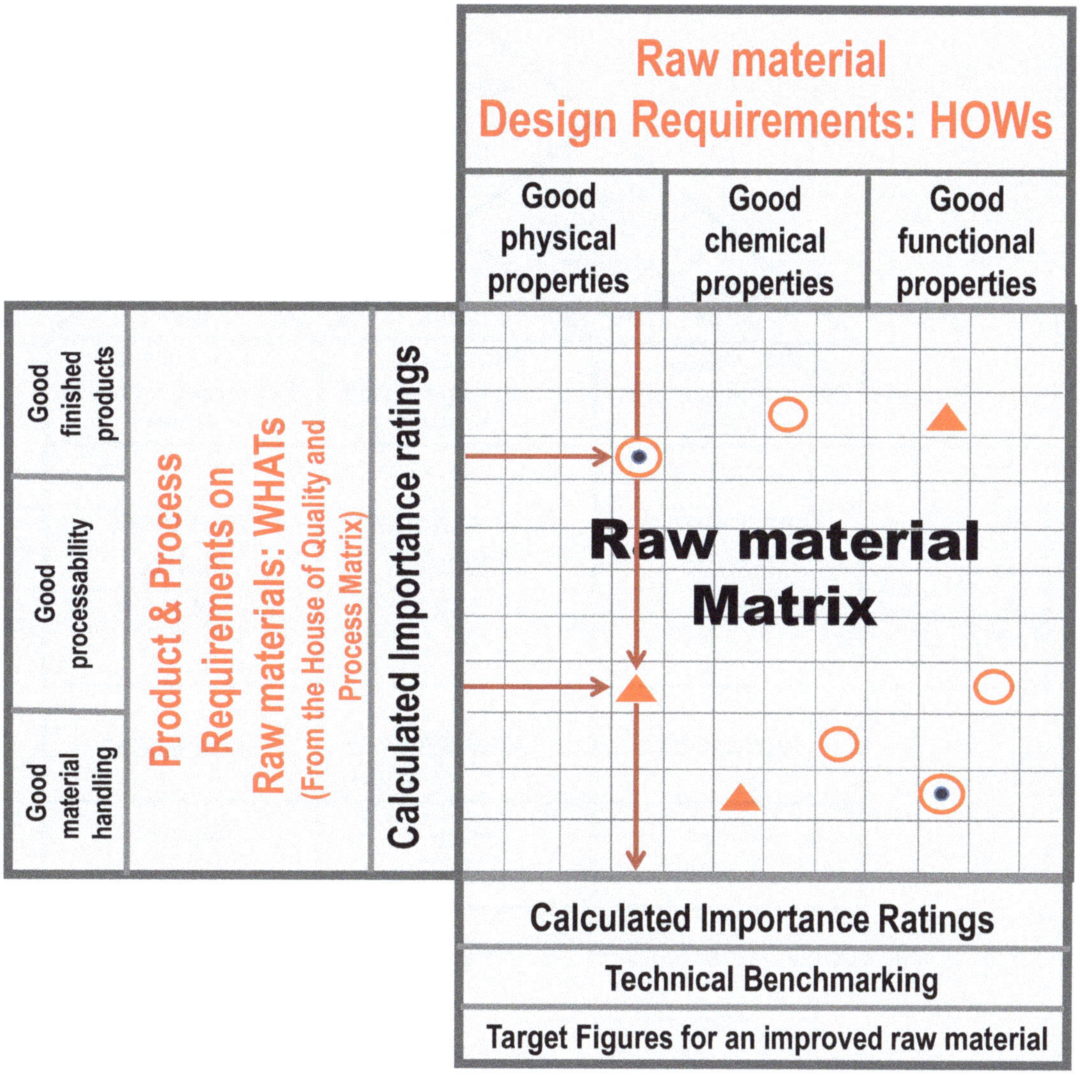

Figure 15.1 A general structure of the Raw material Matrix and associated default matrix rooms.

requirements are not derived from the *House of Quality* but rather are developed by the company's own production organization. The *Raw material Design Requirements* usually include many of the specifications that are given to the company procurement organization to be negotiated with external or internal raw material suppliers. The development of the Raw material Relationship matrix is rather similar to the development of the Product Relationship Matrix in the *House of Quality*, and the same goes for many of the associated complementary rooms used in this matrix.

In the process industries, the production process is usually supplied not only with a number of raw materials but also with a number of reagents or modifying ingredients. However, in developing the Raw material Matrices, it is initially recommended to select only one important raw material to start with and to develop other supplementary Raw material

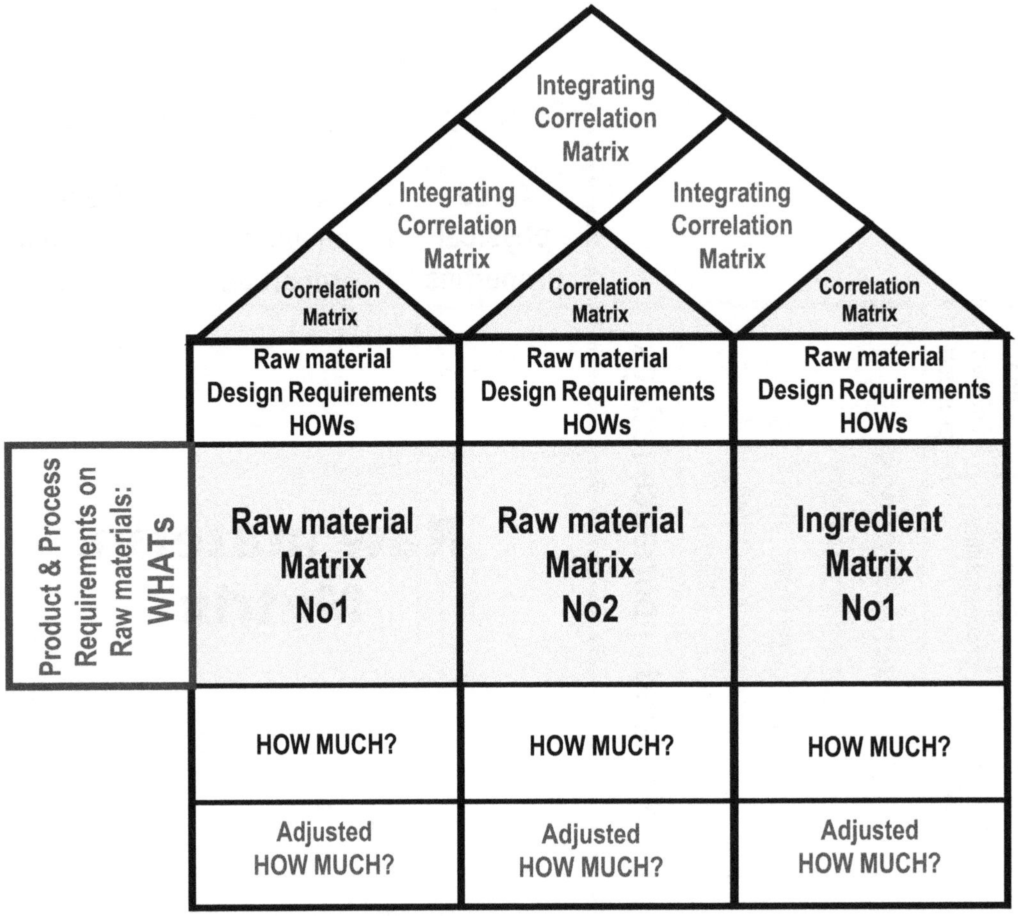

Figure 15.2 Using correlation matrices for the integration of multiple Raw material Matrices.

Matrices later on. In Fig. 15.2, this situation is illustrated; note that the selected *Product and Process Requirements* can also be used later on in the development of all raw materials and ingredient matrices.

It is generally advisable to develop each Raw material Matrix as a separate "sub-chart", which often is facilitated by QFD software like the **QFD Designer**™ software. The use of sub-charts will simplify the further development and discussions with individual raw material or ingredient suppliers. Figure 15.2 also shows how interactions between individual raw material matrices can be achieved using integrating Correlation Matrices in the combined roof. After the development of multiple Raw material Matrices, provisional Target Figures (How Much) may need to be adjusted, which is also illustrated in Fig. 15.2.

15.1.1 *The overall work process for the Raw material Matrices*

The overall work process for the development of a Raw material Matrix is shown in Fig. 15.3. The recommended work process is shown briefly here and will be discussed in more detail

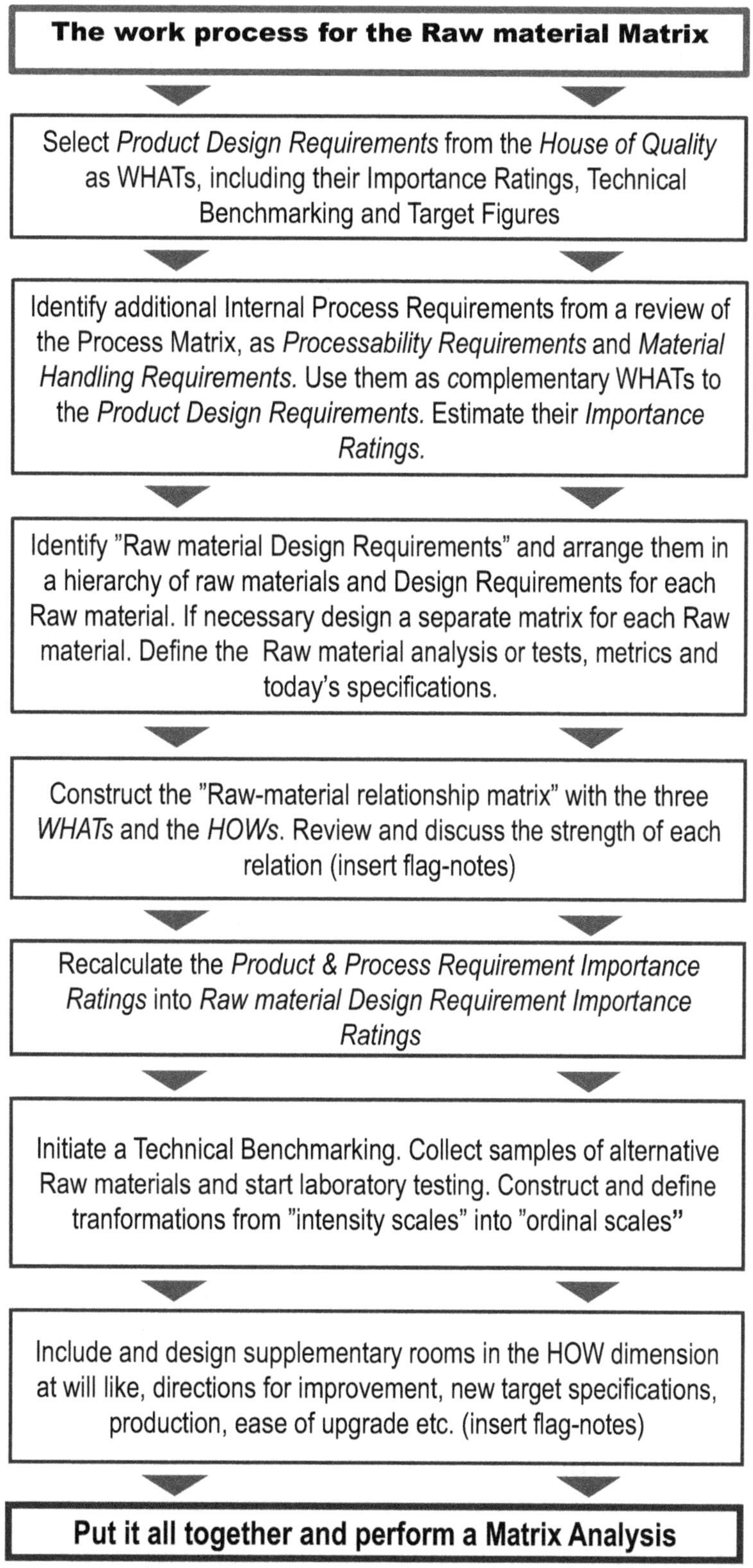

Figure 15.3 The overall work process for the Raw material Matrix.

in the following sections. Since Pilsner Malt is an important raw material in the manufacturing of beer, it has been chosen as the illustrative case, which is later presented in Fig. 15.9.

15.2 Select product Design Requirements from the House of Quality

In Fig. 15.4, the design of the Raw material Matrix is illustrated using information from both the *House of Quality* and the Process Matrix. Selected *Product Design Requirements* from the *House of Quality*, along with their Importance Ratings and Target Figures, can be progressed directly into the Raw material Matrix. In the development of this matrix, the previously developed HOWs in the *House of Quality* must be scrutinized and assessed for possible use as WHATs in this matrix. If the number of HOWs is below 20, all of them should generally be progressed into the Raw material Matrix. On the other hand, if the number of HOWs are larger than 20, it is recommended only to progress a select number of HOWs into the Raw material Matrix, and the following criteria can then be used:

- *Select "functional product attributes" with high Technical Importance Ratings from the* House of Quality*; mark those WHATs in a normal black font.*
- *For consumer products and product specifications generally communicated with customers, including tables of product contents, mark those WHATs with a green font.*
- *For products intended for B2B customers, focus should be on product attributes which are communicated with the customers; mark those WHATs with a green font.*
- Product Design Requirements *that during the Matrix Analysis of the* House of Quality *have been identified for improvement in the development of a new or improved product are naturally important to progress into the Raw material Matrix; mark those WHATs with a bold font.*
- *If, in the development of the HOWs in the* House of Quality, *new* Product Design Requirements *have been identified, including new test methods, such HOWs are truly important to progress into the Raw material Matrix; mark those WHATs with an italic font.*

The QFD team must consider all of the above criteria and decide on which *Product Design Requirements* should be selected for phase progression into the Raw material Matrix. However, the overall guiding principle for the selection of *Product Design Requirements* and the use of the Raw material Matrix is to assist the development of raw materials for improved product and production performance.

15.3 Identify internal process requirements in a review of the process matrix

In a similar way to the selection of *Product Design Requirements*, internal process requirements on raw material properties must also be identified and selected. Since such

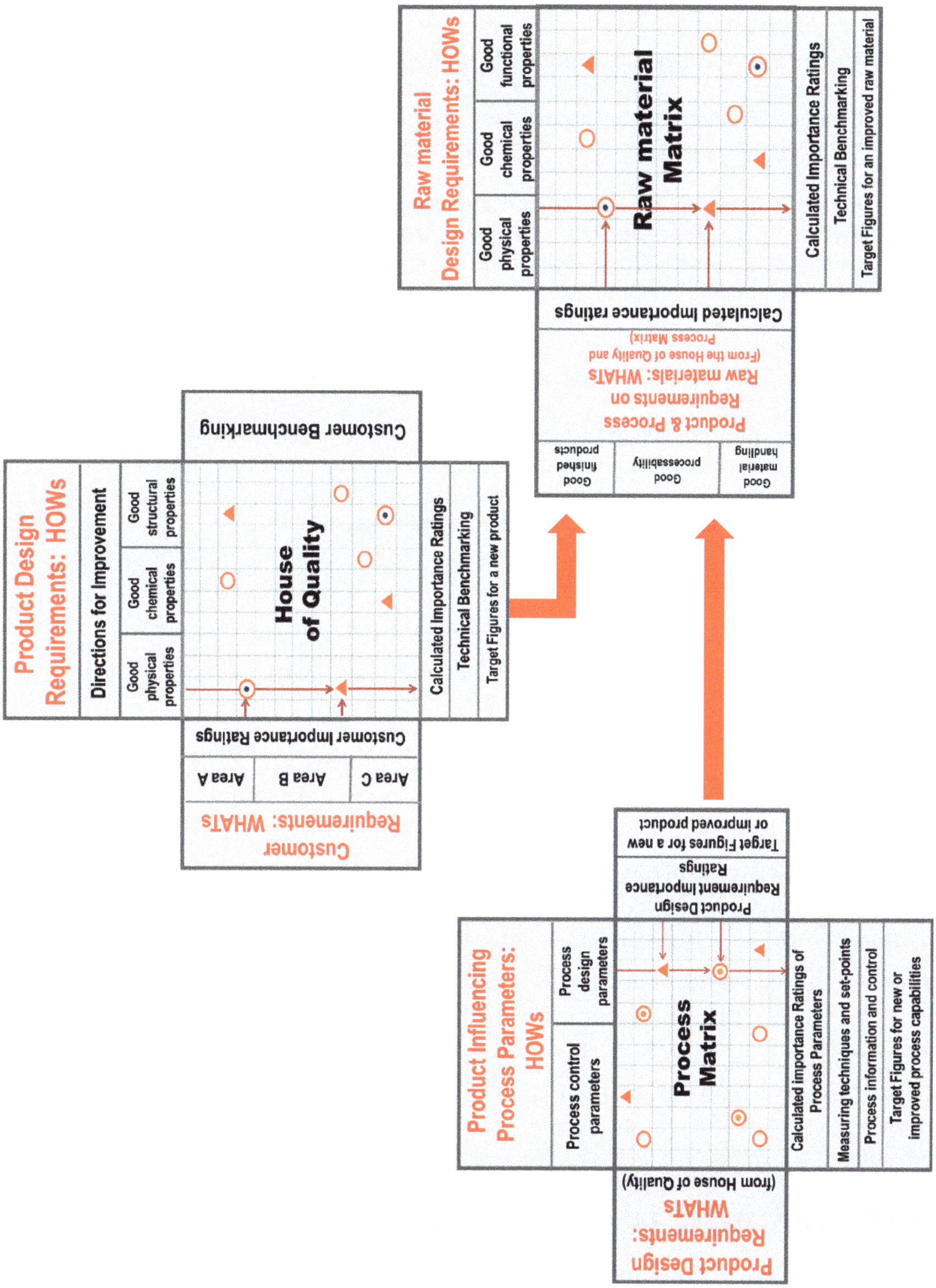

Figure 15.4 Designing the Raw material Matrix using information from both the *House of Quality* and the Process Matrix.

An explanatory example

Selecting internal Process Requirements on malt raw material – the Voice of the Production Process

The production system imposes a number of requirements on the raw materials used in the production process. In Fig. 15.1, such Process Requirements have been grouped as "Good material handling properties" and "Good processability properties".

In the production of beer in the brewery, there are a number of Pilsner malt properties that influence the different stages of the brewing process. When the malt is introduced in the brewing process, dusting should naturally be avoided and it is also desirable that the malt has good storing properties and "Grindability" properties. Furthermore, good extract recovery and good fermentation are malt properties that the brewmaster is searching for.

Handling and processability requirements on raw materials are often related to process efficiency and production economy and are thus of a different character compared to the *Product Influencing Process Parameters* that are used for production process control.

Figure 15.5 Selecting internal Process Requirements from the brewery on raw materials for beer production.

requirements are associated with production process performance, they are denominated "processability" requirements; that is, they are production process demands on raw material properties. Such processability demands are not similar to the previously developed *Product Influencing Process Parameters* in the Process Matrix; as an analogy to the *Voice of the Customer*, they can be called the "Voice of the Production Process" (further explained in Fig. 15.5). The importance ratings of such "processability" demands must be estimated together with production plant operational staff, and they must be set in a proper relation to the importance rating of the *Product Design Requirements* from the *House of Quality*.

15.4 Identify raw material design requirements

In the development of *Raw material Design Requirements* to be used as HOWs in the Raw material Matrix, existing test methods and chemical and sensory analysis for the selected raw material usually represent a good point of departure. However, experience tells that existing *Raw material Design Requirements* and related raw material specifications used for the procurement of raw materials often must be complemented with a number of more sophisticated functional requirements in order to better comply with the development of new or improved company products. Moreover, definition of test methods and their related metrics for all *Raw material Design Requirements* must also be included in the matrix, along with raw material specifications and tolerance levels. In the "full version" of the Raw material Matrix, the *Raw material Design Requirements* for Pilsner Malt are used as the HOWs.

15.5 Construct the Raw material Relationship Matrix

The Raw material Relationship Matrix can be developed in a manner analogous to the development of the Process Matrix, which was previously explained in Section 14.5.

15.6 Recalculate Product and Process Requirement Importance Ratings into Raw material Design Requirement Importance Ratings

Using the calculated Product Design Requirement Importance Ratings from the *House of Quality* and the Estimated Importance Ratings of Processability and Material Handling from the Process Matrix, the relationship matrix can now recalculate those figures into Calculated Raw material Design Requirement Importance Ratings. They can be presented as figures, but it is generally advised to illustrate the results with bar charts at the bottom of the matrix. In the "full version" of the Raw material Matrix for Pilsner malt, this is further illustrated.

15.7 Develop the room Directions for Improvement

The development of the room Directions for Improvement is similar to the development of this kind of room in the *House of Quality*. This is a room where the direction for the improvement of each *Raw material Design Requirement* is reviewed, from the perspective of desired properties of a finished product and production processability. An upward pointing arrow symbolizes "more is better", an arrow pointing downward symbolizes "less is better", and a ring symbolizes "nominal is best". When the direction for improvement is to be decided on for an individual raw material property, one must only consider this individual requirement in a "sterile environment"; furthermore, one must not take a stand regarding whether or not improvements are technically possible or not, or whether or not a change is desirable. In Fig. 15.6, the Direction for Improvement of a few Pilsner Malt properties is illustrated. Completion of this room is generally simple and is usually based on company representatives' "gut feeling" of desired directions for improvement. In the "full version", Directions for Improvement of Pilsner Malt are presented in a room above the raw material relationship matrix.

An explanatory example

Directions for Improvement of malt raw material properties

In the "full version" of the Raw material Matrix, you will find the room Directions for Improvement on top of the Raw material Relationship Matrix. Among the physical properties of malt, you will find that "top size" and "moisture content" of the malt are generally desired to be low. The chemical property "Beta-glucan" should also be low, while "Free amino nitrogen content" should be high.

The functional property "Viscosity in Wort" should be low, while it is desirable that the "Diastatic power" should be high.

Figure 15.6 Directions for Improvement of some Pilsner Malt *Raw material Design Requirements* for beer manufacturing.

15.8 Initiate a Technical Benchmarking of raw materials

Technical Benchmarking in the laboratory of raw materials can be carried out in a fashion corresponding to the Technical Benchmarking in the *House of Quality*, which was explained in detail in Chapter 11. The Technical Benchmarking is recommended to await the development of the full Raw material Matrix in order to give guidance for the acquisition of alternative raw materials.

The selection of raw materials of interest for the Technical Benchmarking from external (or sometimes internal) raw material suppliers should preferably be carried out in close interaction or even intimate collaboration with alternative supplying companies. The information from the completed Raw material Matrix will then provide important advice on which *Raw material Design Requirements* are of interest to improve. In discussions with individual raw material suppliers on specifications for presently used or potential alternative new raw materials, the Raw material Matrix will be an excellent framework and guiding instrument for fact-based raw material development at the supplying company for new or improved product innovation at the customer company.

15.9 Develop and design supplementary rooms of interest for the Raw material Matrix

There are a number of optional supplementary rooms in the HOW dimension that could be of interest to develop and use in the Raw material Matrix, depending on its intended purpose and future use:

A sustainability room

The sustainability room is of growing importance because of the increased company focus on sustainable production and products and on environmental issues in general.

A room indicating the difficulty of changing or improving each requirement

This room provides an assessment of how difficult each *Raw material Design Requirement* is to change or improve. Such assessments can be from either a technical or an economic perspective.

A room for the evenness of Target Figures

In the process industries, a radically improved raw material may be a raw material with similar product specifications to the original material but radically improved and lowered tolerance levels.

A room indicating how important it is to improve the requirements for the development of a new or improved product

This room provides an assessment of how important each *Raw material Design Requirement* is to improve as a conclusion from the first Matrix Analysis exercise and after setting provisional Target Figures for a new or improved raw material.

A Correlation Matrix as a top roof

The individual *Raw Material Design Requirements* in the Raw material Matrix usually interact, since a change in the intensity of one HOW sometimes will influence another HOW. This situation is quite similar to the Correlation Matrix in the *House of Quality.* In more in-depth discussions with raw material suppliers and in search of improved Target Figures for a select raw material, the development of such a correlation roof on top of the Raw material Matrix is a worthwhile activity. In Fig. 15.7, this is further exemplified.

An overall QFD "system perspective" on the use of multiple raw materials in the production process

If a product is intended for B2B customer use, the *House of Quality* is really, from the perspective of the customer, a Raw material Matrix, and then often only one out of a number of raw materials and Raw material Matrices. A raw material can be a singular raw material interacting with similar kinds of raw materials from other suppliers or with different kinds of raw materials. In Fig. 15.8, an example is shown of how multiple Raw material Matrices for beer manufacturing can be developed and interact in different stages of the beer

An explanatory example

Correlation between Raw material Design Requirements

The Correlation roof is not inserted in the full Raw material Matrix. However, the functional property "Viscosity in wort" is correlated with particle size of the crushed malt. The particle size is, in reality, a screen curve, but two points on this curve have been selected as the requirements "Top size" and "Fine size". Moreover, the viscosity is correlated with the "pH in wort" but also with the "Beta-glucan content".

Furthermore, the "Viscosity of wort" has a strong relation to the Process Requirement "No filtering problems" and a medium strong relation to the Process Requirement "Short mashing time".

Figure 15.7 The correlation between Raw material Design Requirements for malt.

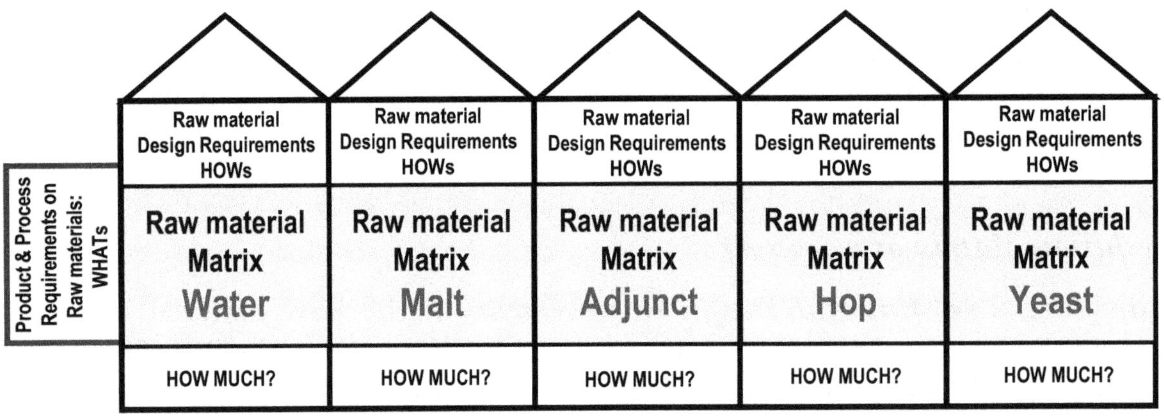

Figure 15.8 An example of Multiple Raw material Matrices for beer manufacturing. Interactions between individual matrices can be analyzed using interacting Correlation Matrices on top of the selected correlation matrices.

manufacturing process. Such interactions will create an important "system aspect" on the use of multiple raw materials in the production process that may generate requirements on the individual supplied raw materials. A similar situation is when components interact in modular designed products in other manufacturing industries. This kind of "system interaction" between individual raw materials in the customers' production process is still often difficult to simulate and understand. However, the use of interacting correlation roofs may be a supporting facilitating tool in such an advanced analysis and in the use of multiple Raw material Matrices.

15.10 The illustrative case of development and manufacturing of beer — The Raw material Matrix for malt

You have previously been introduced to the "full version" of the Raw material Matrix located in Appendix A of the book and in the previous sections of this chapter. Now, you will find the "mini version" of the Raw material Matrix presented in Fig. 15.9. Please observe that recalculated Importance Ratings from the *House of Quality* and the Process Matrix presented at the bottom of the matrix in the "mini version" will not correspond to the figures in the "full version", since additional Product and Process Requirement relations are included in the calculation of the figures in the "full version".

In the Raw material Matrix, we move from a product and process dimension into a Raw material dimension. Only eight WHATs have been selected from the "full version", despite the fact that there are additional such requirements that also relate to the *Raw material Design Requirements*. In the Raw material dimension, only seven Raw material attributes have been selected for the construction of the "mini version" of the Raw material Matrix.

The system of Multiple Progression QFD

The *Product and Process Requirement Importance Ratings* presented in the "mini version" are recalculated with the Raw material Relationship Matrix into *Raw material Design Importance Ratings.* For example, the importance of the Raw material attribute "Total Protein content" is as follows: 230*3 + 260*1 + 54*3 = 1112.

Product Influencing Process Parameters

In a review of the "full version" of the Process Matrix, the importance of the Raw material attributes "Extract yield", "Diastatic power" and "Saccharification time" stand out.

Please remember that Pilsner malt is only one out of many raw materials in the manufacturing of beer, and thus the *Product Design Requirements* (Product attributes) of the beer product like "Bitterness", "Hop aroma" and "Hop taste" are consequently not affected by the Pilsner malt product attributes. The importance rating figures for individual Pilsner malt attributes are, however, to a large extent influenced by the processability requirements.

Supplementary rooms in the Raw material Matrix

There are some supplementary rooms inserted in the "full version" of the Raw material Matrix, but they have not been filled in, since the focus of this book is not on brewing science but on the use of the methodology.

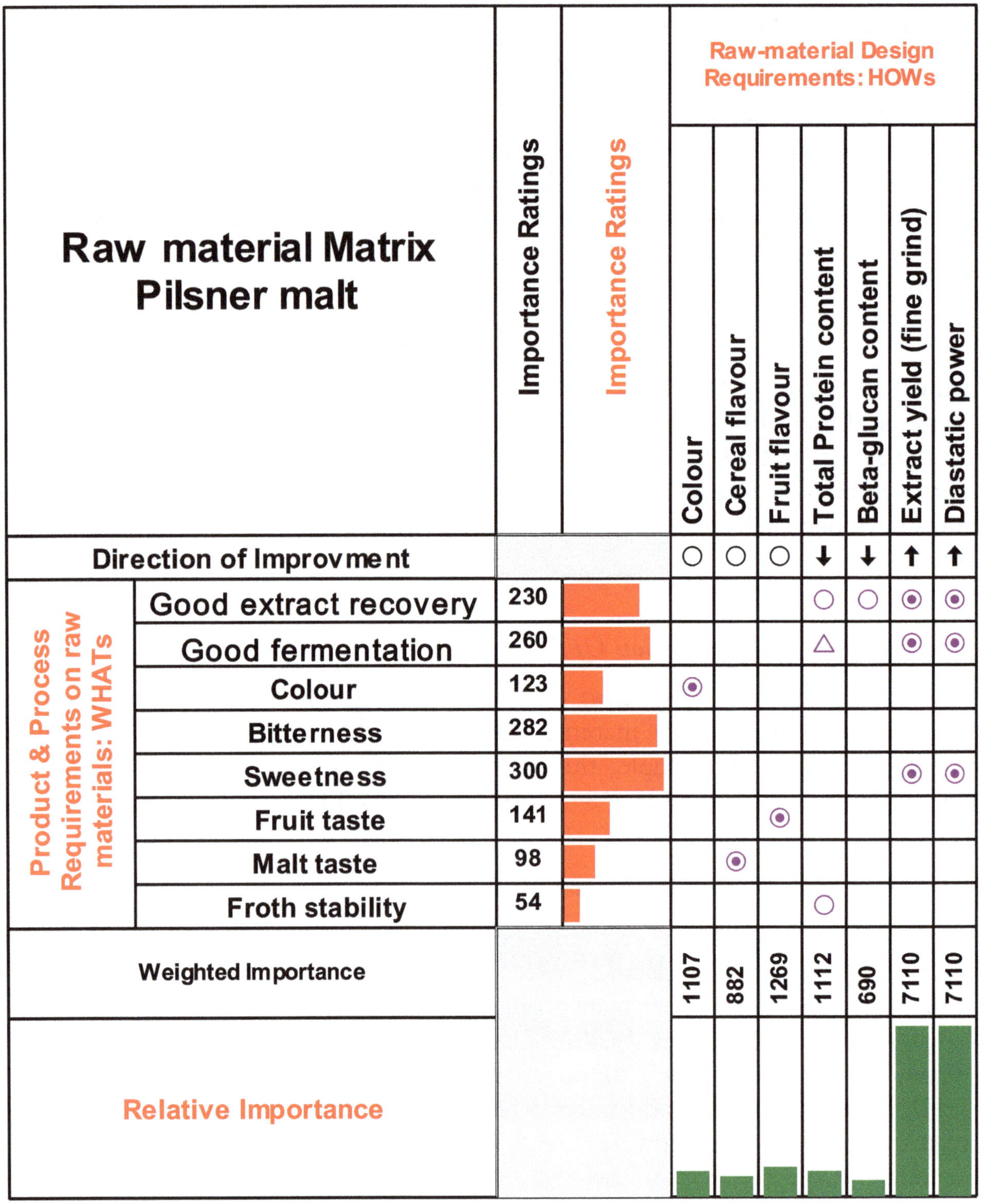

Figure 15.9 The simplified "mini version" of the Raw Material Matrix

15.11 Put it all together and perform a Matrix Analysis and provisional Target Setting for the Raw material Matrix

15.11.1 *Matrix analysis*

A Matrix Analysis of the Raw material Matrix can, in many respects, follow and be guided by the previously presented Matrix Analysis of the *House of Quality*. In an "open innovation" approach, it could be of interest to invite external experts or even a select raw material supplier for this exercise.

Day 1 Review the total Raw material Matrix and its individual parts

Part I (Morning session) Take a "guided tour of the Raw material Matrix" and review the information together

The first activity after "cooling out" from other kinds of work is the introduction of all members and a review of the program for the Matrix Analysis. Afterwards, the first activity is to review the total Raw material Matrix together. During this update, questions and open discussions should be allowed, but the analysis should wait until group work has been carried out.

Part II (Afternoon session) Analyze the different parts and content of the rooms in separate group work

Group A — Analyzing *Product & Process Requirements on Raw materials* and the raw material relationship matrix

- *Review the total structure of* Product & Process Requirements on Raw materials *according to the previously presented list in Section 15.2. Discuss the different kinds of* Product and Process Requirements on Raw materials *in the perspective of the selection criteria (fonts and colours).*
- *Do any horizontal parts of the matrix lack "fill" or contain only weak relations, indicating that such* Product and Process Requirements on Raw materials *do not really relate to any Raw material Design Requirements? If so, how could such WHATs be measured?*
- *How much "fill" is there in the matrix (too much or too little)? In consequence, should some WHATs or HOWs be expanded for clarification or possibly removed?*
- *Review the total matrix structure and discuss where in the matrix the various product functionalities are influenced by the raw material properties. The information is partly on display as strong symbol relational patterns in each matrix row in the relationship matrix (use a marker pen).*
- *Review the matrix structure and discuss where in the matrix production processability is influenced by the raw material properties. The information is partly on display as strong symbol relational patterns in each matrix row in the relationship matrix (use a marker pen).*

- *Are there any areas of significant "weak knowledge" or even important areas as "white spots on the map" indicating potentially interesting areas for further experimental work?*
- *Review the "flag notes" (focus on strong relations) and discuss the overall content of available information.*
- *Should the Raw material Matrix be split up into several sub-matrices, and which areas should be further discussed, reviewed and validated in different internal company forums and which in collaboration with external experts?*

Group B — Analyzing in the *Raw material Design Requirement* dimension of the Process Matrix

- *How good is the total structure of* Raw material Design Requirements? *Have any requirements been forgotten, or should any be removed?*
- *Mark the 20% most important and the 20% least important* Raw material Design Requirements *with different colors. Discuss and reflect on these. Should the company improve some test methods, develop new ones or make any existing ones obsolete?*
- *Do any vertical parts of the matrix lack "fill" or contain only weak relations, indicating that such* Raw material Design Requirements *do not really relate to any* Product & Process Requirements on Raw materials. *Should they be removed from the Raw material Matrix?*
- *Review the information in all supplementary rooms. If there are a large number of such rooms inserted in the Raw material Matrix, it is recommended that a third group C should focus only on that part of the matrix and that the second day for Matrix Analysis is extended to a full day.*
- *Review the results from the Technical benchmarking of presently used raw materials, alternative raw materials or substitute products.*
- *Should the total Raw material Matrix be split up into complementary matrices?*

Day 2 Presentation of group work and a final synthesis of the results

Part I (Morning session, 2 hours) Presentations of the results from the group work

The presentation of the group work should be carried out by the selected work group leaders using prepared slides and in a complementary view of the large printed paper version of the matrix.

Part II (Morning session, 2 hours) Discussion and synthesis of the results in a holistic fashion and decisions on further actions

The following list of questions to be answered and actions to be taken are only tentative, since each project may require separate analyzing activities:

- *Are existing raw material specifications (internal or external) usually contained in the presently used raw material? If not, which specifications are generally not achieved?*

- *How good are the individual characteristics (specifications) of raw materials used in the production process today, and how do they compare with raw materials from competitive suppliers or possible substitute materials?*
- *What kind of internal, theoretical or experimental work should be carried out? If possible synchronize with already on-going activities or programs.*
- *What kind of external, theoretical or experimental work should be carried out? Are there any suggestions for collaborative partners?*
- *What kind of input is especially of interest for Purchasing? Are there any actions to be taken?*
- *What kind of input is especially of interest for Production? Are there any actions to be taken?*
- *Should the Raw material Matrix be enlarged or possibly reduced?*
- ***What are the most important Raw material Design Requirements to improve in order to achieve provisional product Target Figures for a new or improved product?***
- *Discuss and try to answer the three important questions initially presented in Section 15.1.*

15.11.2 *Provisional target setting*

After a Matrix Analysis and in discussions and collaboration with raw material and ingredients suppliers, the first objective should be to decide on optimum specifications for presently used ingredients, and the second should be to deliver feedback to the ingredient supplier on desirable improved raw material specifications for company development of new or improved products. In the use of the Raw material Matrix, the possibility of testing and finding substitutes for new raw materials or cheaper raw materials from alternative suppliers must not be overlooked. The combination of functionality and price will be the key selection criterion here.

The Raw material Matrix as only one part of an Integrated Knowledge Platform

The Raw material Matrix and the results from the Matrix Analysis should be included in the *Integrated Knowledge Platform (IKP)* for the product or product family to be deployed in further company product innovation. IKPs will be presented and further discussed in the following Part V in this book.

Chapter 16

The Product Matrices — The Facilitating Instruments for an Improved "Understanding of Product Functionalities"

> **"The Product Matrix proved to develop into a very interesting instrument for organizational learning and collection of tacit LKAB knowledge that already existed to some extent in the R&D organization. As one researcher pointed out, this matrix will provide a platform for the development of 'conceptual explanatory models' that describe how product functionalities are related to more or less measurable inherent product characteristics".**
>
> **(Tottie et al., 2016)**

It has already been mentioned that the largely homogenous products produced in the process industries appear "low tech" for the superficial onlooker but in reality often have a high inherent material complexity. It was thus before pointed out in Chapter 8 that a QFD system that is well adapted to process-industrial product innovation must have an ability to relate *Product Design Requirements* to such inherent product characteristics. In other words, the system must facilitate an understanding of how internal product characteristics cause improved product functionalities in a customer's use of a product. In this book, such product characteristics have been denominated *Explanatory Product Characteristics*; in Fig. 16.1, this concept is further discussed.

Advanced products from the Steel Industries are examples of complex product inherent structures that strongly influence product functionalities; advanced paperboards from the Forest Industries likewise. The different sensory properties of products from the Food and Beverage Industries are often created by very structurally complex chemical entities, which are of no interest to consumers. Nevertheless, for the producer of such products, an improved knowledge of which inherent product characteristics generate such improved product functionalities is of vital interest in advanced product innovation and renovation. **Product**

An explanatory example

What is an "Explanatory Product Characteristic"?

Explanatory Product Characteristics are defined in this book as the inherent product properties that create product functionality in a product attribute in a customer's use of a product. To give an example of such a characteristic, the following case may serve:

One *Customer Requirement* WHAT on a beer is that it should have a "Good head", and related *Product Design Requirements HOWs* are "Good frothability" and "Good froth stability", the latter measured with the Rudin test method (see further in the "full version" of the *House of Quality*). However, these HOWs do not explain how such functional properties are made, nor how to proceed if one wants to improve them?

Apart from other correlated *Product Design Requirements* that cause a "Good head", like the "Content of carbon dioxide" in the beer, there are a number of inherent *Explanatory Product Characteristics* that will influence the frothability and froth stability. Most important for improved frothability is the content of hydrophobic proteins from malt and alpha acids from hops.

These compounds are produced by and dependent on brewing process conditions, but they are also influenced by malt and hop properties, thus illustrating the linkage between raw materials properties and production conditions for the creation of functional properties of manufactured products in the process industries.

Figure 16.1 Clarifying the concept of *Explanatory Product Characteristics.*

Matrix A serves this purpose and thus uses the *Product Design Requirements* from *The House of Quality* as HOWs and *Exploratory Product Characteristics* as WHYs in the matrix design.

If customers and consumers, the product users, are not very interested in what causes product functionalities, they are even less interested in how such product functionalities are created in the supplier's production process. Conversely, the manufacturing company searches eagerly for a better understanding of how to operate their production processes for improved inherent product characteristics and related product functionalities. **Product Matrix B** serves this purpose, using the previously presented *Exploratory Product Characteristics* from Product Matrix A as WHYs and selected *Product Influencing Process Parameters* from the Process Matrix as HOWs in the development of this matrix.

Product Matrix A and Product Matrix B are two novel kinds of matrices specially developed and integrated in the Advanced Multiple Progression QFD system (mpQFD). Experiences from the industrial use of those new matrices tells that an improved understanding of the relationships through those matrices sometimes even makes it possible to develop new breakthrough product functionalities. It has previously been illustrated in Fig. P4.2 that Product Matrix A can only be developed after the development of the *House of Quality*, as well as that Product Matrix B also must rely on a previous development of the Process Matrix. You will learn in the forthcoming chapters that the development of these matrices often involves also applied research, and thus they naturally "belong" to the company's R&D department. However, they will certainly substantially benefit from input from both internal and external experts in a truly "open innovation" approach.

16.1 Introducing Product Matrix A and the guiding work process for matrix development

Product Matrix A is the facilitating instrument for an improved understanding of product functionalities and will help to answer the question: "How do *Product Design Requirements* in the *House of Quality* relate to inherent *Explanatory product Characteristics*?"

How individual *Product Design Requirements* (product attributes) can be measured is usually well known to companies in the process industries. However, which inherent product characteristics that cause such product properties is often less well known and generally much more difficult to measure. The more structurally heterogeneous a product is, the greater is the need for Product Matrix A, which is a complementary matrix to the *House of Quality*. This matrix thus provides an in-depth understanding of WHY a product acquires certain functionalities and serves as an instrument for a more systematic investigation of the underlying mechanisms that create such properties.

The above question could be expanded into the following three detailed questions:

- *Selecting the most important functional* Product Design Requirements *(product attributes) to improve in the development of a new or improved product, which are the most important related* Explanatory Product Characteristics *to improve?*
- *In a review of those important* Explanatory Product Characteristics, *how do they compare with the* Explanatory Product Characteristics *of competitive products (i.e., the results from the competitive Technical Benchmarking of such product characteristics)?*
- *What can be learned from a review of the information within Product Matrix A and the state-of-the-art knowledge of* Explanatory Product Characteristics *in the development of new or improved product functional attributes?*

Product Matrix A

The general structure and design of Product Matrix A is depicted in Fig. 16.2, including the associated default matrix rooms. Selected *Product Design Requirements* from the *House of Quality* are used as the HOWs in this matrix. The HOWs also include potential new *Product Design Requirements* and their related test methods, which have been discovered and delineated during the first matrix analysis of the *House of Quality*. Related Technical

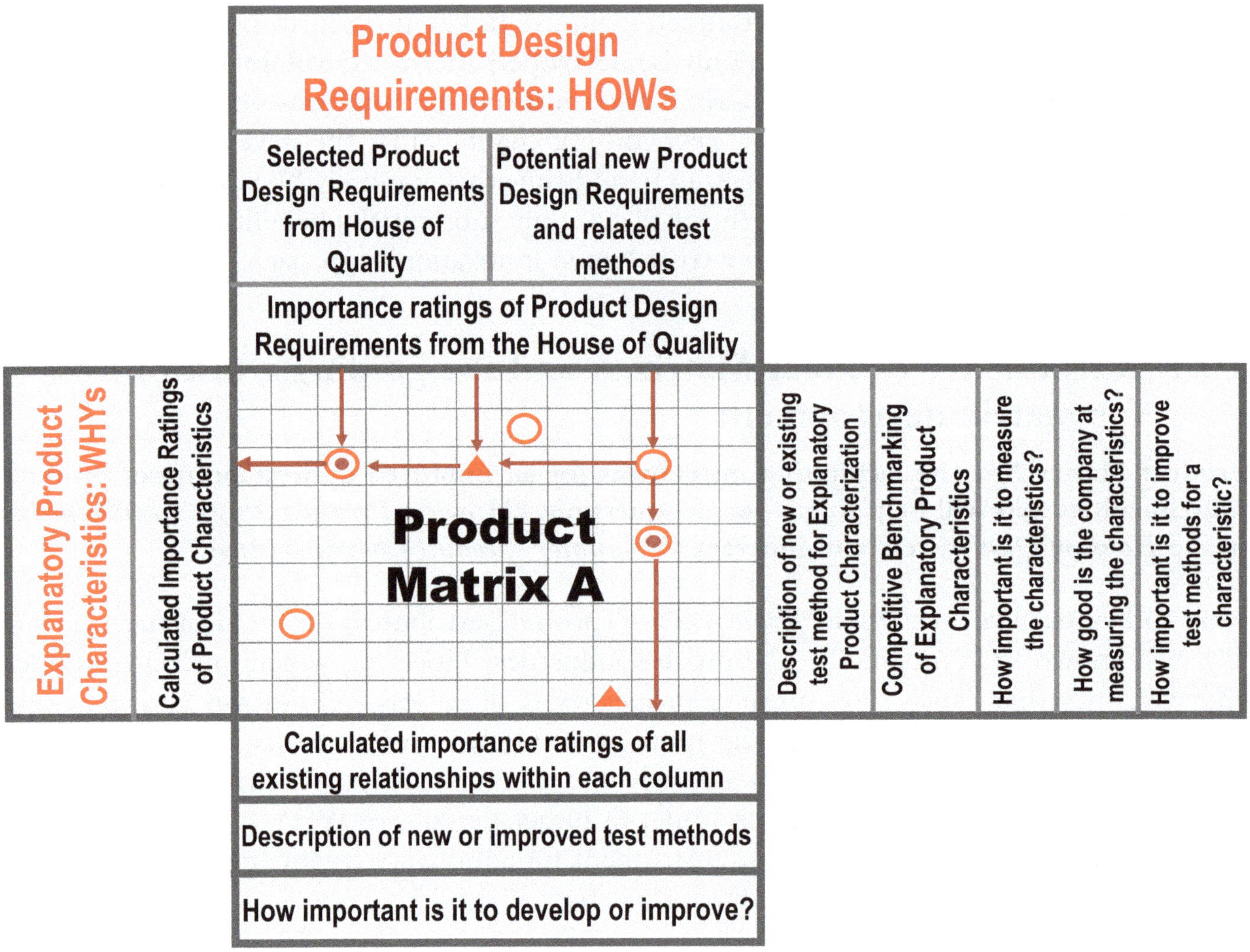

Figure 16.2 A general structure of Product Matrix A and associated default matrix rooms.

Importance ratings for both kinds of requirements could also be progressed into this matrix. Selected *Product Design Requirements* are introduced and connected through the relationship matrix in the other matrix dimension to *Explanatory Product Characteristics*, the WHYs. The *Product Design Requirements Importance Ratings* are translated by the relationship matrix into *Importance Ratings of Explanatory Product Characteristics*. In the following sections and in the presentation of the overall work process, the development of the individual rooms will be further discussed and explained.

16.1.1 *The overall work process for Product Matrix A*

The overall work process for the development of Product Matrix A is shown in Fig. 16.3. The recommended process steps are summarized here and in the following sections they will be further explained in more detail.

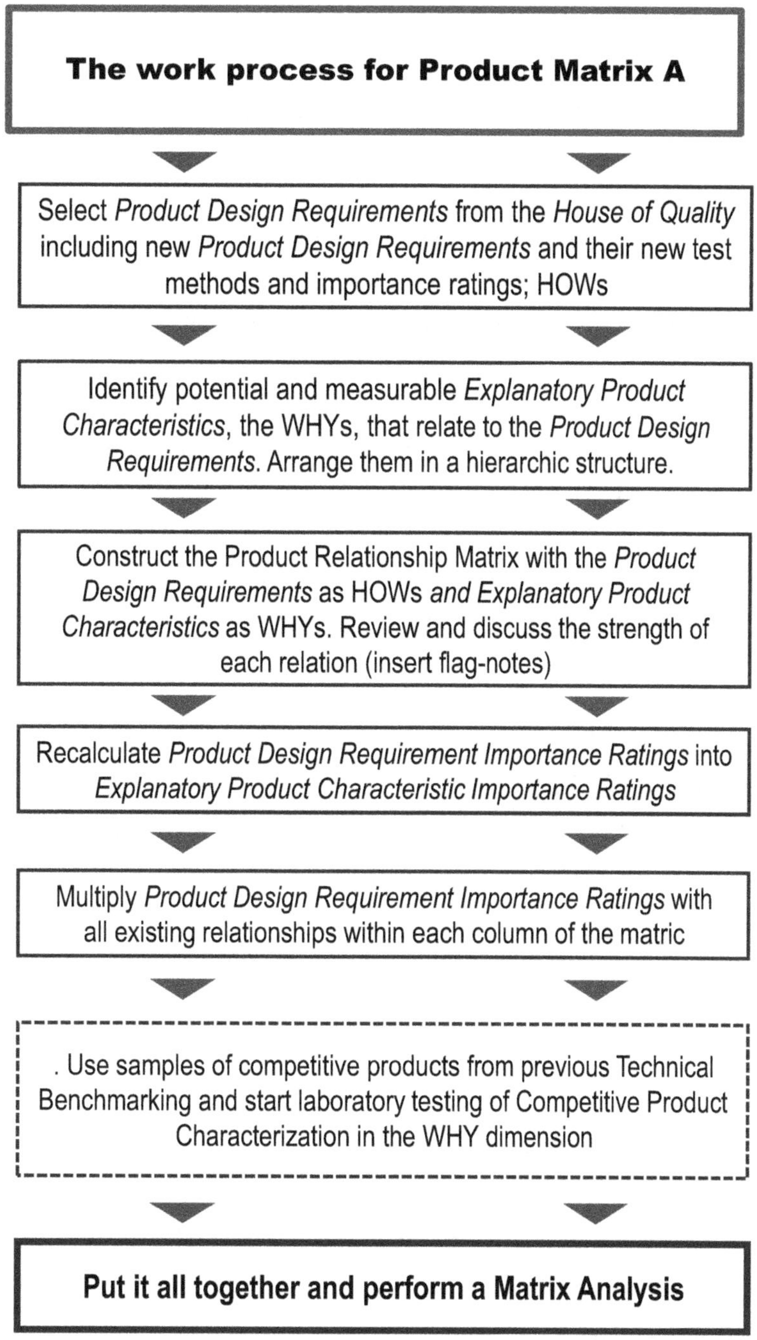

Figure 16.3 The overall work process for Product Matrix A.

16.2 Select Product Design Requirements from the House of Quality

When *Customer Requirements*, the WHATs, in the *House of Quality* are reviewed, if there is a need to introduce new *Product Design Requirements* to measure un-answered *Customer Requirements*, those complementary HOWs should be included in Product Matrix A. It is further necessary to conceptually develop associated, and often sophisticated, test methods and metrics that are intended to capture and measure such requirements. If there are *Product Design Requirements* that are not measured on a continuous basis in the production process, or if they involve research and development, this is the matrix where they should be introduced.

An enhanced product understanding can be obtained through the development of Product Matrix A for products that have complex functionalities or that are internally structurally designed. Figure 16.4 illustrates how selected *Product Design Requirements* from the *House of Quality*, the HOWs, are progressed into *Product Matrix* A, relating them to the products' inherent (structural) product properties or other explanatory properties, called the WHYs.

In the development of this matrix, previously developed HOWs from the *House of Quality* must be scrutinised and assessed if they are to be included as HOWs in this matrix. Consequently, if the number of HOWs is below 20, all of them should generally be progressed into Product Matrix A. If the number of HOWs in the *House of Quality* is greater than 20, only a selected numbers of HOWs should be into the Product Matrix, and the following criteria can be used:

- *It is first of all recommended to choose "functional product attributes" with high Technical Importance Ratings from the* House of Quality; *mark those HOWs in a normal black font.*
- *If, in the development of the HOWs in the House of Quality, new Product Design Requirements have been identified including new test methods, such HOWs are truly important to progress into the Product Matrix; mark those HOWs with an italic font.*
- *The* Product Design Requirements *that, during the Matrix Analysis of the* House of Quality, *have been identified for improvements in the development of a new or improved product are naturally important to progress into the Product Matrix; mark those HOWs with a bold font.*

The QFD team must consider all of the above criteria and decide which *Product Design Requirements* should be selected for phase progression to Product Matrix A. However, the overall guiding principle for selection of *Product Design Requirements* and the use of the Product Matrix A is to get an improved product understanding in order to achieve enhanced product innovation performance.

16.3 Identify Explanatory Product Characteristics

In order to create Product Matrix A, brainstorming sessions about which features that create the functionality of a product must be held together with company representatives from applied research and product development. These inherent product characteristics can then be classified and organized in a hierarchical structure (as previously explained in Section 10.3) and denominated the WHYs in the matrix; see Fig. 16.2. Since company test methods

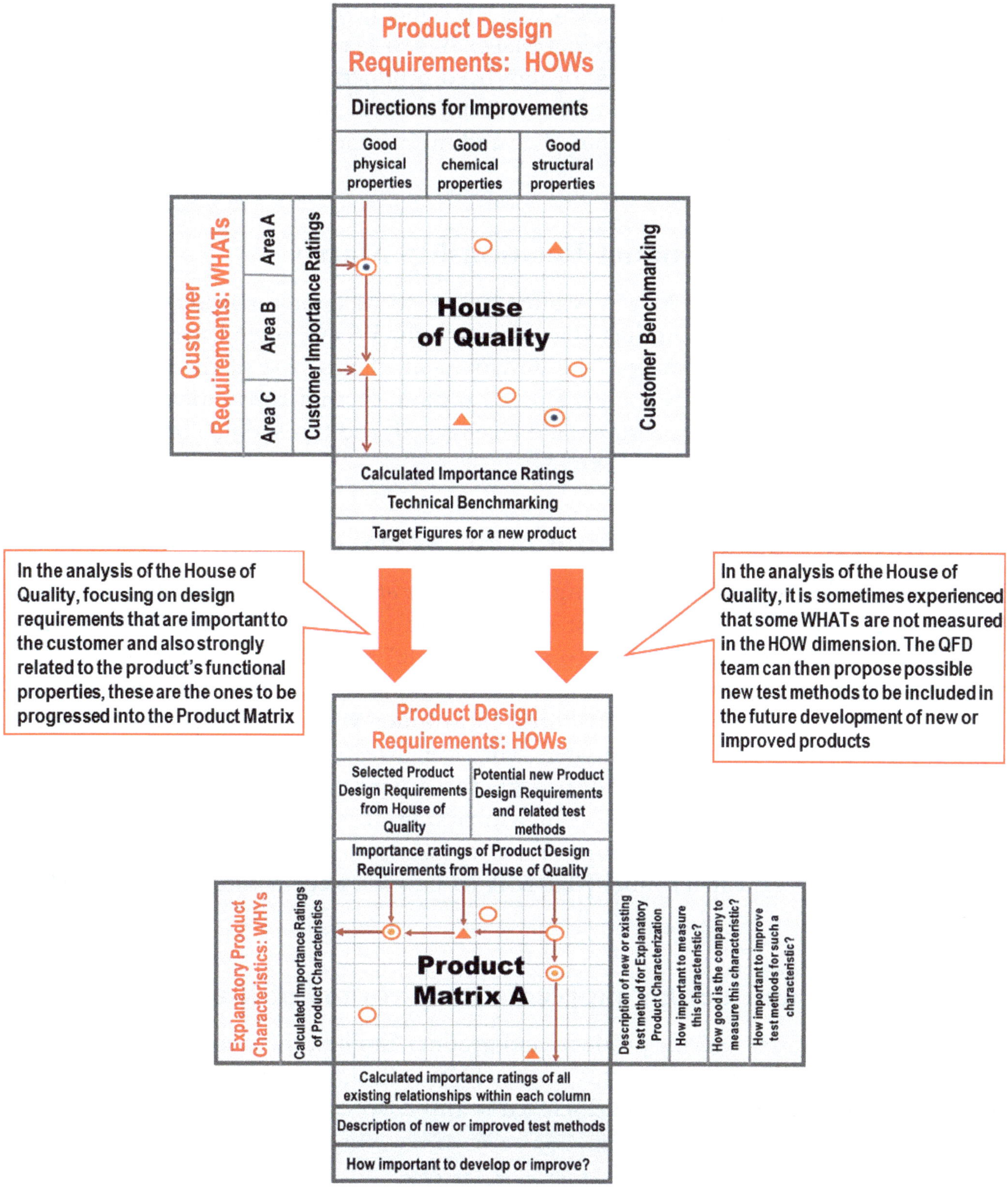

Figure 16.4 Designing Product Matrix A from the *House of Quality* (*Tottie et al., 2016*).

sometimes do not correlate well with how well a product behaves in the customers' production processes, one objective with Product Matrix A is to examine potentially better internal test methods in order to develop new or improved products. Thus, Product Matrix A is a tool not only for relating the *Product Design Requirements*, the HOWs, to the *Exploratory Product Characteristics*, the WHYs, but also, and most importantly, to guide the development of better test methods.

16.4 Construct the Product Relationship Matrix

The development of Product Matrix A often proves to be a long and iterative journey of learning. Since many of such product characteristics, the WHYs, often have never been collected and systematically structured, the development of this part of the matrix can take considerable time and effort by the R&D organization. During consecutive meetings, different internal experts and specialists should be engaged, and this should continue throughout the development of Product Matrix A.

16.5 Initiate laboratory testing in the WHY dimension

In the previous Technical Benchmarking in the *House of Quality* company *Product Design Requirements* and their associated test methods and metrics were used. The "intensity results" were afterwards translated into a "value scale" illustrating the different product performances in the technical dimension. The results pinpoint how good the company product is for each of its product attributes. In Product Matrix A, the product functionality for the customer is complemented with the investigation of how such product attributes relate to individual *Explanatory Product Characteristics*, the WHYs. Using the same samples from previous Technical Benchmarking, it is now possible to start laboratory testing of competitive *Exploratory Product Characteristics* in the WHY dimension. The results are inserted in Product Matrix A to be reviewed in the subsequent Matrix Analysis.

16.6 Develop and design supplementary rooms of interest

Depending on company-specific products and product families, there are often a number of alternative rooms to be inserted in this matrix. However, since such rooms are often product-specific, they will not be further discussed in this chapter.

Insert a side roof in the matrix correlating the WHYs

Since a number of *Exploratory Product Characteristics* often interact, it is sometimes of interest to develop and design a correlation matrix as a side roof for the Product Matrix A.

16.7 Put it all together and perform a Matrix Analysis

Process Matrix A provides an outstanding instrument for "organizational learning" and a collection of "tacit" knowledge about an existing product in the R&D organization. A well-managed Product Matrix A should thus be in a constant state of flux during on-going product development activities and in future development cycles. As previously noted, this matrix will provide a platform for the development of "Conceptual Explanatory Models" that describe how product functionalities are related to more or less measurable inherent product characteristics. Depending on the size and complexity of Product Matrix A, the Matrix Analysis may take only one day, though it may take two days. It is of utmost importance that this event is carried out in a relaxed creative atmosphere and that sufficient time is allowed for discussions. The use of flag-notes for capturing tacit information during discussions is strongly recommended!

Day 1 Review the total Product Matrix A and its individual parts

Part I (morning session) Take a "guided tour of Process Matrix A" and review the information together

Since the development of Process Matrix A is generally a temporally lengthy process with many iterative development loops, it is advised that half a day should be allocated to recapitulate the previous steps in matrix development, refamiliarise all participants with the different rooms, and then decide on an agenda for the analysis. If the development of Process Matrix A has been related to an introduction of the QFD methodology as such, it is also advisable that a brief review of the mpQFD system is initially provided. Moreover, if some members have not been directly involved in the specific product development project, it is necessary to introduce the overall project and project objective. Most of the general guidelines and recommendations previously mentioned in Chapter 12 are also applicable for the Product Matrix A analysis.

Part II (Afternoon session) Analyze different parts of the matrix in separate group-works

Process Matrix A is generally well-structured in the *Exploratory Product Characteristic* dimension. The hierarchical structure of the characteristics will usually provide good advice on how many different groups should be deployed in group work (e.g., for a homogenous product like a paperboard, one group could work with surface characteristics, one with inherent structural characteristics, etc.). Before the start of the group works, mark the 20% most important and the 20% least important *Explanatory Product Characteristics* with different colors. Each group should review the following topical areas:

Group A — Analyzing one section of Explanatory Product Characteristics and related test methods

- *Review the* Explanatory Product Characteristics *in the perspective of their relations to* Product Design Requirements. *Do any horizontal parts of the matrix lack "fill" or contain only weak relations, indicating that such* Explanatory Product Characteristics *do not really relate to any* Product Design Requirement*? Should they be removed from the matrix, or are the relations not well-understood?*

- *Review and analyze the selected part of the relationship matrix and focus especially on the most important* Explanatory Product Characteristics. *Are proper symbols for relationship strength selected for them? Make notes and comments for insertion into the software as "flag notes".*
- *How can the important* Explanatory Product Characteristics *be measured today, and which "unmeasurable characteristics" would you like to be able to measure in the future? What new test methods need to be developed, or what kinds of research facilities and equipment are needed for such measurements?*
- *In order to better understand individual relations in the relationship matrix, what kind of theoretical or experimental investigations should be carried out? Suggest proper activities and suggest potential collaborating National of International research bodies.*

Group B — Analyzing one section of *Explanatory Product Characteristics* and related test methods

See the list of previously presented topics for Group A.

Day 2 Presentation of group work and final synthesis of the results

Part I (Morning session, 2 hours) Presentations of the results from the group work

The presentation of the group work should be carried out by the selected work group leaders using prepared slides and in a complementary view of the large printed paper version of the matrix.

Part II (Morning session, 2 hours) Discussion and synthesis of the results in a holistic fashion and decisions on further actions

The following list of questions to be answered and actions to be taken is only tentative, since each project may require separate analyzing activities:

- *Review and answer the three questions previously articulated in Sub-section 16.1.*
- *How do the important* Explanatory Product Characteristics *of the company's own products compare with such characteristics of competing company products in a Technical Benchmarking perspective? What can be learned about the importance of individual* Explanatory Product Characteristics?
- *What kind of internal or external applied research activities of theoretical or experimental character should be initiated for a further development of the knowledge contained in Product Matrix A? If possible, synchronize with already on-going activities or programs.*

Delineate an action plan for the company's applied research activities

The experiences and conclusions from the Matrix Analysis should be developed into an action plan and program for applied research by the company. Such an action plan should partly follow the previously presented topics and structure for the Matrix analysis. The program may include:

- *An overall research program aimed at an improved understanding of selected important areas of Product Matrix A delineated during the Matrix Analysis.*
- *A research program focusing on how selected important Explanatory Product Characteristics can be better understood and improved.*
- *A program for the development of important test methods and research facilities for selected important Explanatory Product Characteristics.*

Such a research programs can benefit from a previous development of Product Matrix B, which will be further explained in the following chapter.

16.8 The Product Matrix A as only one part of an Integrated Knowledge Platform

The Product Matrix A and the results from the Matrix Analysis should be included in the *Integrated Knowledge Platform (IKP)* for the product or product family to be deployed in further company product innovation. IKPs will be presented and further discussed in the following Part V in this book. Product Matrix A is generally so technically complex that it is not suitable with an illustrative case.

16.9 Introducing Product Matrix B and the guiding work process for matrix development

Product Matrix B is the facilitating instrument for an improved understanding of product functionalities and will help to answer the following question: "How do inherent *Explanatory Product Characteristics* from Product Matrix A relate to *Product Influencing Process Parameters* from the Process Matrix?"

The previously presented Product Matrix A relates *Product Design Requirements* to measurable *Explanatory Product Characteristics*, and both of these product parameters are measured in the laboratory. In Chapter 14, the Process Matrix was introduced, relating the *Product Design Requirements* to *Product Influencing Process Parameters* in the production process. Product Matrix B is a combination of the contents of those two matrices that relates *Explanatory Product Characteristics,* the WHYs, to *Product Influencing Process Parameters*, the HOWs, and thus attempts to explain how such inherent product characteristics are made in the production process. The more structurally heterogeneous a product is, the greater is the need for Product Matrix B as an aid to systematically investigate the dependencies of inherent product characteristics on process conditions.

The above question could be expanded into the following three detailed questions:

- *Selecting the most important* Explanatory Product Characteristics *to improve in the development of a new or improved product (see Product Matrix A), which are the most important related* Product Influencing Process Parameters *to control and improve?*
- *What can be learned from a review of the information within Product Matrix B and the state-of-the-art knowledge of how* Explanatory Product Characteristics *are made in the production process?*
- *What kind of internal or external applied research activities are necessary for further development of the knowledge contained in Product Matrix B? If possible synchronized with already on-going activities or programs. Are there any suggestions for collaborative partners?*

Product Matrix B

In Fig. 16.5, the general structure and design of Product Matrix B is depicted, including some of the most commonly used matrix rooms.

16.9.1 *The overall work process for Product Matrix B*

The overall work process for the development of Product Matrix B is shown in Fig. 16.6. The figure outlines the recommended process steps, which in the following sections will be explained in further detail.

16.10 Select Explanatory Product Characteristics from Product Matrix A, including their calculated importance ratings

All *Explanatory Product Characteristics* from Product Matrix A should usually be selected for the development of Product Matrix B, as illustrated in Fig. 16.7.

16.11 Identify and select Product Influencing Process Parameters from the Process Matrix

In the selection of *Product Influencing Process Parameters*, the general advice is to select the most important Process Control Parameters from the previously developed Process Matrix; see Fig. 14.1. If phase progression has been properly administered, those parameters should

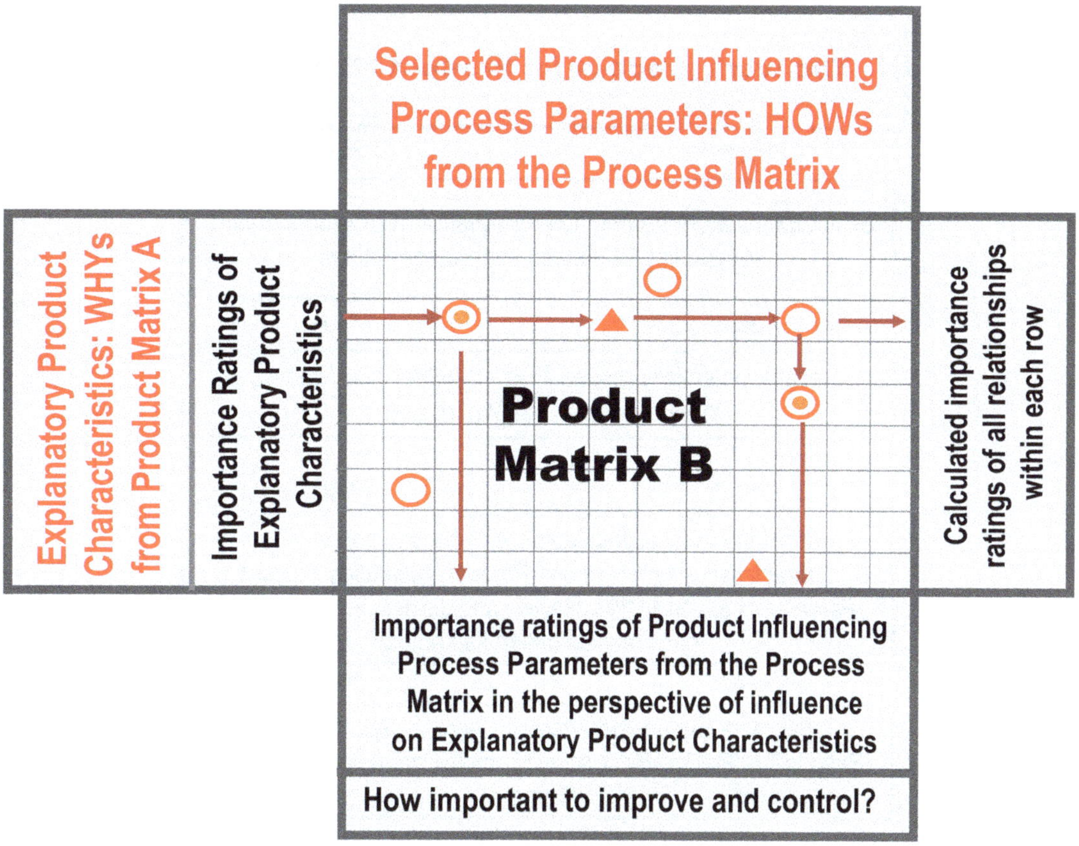

Figure 16.5 A general structure of Product Matrix B and associated default matrix rooms.

reflect the most important *Product Design Requirements* preferences from the previously developed *House of Quality*.

16.12 Construct the Relationship Matrix

Construct the Relationship Matrix with the *Explanatory Product Characteristics* as the WHYs and *Product Influencing Process Parameters* as the HOWs. Review and discuss the strength of each relation (insert "flag notes").

16.13 Recalculate the Explanatory Product Characteristic Importance Ratings into Product Influencing Process Parameter Importance Ratings

Recalculate the Explanatory Product Characteristic Importance Ratings into Product Influencing Process Parameter Importance Ratings.

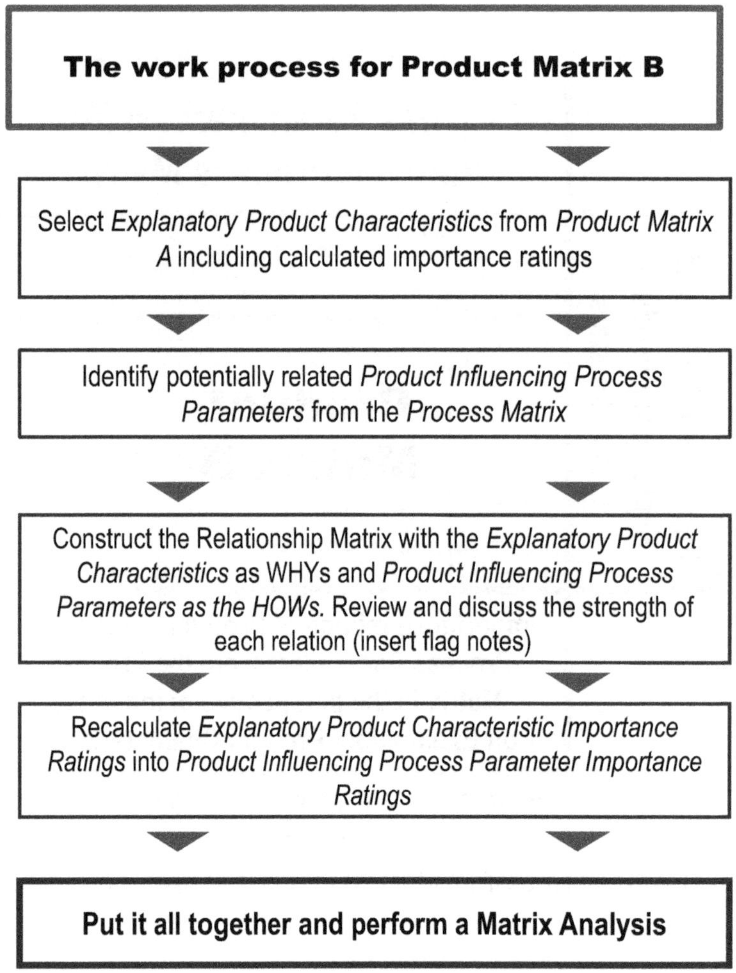

Figure 16.6 The work process for Product Matrix B.

16.14 Develop and design supplementary rooms of interest

Depending on company specific products and product families, there are often a number of alternative rooms to be inserted in this matrix. However, since such rooms are often product-specific, they will not be further discussed in this chapter.

16.15 Put it all together and perform a Matrix Analysis

A well-managed Product Matrix B should be in a constant state of flux during product development activities and in future development cycles. As previously noted for Product Matrix A, this matrix will also provide a platform for the development of "Conceptual Explanatory Models" that describe how *Explanatory Product Characteristics* relate to the production process and its *Product influencing Process Parameters.* Depending on the size and complexity of Product Matrix B, the Matrix Analysis may take only one day, though in some cases it may take two.

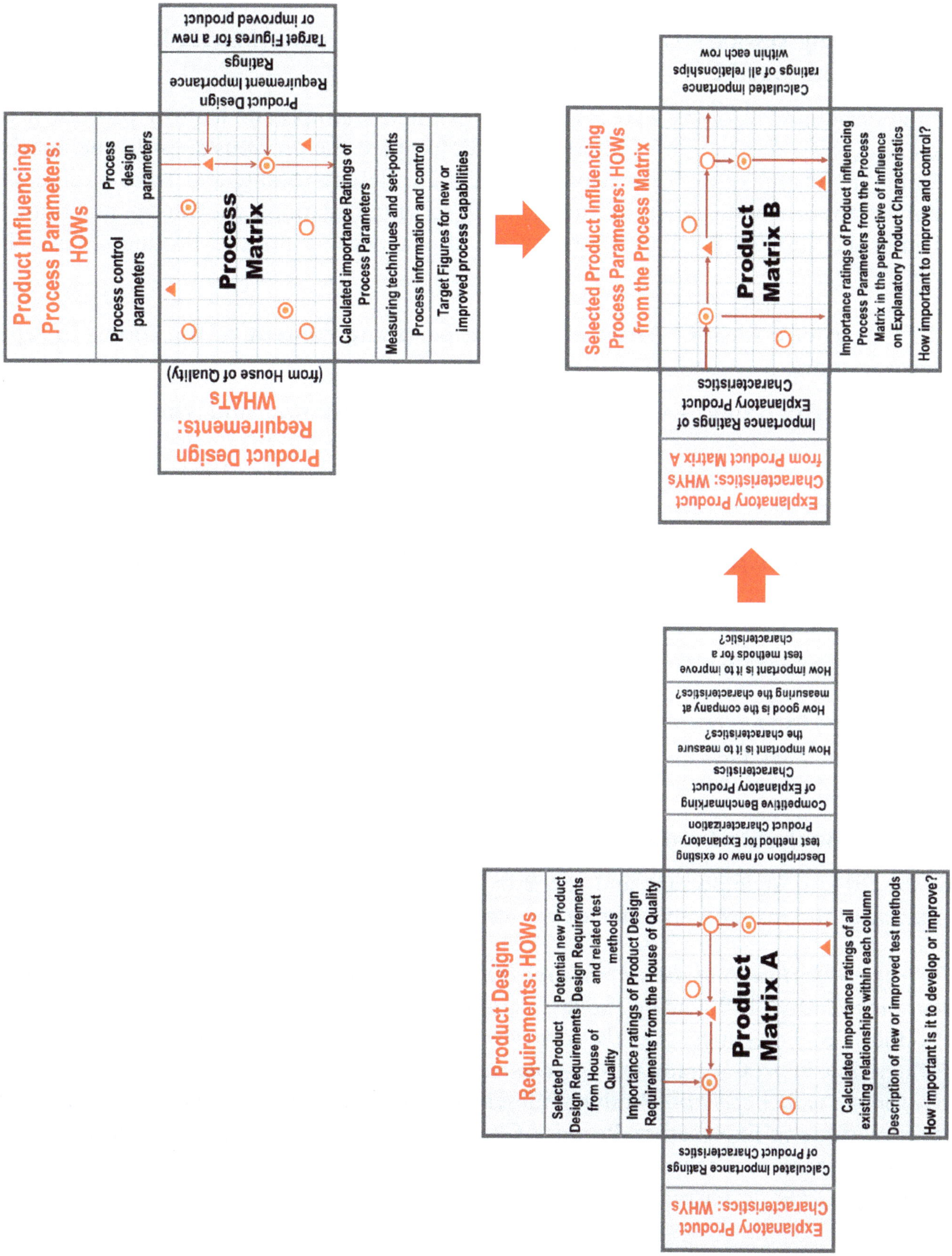

Figure 16.7 Designing Product Matrix B using the Product Matrix A and the Process Matrix.

Day 1 Review the total Product Matrix B and its individual rooms

Part I (Morning session) Take a "guided tour of the Process Matrix B" and review the information together

Since the development of the Process Matrices is often a temporally lengthy process, it is advised that half a day should be allocated to recapitulate the previous steps in the matrix development, refamiliarise all participants with the different rooms, and then outline an agenda for the analysis. Most of the general recommendations previously mentioned in Chapter 12 and for Product Matrix A are also applicable for the Product Matrix B analysis.

Part II (Afternoon session) Analyze different parts and content of the rooms in separate group work

Group A — Analyzing *Explanatory Product Characteristics* and the relationship matrix

Mark the 20% most important and the 20% least important *Explanatory Product Characteristics* in total matrix with different colours

- *Review the most important* Explanatory Product Characteristics *in the perspective of their relations to* Product Influencing Process Parameters. *Do any horizontal parts of the matrix lack "fill" or contain only weak relations, indicating that such* Explanatory Product Characteristics *cannot really be controlled in the production process? Should they be controlled?*
- *In order to better understand individual relations in the relationship matrix, what kind of theoretical or experimental investigations should be carried out? Suggest proper activities and potential collaborating National of International research bodies.*

Group B — Analyzing Product Influencing Process Parameters

Mark the 20% most important and the 20% least important *Product Influencing Process Parameters* in the matrix with different colours.

- *Do any vertical parts of the matrix lack "fill" or contain only weak relations, indicating that such* Product Influencing Process Parameters *do not really relate to any* Explanatory Product Characteristics*? Are the relations not well understood, or should such* Product Influencing Process Parameters *be removed from the matrix?*
- *Review the most important* Product Influencing Process Parameters *in the perspective of their relations to* Explanatory Product Characteristics *(recalculated importance ratings from the importance ratings of the* Explanatory Product Characteristics*). Analyze the relationship patterns for the individual parameters.*

Day 2 Presentation of group work and a final synthesis of the results

Part I (Morning session, 2 hours) Presentations of the results from the group work

The presentation of the group work should be carried out by the selected work group leaders using prepared slides and, if desired, a view of the large printed paper version of the matrix.

Part II (Morning session, 2 hours) Discussion and synthesis of the results in a holistic fashion and decisions on further actions

The following list of questions to be answered and actions to be taken are only tentative, since each project may require separate analyzing activities:

- *Review and answer the three questions previously articulated in Section 16.9.*
- *Which are the most important* Product Influencing Process Parameters *for influencing the most important* Explanatory Product Characteristics?
- *How, and under what process conditions, can the important* Explanatory Product Characteristics *be improved?*

Delineate an action plan for the company's applied research activities

The experiences and conclusions from the Matrix Analysis should be developed into an action plan and program for the company's applied research. It is recommended that such an action plan should partly follow the previously presented topics and structure for the Matrix analysis. The programme may include:

- *An overall research programme aiming at an improved understanding of selected important areas of Product Matrix B, delineated during the Matrix Analysis.*
- *A research programme focusing on a better understanding of how Explanatory Product Characteristics can be better understood and produced in the production process; in search of explanatory conceptual models for the manufacturing of desired Explanatory Product Characteristics.*

Product Matrix B as only one part of an Integrated Knowledge Platform

Product Matrix B and the results from the Matrix Analysis should afterwards be included in the *Integrated Knowledge Platform* for the product or product family to be deployed in further company product innovation. This will be presented and further discussed in the following Part V in this book. Product Matrix B is generally so technically complex that it is not suitable with an illustrative case.

Chapter 17

Multiple Progression Within and Outside Company Demarcations — Addressing the Total Supply/Value Chain in Product Innovation

A QFD system must be able to cope with the long and often complex supply/value chain of multiple customers and suppliers, particularly if the company is located in the middle of such a chain of suppliers and customers and may have no direct contact with the end users.

In Chapter 8, the above issue was identified as an important problem that a QFD system must address in process-industrial product innovation of excellence. If the company products are intended for consumers and B2C customers, the *House of Quality* is the final interface between the company and its end-users. However, there is usually also a chain of upstream suppliers of raw materials and ingredients to whom a progression of such *Product Design Requirements* from the *House of Quality* should be further progressed; this is referred to as a **upstream multiple progression**.

For companies delivering products to B2B customers, there is also often a chain of downstream industrial customers until a final product reaches an end-user. In such an instance, the *House of Quality* is only a primary interface, and the company's *House of Quality* is in reality a Raw material Matrix. This is another important issue with which the QFD system must be able to cope. A **downstream multiple progression** using the mpQFD system will thus be further discussed in this chapter.

17.1 Upstream multiple progression of *Raw material Design Requirements* — in search of an improved upstream supply chain understanding

It has been previously mentioned that the final consumer of a beer is only the last link in a chain of suppliers from barley production, malting and manufacturing of beer. In an

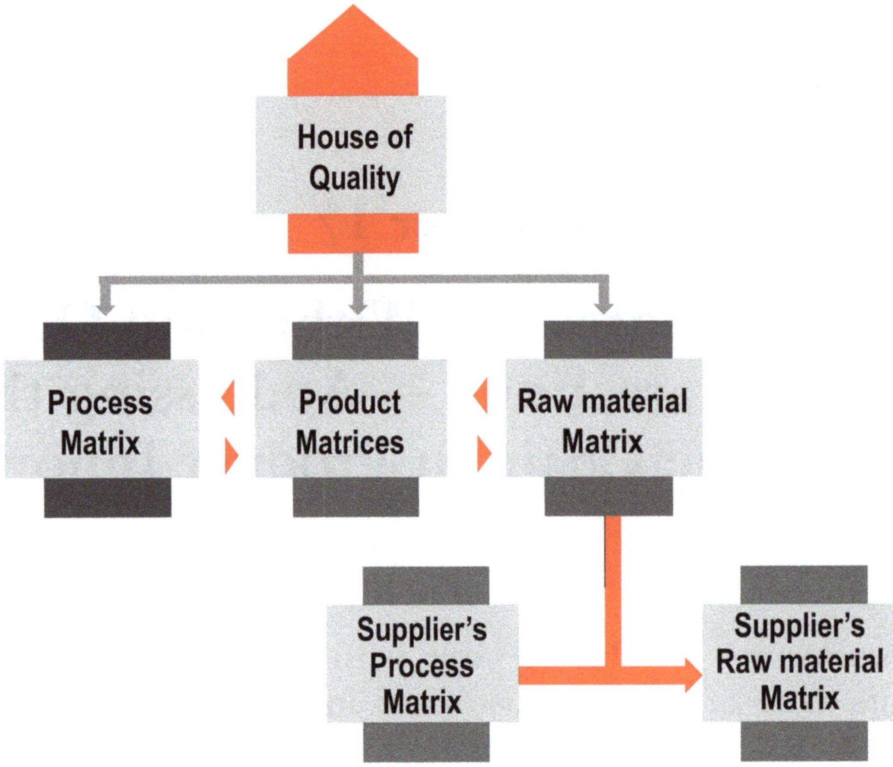

Figure 17.1 Upstream multiple progression.

upstream progression of *Customer Requirements* from the *House of Quality* and its related Raw material Matrix, Raw material Design Requirements could thus be further progressed to the raw material's suppliers, as illustrated in Fig. 17.1.

Since a Raw material Matrix is dependent both on the input from both *Product Design Requirements* and on input from a related production process, as previously shown in Figure P4.2, it is beneficiary in the development of the Supplier's Raw material Matrix if the Supplier's Process Matrix is also outlined (see Fig. 17.1). If not, demands from the production process must be provisionally estimated.

Transcending corporate internal demarcations — from finished products to "in-situ" raw materials

The often long company internal production chains in the process industries were depicted in Figure 14.4, and within a company there may thus be a number of internal production plants connected in an often long internal supply chain. As previously discussed, it is sometimes recommended to treat important internal production interfaces in a similar fashion to external interfaces and thus build a small internal Raw material Matrix (or *House of Quality*).

Transcending corporate external demarcations — from primary raw material suppliers to "in situ" raw materials

The *Raw material Design Requirements* that are expressed in the company Raw material Matrix can be further progressed backwards into deeper levels of the supply chain. The internal or external suppliers of raw materials can view these requirements as the WHATs in their own Raw material Matrix. The raw material supplier to the customer not only has the option to develop its own Raw material Matrix but also can develop a Process Matrix, as illustrated in Fig. 17.1. The *Raw material Design Requirements* HOWs, their related Importance Ratings, and associated Target Figures from the supplier's own Raw material Matrix can be cascaded further into even deeper levels of the supply chain until they become the specifications for raw materials that are available as "in situ natural resources".

17.2 Downstream multiple progression in a B2B supply chain — in search of an improved end-user understanding

If a company is an "in-between" supplier in an downstream supply chain, the *House of Quality* is only the first interface between several B2B customers. Downstream multiple progression is then the instrument for an improved end-user understanding.

Different kinds of test methods and metrics of the performance of supplied products are presented in Fig. 17.2. From measuring *Explanatory Product Characteristics* in the Product Matrices and/or to measuring *Product Design Requirements* in the *House of Quality* and/or in further pilot planting, it is argued that the "real McCoy" is the supplied product's operational behaviour in the customer's production process.

The customer's product properties may also be influenced by the supplied product properties — a fact that has already been thoroughly discussed in Part II of the book. Product innovation of excellence in the process industries must thus be supplemented with application development, which can be defined as follows:

> Application development is not product development, but the significant development of the customer's use of the supplying company's products. The development is primarily intended to optimize and to improve the customer's products and/or production system or to give additional opportunities for other cost savings.
>
> (Storm and Lager, 2014)

The significant and important conclusion is that an improved customer understanding in a B2B context may require the development of the associated customer's Process Matrix, as illustrated in Fig. 17.3.

The development of the customer's Process Matrix is naturally best done in collaboration with the customer's production organization; however, in a less optimal scenario it may

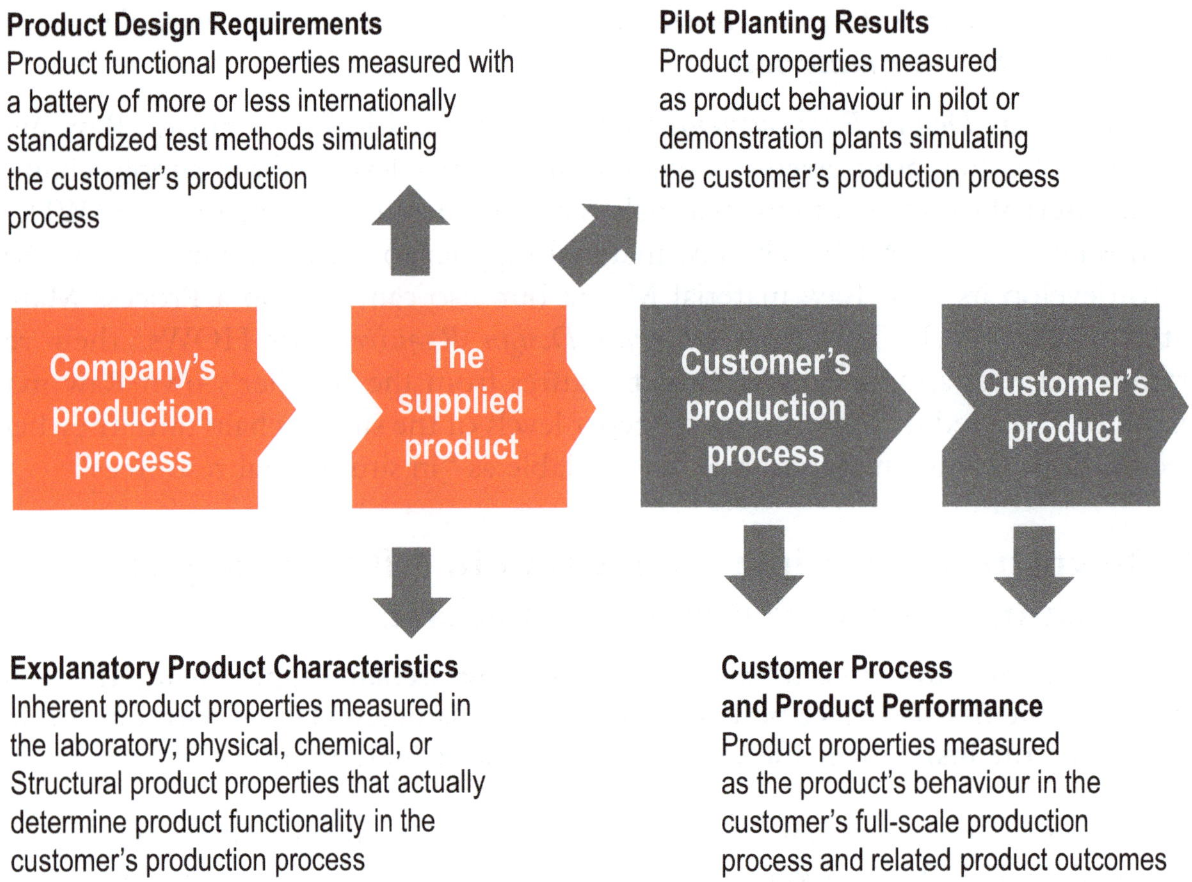

Figure 17.2 In a Business-to-Business (B2B) company relation with its customer, the ultimate *Voice of the Customer* is the product's measurable behaviour in the customer's production system. All other metrics presented in the figure only offer guidance and support for the product developers.

also be carried out by the company's internal experts. ***B2B customers' development of a Process Matrix in collaboration with the product supplier is a truly "open innovation" activity, which has a strong potential to develop into a win-win collaborative exercise.*** The development of such a Process Matrix can follow the presentation in Chapter 14 and develop into a truly important part of an improved end-user understanding.

17.3 The illustrative case of beer development and manufacturing — The Raw material Matrix: barley for malting

The "full version" of the Raw material Matrix: barley for malting is located in Appendix B of the book. You will find here the "mini version" of this Matrix presented in Fig. 17.4. Please observe that recalculated Importance Ratings presented at the bottom of the matrix in the "mini version" will not correspond to the figures in the "full version", since

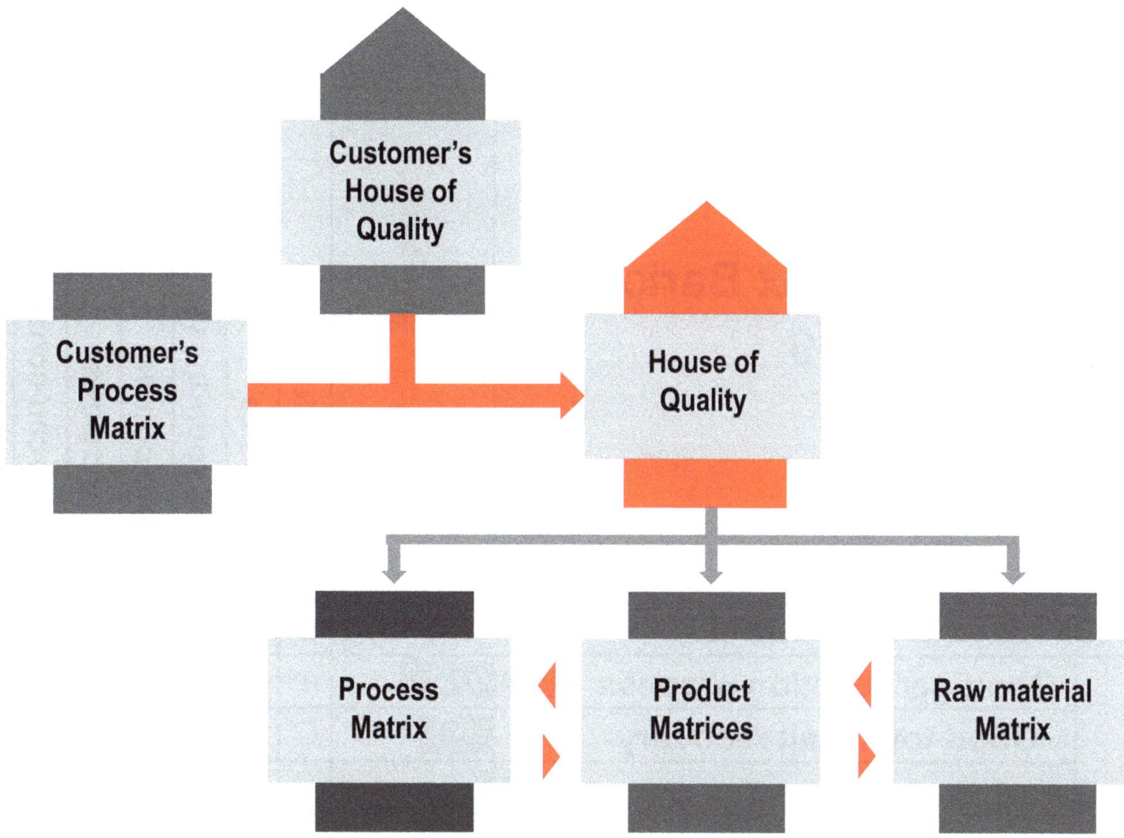

Figure 17.3 Downstream multiple progression.

additional Product and Process Requirement relations are included in the calculation of the figures in the "full version".

In the this matrix, only eight Process and Product Requirements on the raw material barley have been selected for the construction of the "mini version" of the Raw material Matrix. Corresponding eight Raw material Design Requirements on barley have been selected.

The system of Multiple Progression QFD

The *Product and Process Requirement Importance Ratings* presented in the "mini version" are recalculated with the Raw material malt Matrix into *Raw material Design Importance Ratings* for barley. For example, the importance of the barley Raw material attribute "Germination energy" is as follows: 5000*9 + 4000*9 + 8100*9 = 153900.

Product Influencing Process Parameters

In a review of the "full version" of the Raw material Matrix Barley for malting Matrix, the importance of the Raw material attributes "Type of barley" and "Germination energy" stand out.

Raw material Matrix Barley for malting	Ratings	Importance Ratings	Type of barley	Kernael size	Kernel roundness	Protein content	Diastatic power	Beta-glucan content	Germination energy	Pre-germination
			Raw material Design Requirements: HOWs							
Direction of Improvment			○	○	↑	↓			↑	↓
Process & Product Requirements on raw materials: WHATs										
Good germination process	5000		⊙	○		⊙			⊙	△
Good total malt recovery	4000		○	○		○	⊙		⊙	⊙
Colour	1107									
Fruity flavour	4257		○							
Total protein content	2382					⊙				
Beta glucan content	2744		⊙			△		⊙		
Extract yield (coarse grind)	8100			△	△	△		○	⊙	○
Diastatic power	8100		⊙				⊙	⊙		
Weighted Importance			167367	35100	8100	89282	108900	121896	153900	65300
Relative Importance										

Figure 17.4 The Raw material Matrix — Barley for malting.

Chapter 18

Summing Up and Some Issues to Reflect Upon

In this part of the book, you have become acquainted with the upstream and downstream matrices in the Multiple Progression QFD system especially adapted to process-industrial product innovation. It is argued that future company product innovation of excellence must have a stronger focus on "cost innovation" in an overall price/performance product development perspective. Since in the process industries product costs are totally dependent on production process efficiency and the use of selected, adapted and optimized raw materials, the use of phase progression in the QFD methodology is strongly advised.

You have previously learned that the use of the *House of Quality* secures a strong customer relevance in the development of interactive Target Figures for a new or improved product. However, how such new or improved product attributes are achieved is a completely different matter. The development of the upstream matrices will ensure that a complementary knowledge base to the *House of Quality* is developed in a production system dimension. It has been previously advocated that, in the process industries, "the process embodies the product" and that, consequently, neglecting a development of the upstream matrices is a fundamental product innovation error.

In this part of the book, you have further become aware of the fact that, in advanced process-industrial product innovation, it is essential to understand how product functionalities are made. The concept of *Explanatory Product Characteristics* has been introduced, and it has been further explained how the Product Matrices can create an enhanced product understanding. Moreover, it is advocated that the information from a developed *House of Quality* and the information from the upstream and downstream matrices should be further integrated and deployed in *Integrated Knowledge Platforms.* The development and use of such platforms will be further discussed in the next part of this book.

After digesting the theoretical and methodological use of the upstream and downstream matrices, it is now important that you try to relate this information to the perspective of your company's present product innovation work process and philosophy. In order to facilitate such an activity, please reflect on the following questions:

- The price/performance ratio for products is to a large extent dependent on present and future product production cost structure. How does your company's product innovation address the need for reduced manufacturing costs in product innovation and renovation? Could the use of the Process Matrix and Raw Material Matrix be facilitating instruments?
- In your review of a select product and its multiple *Product Design Requirements*, how do they relate to the large number of *Product Influencing Process Parameters* in your company production system? Could the Process Matrix be a facilitating tool for the development of improved explanatory conceptual models for process innovation?
- When Target Figures (product specifications) have been set for new or improved products, how well are related necessary production process capabilities for the manufacturing of a product with such target figures delineated and secured in your company?
- Since the evenness (Target Figures and related upper and lower tolerance levels) of new or improved products is an important dimension in process-industrial product innovation, how are such figures derived and traced back to *Customer Requirements*? How well are such figures related to necessary improved production process capabilities?
- The identification of *Product Design Requirements* and associated test methods to capture individual product attributes and functionalities is essential for interactive target-setting in product innovation. How well do presently used test methods for process innovation capture customers' product user preferences?
- Does your company systematically compile and structure *Explanatory Product Characteristics* for individual products and understand their influence on functional properties?
- Select one important company product and its three highest ranked *Product Design Requirements*. List the five most important *Explanatory Product Characteristics* for those three requirements.
- Is it generally well understood in your company how product functionalities are created, and could the Product Matrices A and B help to establish and delineate such a platform of knowledge?
- Searching for end-user interaction in company product innovation has been a well-recognised and rewarding development strategy since long. Could the presented mpQFD downstream matrices support and invigorate discussions and collaboration with your company's product end-users?
- In the perspective of company B2B customers, it is argued in this book that the customer is, in reality, the customer's production process. Do you think that delineating such customers' *Production Process Requirements* using the customers' Process Matrix not only could benefit company Sales and Marketing activities but also could invigorate company Application Development (improved solutions for customers' use of supplied products)?

- In an analogy to the importance of understanding the end-users' product preferences, an understanding of "Primary Raw material suppliers'" product innovation capabilities is of equal importance in process-industrial innovation. Could the Multiple Progression QFD system and related upstream matrices assist in reaching and collaborating with primary raw material suppliers to your company?
- Since *Raw material Design Requirements*, and their related product specifications and tolerance levels, influence not only company production process performance but also finished product attributes, how does your company's product innovation work process systematically address this issue?
- How often does your company benchmark alternative externally or internally supplied raw materials and ingredients used in the production process, and how is such a benchmarking process administered and communicated within your company?
- In other manufacturing industries, suppliers are generally heavily involved and integrated in product innovation, which unfortunately sometimes is uncommon in many sectors of the process industries. Do you think that well-developed Raw material Matrices could strengthen such interaction with company raw material suppliers?
- It is argued in this book that the Raw material Matrix also can facilitate interaction and dialogues with internal suppliers of intermediate products within a company internal supply chain. If your company sometimes experience barriers between production units, could intermediate Raw material Matrices serve as facilitating instruments for improved integration?

Part V

Consolidating Matrix Information into Integrated Knowledge Platforms (IKPs) — New Perspectives on the Industrial Use of the QFD Methodology

Companies cannot afford to start from a "clean slate" in the future if they want to develop products in a cost-efficient manner and a reasonable timeframe. In order to rectify such a state of affairs, product innovation work processes and projects must start from easily retrievable and previously accumulated company platforms of knowledge. Adoption of the QFD tool in building such well-structured platforms related to individual products or product families is the heart of the matter and the "hidden gem" of QFD methodology use.

Expected company outcomes from the use of QFD

In this part of the book, you will first learn about general industrial experiences from QFD methodology use and expected company outcomes. A number of outcomes from company use of the QFD methodology are generally acknowledged and supported by scientific evidence. Such important outcomes will be highlighted; however, you will also discover that a number of usually advocated outcomes of QFD have no scientific support at all. You will further learn that the most important outcome from the industrial use of the methodology is related to its strong capability as an instrument for knowledge creation and utilization. Unfortunately, this ability is generally unrecognised or overlooked, and for this reason it is advocated in this book to be the "hidden gem" of QFD methodology use.

QFD in a company knowledge management perspective

You will be advised of the importance of capturing and using both "explicit" and "tacit" knowledge in product innovation and of the superior capability of the mpQFD system as a tool for organizational learning. You will be introduced to the concept of "transactive memory system" (TMS), which is usually defined as a collective system for encoding, storing, retrieving and communicating group knowledge; if properly acknowledged and used in the R&D organization, the QFD methodology and its matrices can be seen as such TMS systems. Securing the recovery and use of "tacit knowledge" in innovation is surprisingly still a vastly overlooked area. An area which is not the sole responsibility of the R&D organization, since the recovery of tacit information from customers and collaborative partners by marketers is a product innovation capability that is not seldom overlooked.

Integrated Knowledge Platforms (IKPs) by mpQFD

In Part III, you learned how to build a *Voice of the Customer* and the related *House of Quality*, and in Part IV you further learned how to build the complementary upstream & downstream matrices. You will now learn that the information from all QFD matrices should be well-integrated in order to benefit not only from an improved customer and product understanding but also from an improved understanding of the process-industrial production system. Such a platform of knowledge, integrating the total matrix information, has in this book been denominated *Integrated Knowledge Platforms (IKPs)*.

Such *Integrated Knowledge Platforms* should be related to individual products or product families and serve as supporting platforms for the overall product innovation work process in the development of new or improved products. If such platforms are developed in advance of more project-based product innovation activities, they will overcome the time-consuming methodological disadvantage of QFD methodological use and instead facilitate and speed up the product innovation work process.

It has previously been concluded that company product innovation generally will benefit from a well-structured work process, and that the pre-development part, commonly denominated "the fuzzy front end", is one of the most difficult parts to master. It is advocated in this book that the weakness of the product innovation work process in general, and the difficulties with "the fuzzy front end" in particular, are related to a misconception that the pre-development phase should start from scratch. In order to speed up company product innovation and taking full advantage of product ***methodologies of excellence like QFD, such methodologies should be used to pave the way for subsequent, more project-oriented innovation activities and work processes.***

Such an approach will not only help to take the "fuzziness" out of "the fuzzy front end", but will also take advantage of the diverse number of benefits one should be able to expect from a total "end-to-end" work process use the QFD methodology. Thus, the inefficient and costly use of the QFD methodology in "ad hoc" development projects should be abandoned and substituted by the use of the methodology in the building of company reusable *Integrated Knowledge Platforms* for innovation. Such Platforms are company frameworks and points of departure for faster product innovation; they are not part of the product innovation work process as such, but they serve as launch-pads for the kick-start of subsequent project-based product innovation activities.

Chapter 19

Expected Company Outcomes from the Use of the QFD Development Methodology — Scientific Evidence from the Past

Regrettably, in the past the QFD methodology has often been "marketed" by consultants and academics as a panacea for improving everything in company product innovation. As an analogy to the classical saying "if you are everywhere, you are nowhere", it is time to recognize and focus on the proven advantages and strengths of methodology use while avoiding its inherent disadvantages and shortcomings.

A common mistake in too many QFD textbooks and other publications has been to advertise the industrial use of the QFD methodology as an instrument for the improvement of all aspects of product innovation. In the seminal article, "The Manager's Job: Folklore and Fact", Professor Henry Mintzberg (1990) revealed some myths about general management. In the light of today's scientific evidence, it is now time to reveal some myths about the expected benefits from company use of the Quality Function Deployment development methodology!

Thus, it is important first to identify the prominent and specific expected advantages and outcomes from its industrial use, in order to better capture them in company day-to-day product innovation. Along the same line, and most importantly, the methodological weaknesses must also be identified, addressed and remediated, in order to secure the future efficient and effective industrial use of the methodology. Hence, we must go from folklore to facts in the presentation of the usability of the QFD methodology in order to convince experienced industry professionals that, if properly used, it is a versatile and efficient tool for improved product innovation performance.

19.1 On the industrial usability and expected outcomes of QFD

> The conclusion is that the benefits of the QFD methodology will not be realised in the first project, but mainly in the further improvement of the overall company's product development process.
>
> (Griffin, 1992)

19.1.1 *A review of the usability of the QFD methodology — Lessons from the past*

In one of the first studies of QFD use, the results from a large survey of mainly large manufacturing companies in Japan were presented. These results give a clear impression that the Japanese companies had already then come far along the road of using QFD and closely related quality tools (on average, 14 products had been developed with the QFD methodology in each company). The highest ranked expected outcomes from QFD use were: *easier setup of design quality (design requirements)* and *reduced initial quality troubles* (Akao *et al.*, 1987). In the first study of the use of QFD in the USA, the results from 35 development projects in nine companies were reported (Griffin, 1992). The results showed that it cannot be taken for granted that a company using QFD will necessarily benefit from lower development costs or shorter development times. Surprisingly, however, nearly all projects showed other interesting company benefits, like *more rational decisions in the product development process, improved information flow, melding the team together, forging cross functional relationships* and *capturing knowledge.* The study indicated that a company's first QFD project does not necessarily shorten the time for development; on the contrary, the project may take even longer than it otherwise would.

In a qualitative study of four Swedish companies, the company use of the QFD methodology was examined (Norell, 1992). Only the *House of Quality* was used, the methodology was adapted to company needs, and the work was carried out in cross-functional teams. The following (non-ranked) problems with the methodology were identified: *difficult to distinguish between WHATs and HOWs, the methodology is sometimes subjective and not precise, very time-consuming*, and *difficult to estimate the time frame.* The mentioned benefits were: *well retrievable information* and *a good group consensus.*

Other advantages reported were: *improved contacts with the marketing department and within the cross functional team and a well-structured work process with no irrelevant chatting.* Another QFD experience of 31 Swedish manufacturing companies was investigated in an early survey (Ekdahl and Gustafsson, 1997). The identified problems encountered in using QFD were as follows, in ranking order: *lack of management support, group member commitment, lack of resources, lack of QFD experience.* The ranking of benefits achieved after implementing QFD, were as follows, in ranking order: *better communication, preservation of knowledge, unity in the group, more rational decisions, and better designs.*

A study of the use of QFD in mainly manufacturing Fortune 500 companies assessed the results from 80 QFD projects (Vonderembse *et al.*, 1997). The study showed, rather convincingly, that all companies concluded that QFD had improved their products. The methodology created an *information-intensive climate, where communication is improved and ideas are freely communicated.* Nearly all companies were of the opinion that project documentation was improved, and so was organizational learning. The results, however, do not indicate either lower product costs or significantly reduced time to market. Two case studies focused more on the implementation of the methodology and the selection of the right application (Comstock and Dooley, 1998). However, it is an interesting study for many reasons, one of which is the unconventional approach in both projects. In one of the projects, aimed at the development of a completely new consumer product, a market research company was used as a consultant not only for market research but also to build the *House of Quality*. In the other project (for an industrial customer in the defence industry), the technical specifications were used as the WHATs in the *House of Quality*.

A comparative study of the use of QFD in 101 companies in the USA and 47 in Japan addressed the QFD experience of mainly large manufacturing companies in both countries (Cristiano *et al.*, 2000, 2001). The results from the studies are presented in two separate papers, of which the first (Cristiano *et al.*, 2000) presents the use of QFD and expected outcomes. The Japanese companies relied to a large extent on customer complaints, while the American ones made extensive use of market research and interviews for their *House of Quality*. The QFD benefits reported are *better overall project results* and *better product design*. Companies in the USA reported *better customer satisfaction*, which was significantly lower in Japanese companies. No findings of improved product design for manufacturing or lower production costs were reported. In the area of better group communication and better decision process, companies from the USA reported significantly higher figures than the Japanese companies. These areas included: *better cross-functional work*, *better knowledge building*, *more rational decisions*, *better group dynamics*, and *better communication with market and design departments*. No evidence of *reduced time to market* was reported from either Japanese or American companies.

Another study presents the results from a large survey in Great Britain including 246 companies (Martins and Aspinwall, 2001). The results from the study showed that 39% of the companies that used QFD considered a better product (reduced costs, time, complaints and defects and better adherence to product specifications) to be the most important result of using QFD. Top-ranked benefits were *an improved and more dynamic decision process* (24%), *improved teamwork skills* (20%) and *better communication internally and with customers* (15%). Finally, a case study reported on the use of QFD in five large manufacturing companies (Baglieri *et al.*, 2002). There is general but unfortunately not very specific agreement among the companies on the organizational and conceptual power of QFD. From an organizational point of view, the main conditions for making QFD more effective were considered to be: *top management's strong commitment*, *short-term orientation*, and *strong investment in training*.

19.2 A meta-analysis of the previously presented studies

In the deployment of the QFD methodology, there appears to be some general misunderstanding about the "cause and effect" of expected outcomes. Let us thus get this sorted out once and for all. Using a "retrievable and accumulated company knowledge base" coupled with "improved company communication", you will get "better decision support", assisting in the development of "better product specifications", resulting in "better products" and ultimately leading to "better overall customer satisfaction".

The research results from a synthesis and meta-analysis of the industrial usability of quality function deployment (Lager, 2005a), using the previously presented studies, are presented in Table 19.1. The expected outcomes that received strong support in the meta-analysis analysis are presented in large bold font, and other supported outcomes are presented in bold font.

The results from the study are somewhat surprising in the light of previously published presentations. One of the most frequently cited arguments regarding why companies should introduce and use QFD is an expected reduction of time to market. For example, the title of one of the first published books on QFD in America was *Better Design in Half the Time* (King, 1987). This slogan regarding the Japanese experience seems to have developed into a myth of huge proportions. There is, however, no scientific evidence for this argument. Not even in the study by Akao *et al.* (1987), which is referred to by King, is there any

Table 19.1 Expected outcomes from the use of the QFD methodology. Expected outcomes that received strong support in the meta-analysis analysis are presented in large bold font, and other supported outcomes are presented in bold font.

Product-related outcomes

- **Better overall customer satisfaction**
- **Better product (product design)**
- Lower manufacturing cost
- Lower design cost

Work-process-related outcomes

- A generally improved product development process
- **Improved cross-functional team work**
- Shorter time to market
- Shorter design time
- **Better product specifications**
- **Better decision support**

Company-related outcomes

- **Improved company communication**
- **Retrievable and accumulated company knowledge**
- Improved manufacturing process

evidence of such an outcome. One could possibly hope that after developing a comprehensive QFD matrix on a product, the development of the next product generation should be faster, but there is only weak support even for this outcome in a few studies. One can only guess that companies that have jumped on the QFD bandwagon in order to speed up product development have been greatly disappointed. So, let us leave this misconception behind once and for all.

Likewise, there is no support at all for an improved manufacturing process. Since all evidence indicates that few companies have tried to go beyond the *House of Quality*, how could QFD support an improved manufacturing process? This seems very unlikely. There is also practically no support for the likelihood of getting lower manufacturing costs or lower design costs with QFD — not even in Japan, where QFD systems often include cost deployment. The same is true for a generally improved product development process, shorter design time and lower design costs.

The very good news from the presented studies is that there are a number of significant and important benefits from company use of the QFD methodology. There is strong scientific support for the outcomes of better overall customer satisfaction, better products, retrievable and accumulated company knowledge and improved company communication. Arranging these results in a logical sequence, the general expected outcomes from the use of the QFD methodology were presented in the introduction to this subsection. ***Such well-proved outcomes from QFD industrial use should certainly incentivize product innovation professionals to introduce and make good use of the QFD methodology, focusing on its knowledge creation and utilization capabilities.***

Chapter 20

A Review of the QFD Methodology in the Perspective of Company Knowledge Creation and Utilization — Discover the "Hidden Gem" of QFD Use in Product Innovation!

QFD appears complicated at first glance, and technical personnel might tend to respectfully ignore it, but the data can be considered as an accumulation of the past that can be added to or improved with each new development cycle and therefore become an important asset to the company.

(Dir. Nakahita Sato, former Director of Toyota Auto Body)

In Chapter 19, the results showed — somewhat surprisingly, but very convincingly — that one of the most important overall benefits from the use of the QFD methodology relates to its outstanding ability as an "organizational learning" and "knowledge creating" instrument in product innovation. In light of these results, the insightful statement quoted above from a representative from one of the pioneering industrial users of the methodology (Toyota Auto Body) acknowledged very early this important and outstanding methodological capability in a knowledge management perspective (American Supplier Institute, 1989).

In an R&D organization, all individual members are "knowledge workers" in that their output, broadly speaking, is new knowledge, and a knowledge-based perspective on R&D organizations and product innovation is consequently always interesting and applicable. However, in order to operationalise this idea, one must provide the product innovation organization with proper instruments, tools and systems. In this book it is argued that the QFD matrices are first-rate such instruments and that the overall QFD methodology can well serve as such an organizational learning system.

20.1 Product innovation in a knowledge management perspective

> **We argue that the success of Japanese companies is not their manufacturing prowess; access to cheap capital; close and cooperative relationships with customers, suppliers, and government agencies; or lifetime employment, seniority systems, and other human resources management practices — although all these factors, of course, are important. Instead we make the claim that Japanese companies have been successful because of their skills and expertise at "organizational knowledge creation". By organizational knowledge creation we mean the capability of a company as a whole to create new knowledge, disseminate it throughout the organization, and embody it in products, services, and systems.**
>
> **(Nonaka and Takeuchi, 1995)**

In reference to the management of knowledge in company R&D organizations, it goes without saying that in a meta-organizational-perspective, building and sustaining knowledge is an activity second to none. In this subsection, however, you will first be introduced to some relevant and important knowledge management concepts, which afterwards will be deployed in the perspective of QFD methodology use.

20.1.1 *Organizational learning and transactional memory systems in R&D — from personal to shared knowledge*

> Transactive memory theory describes how groups develop a shared memory system in which knowledge is distributed among group members and retrieved using personal directories of meta-memory. Such directories support expertise location within the group by identifying 'who knows what.'
>
> (Nevo *et al.*, 2012)

Organizational learning is a fairly old concept. Crossan *et al.* (1999) took the very broad perspective that organizational learning can be conceived as a principal means of achieving the strategic renewal of an enterprise. They further identified three levels of organizational learning: individual (intuiting and interpreting), group (integrating) and organization (institutionalization). Fostering cross-functional organizational learning capabilities in product innovation has long been generally acknowledged to be of utmost importance and a strong success factor for product innovation, and company R&D work processes today are usually supportive and encouraging of such behaviour in product innovation. Successful organizational learning in innovation is an epitome for successful product innovation, and it is thus important to highly rate the ability of a development methodology that can promote and facilitate such an outcome.

The related concept of "transactional memory systems" was first proposed by Wegner *et al.* (1985) as a response to earlier theories of "group mind" and "groupthink." It constitutes stored information in the memories of individuals, in combination with the "meta-memory" of the group's collective expertise. A transactive memory system (TMS) is usually defined as a collective system for encoding, storing, retrieving and communicating such group knowledge (Lewis *et al.*, 2005). It promotes cycles of learning that produce not only knowledge that is relevant for the current task but also transferable knowledge that can be applied to other tasks in the same domain. In this chapter, you will discover that, if properly acknowledged and used in the R&D organization, the QFD methodology and its matrices can truly be viewed as such a TMS system. Locating expertise in large organizations is a pressing problem today, with approaches ranging from social networks and communities and expert directories to a host of expertise location systems aimed at providing support for "expertise seekers." In that perspective, Nevo *et al.* (2012) recognized that members of a group can serve as external memory aids to one another. Industrial product innovation today is, with few exceptions, project-based and team-based, and the countless organizational learning opportunities in product development teams represent a fascinating area to explore and promote in companies. Using the TMS as a mental model and a framework for the improvement of team-based performance appears to be a promising avenue from a knowledge-based perspective. This book argues that a meta-memory view of a product development team's interactive knowledge-building capabilities and the TMS model for guiding information processes are important perspectives, and which motivate company use of the QFD methodology.

20.1.2 *The interplay between tacit and explicit knowledge creation in innovation — a Japanese outlook*

> Western managers need to 'unlearn' their old view of knowledge and grasp the importance of the Japanese view. They need to get out of the old mode of thinking that knowledge can be acquired, taught, and trained through manuals, books or lectures. Instead they need to pay more attention to the less formal and systematic side of knowledge and start focusing on highly subjective insights, intuitions, and hunches that are gained through the use of metaphors, pictures or experiences.
>
> (Nonaka and Takeuchi, 1995)

Central to the concept of organizational learning is the concept of "tacit information," which was originally presented in the book *The Tacit Dimension* (Polanyi, 1983). Polanyi's conclusion was that if problems exist, and discoveries can be made by solving them, one can know important things that one cannot tell others. He called this kind of knowledge "tacit knowledge" and claimed that one can have tacit foreknowledge of yet undiscovered things, something that might be called a "hunch". Tacit knowledge, as opposed to codified or explicit knowledge, can thus be exemplified (an ostensive definition) as the knowledge of how to drive a car or how to bake bread; a kind of knowledge that is difficult to transfer to another person through written text.

Japanese companies often have a rather different understanding of knowledge and organizational learning compared to most Western organizations, since the former recognize that the knowledge expressed in words and numbers represents only the "tip of the iceberg". They view knowledge as primarily "tacit" and thus not easily visible and expressible. The subjective and intuitive nature of tacit knowledge makes it difficult to convert into words and numbers, but it is argued (Nonaka and Takeuchi, 1995) that it is precisely during the time that this conversion takes place (from tacit to explicit and back to tacit) that organizational knowledge is created. You will learn in this chapter that, if supported by a proper QFD software, the QFD matrices are excellent facilitating instruments to capture and transfer tacit information into explicit information and vice versa.

20.1.3 *Socialization: from tacit knowledge to tacit knowledge*

As the name indicates, socialization is the process of sharing experiences between individuals. Knowledge gained through socialization is acquired through observation, imitation and practice. Referring to the previous example, one cannot learn to drive a car solely by reading written information about driving; rather, one must watch others and engage in driving behaviours for oneself.

20.1.4 *Externalization: from tacit knowledge to explicit knowledge*

The externalization mode of knowledge conversion is typically referred to in the process of concept creation and is triggered by dialogue or collective reflection.

20.1.5 *Combination: from explicit knowledge to explicit knowledge*

Combination is a process of systemising concepts into a knowledge system and can be exemplified by formal education and training in schools.

20.1.6 *Internalization: from explicit knowledge to tacit knowledge*

Internalization is the process of embodying explicit knowledge into tacit knowledge and is closely related to "learning by doing." Documents, manuals and verbal stories are examples of the instruments that can be used in this process.

Securing the recovery and use of tacit knowledge in innovation is still a vastly overlooked area in Western industrial innovation, despite the emphasis on this area in many publications. The reason for this oversight may be that R&D organizations do not recognize or understand the importance of this aspect of innovation, but more likely depend on a general uncertainty of how to capture tacit knowledge in innovation. However, recovering tacit knowledge is not the sole responsibility of the R&D organization, since the recovery of tacit information from customers and collaborative partners by marketers is a product innovation capability that has been previously discussed. The capability of the QFD methodology to capture tacit information is thus a most interesting aspect of methodology use.

20.1.7 *Reassessment of the QFD methodology from the perspective of knowledge creation and utilization*

> We argue that while R&D obviously generates innovations, it also develops the firm's ability to identify, assimilate and exploit knowledge from the environment — what we call a firm's "learning" or "absorptive" capacity.
>
> (Cohen and Levinthal, 1990)

In a seminal paper titled "Innovation and learning: The two faces of R&D", quoted above, Cohen and Levinthal introduced the concept of "absorptive capacity". In the light of this dual role of R&D, they suggested that the long-run cost of learning may be substantial, that most of this cost is borne via the development of a stock of knowledge that constitutes the firm's absorptive capacity, and that a significant benefit from R&D is its contribution to this knowledge base (Cohen and Levinthal, 1990). In a later review and reconceptualisation of absorptive capacity, Zahra and George (2002) distinguished between potential and realized absorptive capacity. The original ideas proposed by Cohen and Levinthal (1990) are still valid, highlighting the importance of the R&D organization's "stock of knowledge" as the prime facilitating mechanism for the acquisition of external knowledge. In that respect, the potential and realized absorptive capacities serve well and are thus selected as an overall guiding framework for an assessment of the QFD methodology (Lager, 2017b), as introduced in Fig. 20.1.

The vast scientific evidence on knowledge management, the organization's absorptive capacity (Zahra and George, 2002) and organizational learning (Crossan *et al.*, 1999,

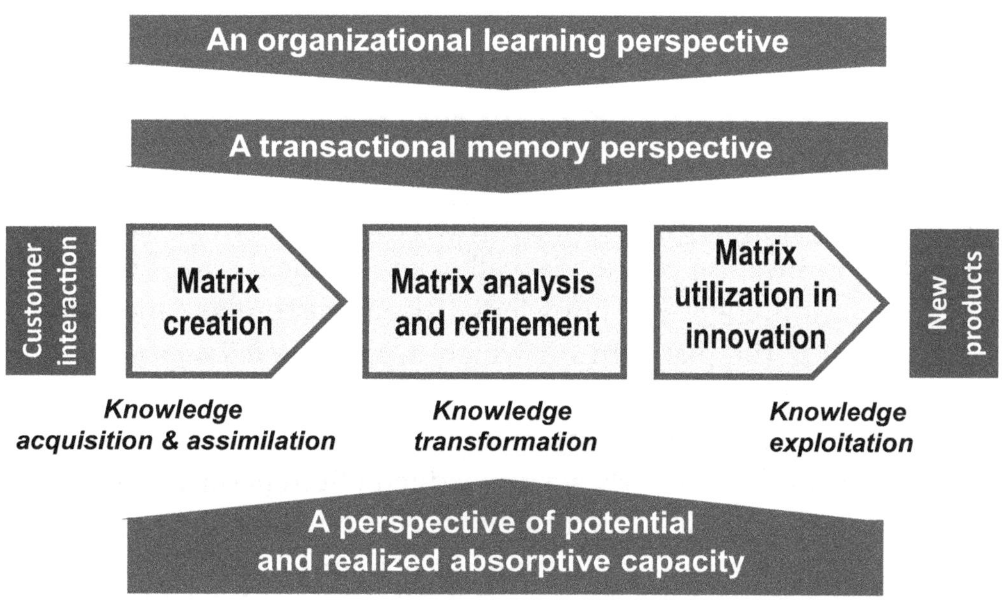

Figure 20.1 A conceptual framework for an assessment of the QFD methodology from the perspective of knowledge creation and utilization (Lager, 2017b).

2011), along with books such as *The Knowledge Creating Company* (Nonaka and Takeuchi, 1995) and *Thinking for a Living* (Davenport, 2005), are challenging companies to improve their knowledge creation and utilization capabilities. It is easy to confess to such ambitions, but it is a different matter to be faithful to such philosophies in the real operations of a company. Despite the importance of these concepts, guidelines and methodologies for operationalising them and ideas for their use in industrial product innovation are scarce.

In the following three subsections, the QFD methodology will thus be reviewed in the perspective of company absorptive capacity and through the lens of the three consecutive activities presented in Fig. 20.1:

- Knowledge acquisition and assimilation
- Knowledge transformation
- Knowledge exploitation

It certainly appears that these areas of importance for reaping the "low hanging fruit" of the industrial use of the QFD methodology have too often been overlooked, possibly misunderstood, and generally ignored in today's QFD publications and in industrial use (Lager, 2017b).

20.2 "Knowledge acquisition and assimilation" in the creation of QFD matrices

> **A firm's capability to identify and acquire externally generated knowledge that is critical to its operations and a firm's routines and processes that allow it to analyze, process, interpret, and understand the information obtained from external sources.**
>
> **(Zahra and George, 2002)**

20.2.1 *Using QFD in a facilitating dialogue with customers and as a framework for market research*

The use of the QFD methodology as an instrument for linking the customer to the company and the product innovation process is well acknowledged and is generally considered the hallmark of the methodology. The knowledge about customers (and, most importantly, also about competitors) is the platform on which a proper *House of Quality* is built. In a business-to-consumer (B2C) context, as well as in the development of consumer products, the growing importance of customer interaction, co-creation and co-development with consumers is today an area that is clearly acknowledged (Bettencourt *et al.*, 2014; Martinez, 2014). However, this is not the whole truth; in discussions with customers and consumers in the development of the *Voice of the Customer*, it is important to recognize that not only explicit knowledge but also customers' tacit knowledge is sought.

The structured information in a well-developed *Voice of the Customer* has been found to be an efficient platform for further discussions with the marketing function and with external marketing research organizations (Lager and Kjell, 2007). Individual well-defined

Customer Requirements can be used as a guide either in contact with individual customers or as a framework for making larger marketing surveys. In the development of the inquiries in surveys or guiding frameworks for focus groups or clinics, the *Voice of the Customer* pinpoints the questions that product developers need to have answered (Day, 1993).

Customers that have been involved in the development of a product using QFD have usually acquired a better understanding of how to use the product (Tottie *et al.*, 2016). In a B2B context, co-development with customers is not only an important approach for the development of products adapted to customer requirements but also a most interesting instrument for "customer relationship building" (Macdonald *et al.*, 2011; Tottie *et al.*, 2016; Vargo and Lush, 2011). Advancing from an "innovation for the customer" development approach to an "innovation with or by the customer" approach, is generally desired in today's company product innovation.

20.2.2 *Supporting the organization and review of customer data and information using QFD*

In the domain of knowledge acquisition, the *Voice of the Customer* as the foundation on which the *House of Quality* is built in the QFD methodology is an acknowledged tool (Bossert, 1991; Cohen, 1995; Day, 1993; Eureka and Ryan, 1988). The *Voice of the Customer* is a well-proven part of the methodology (Miguel, 2013; Terninko, 1997) and, in

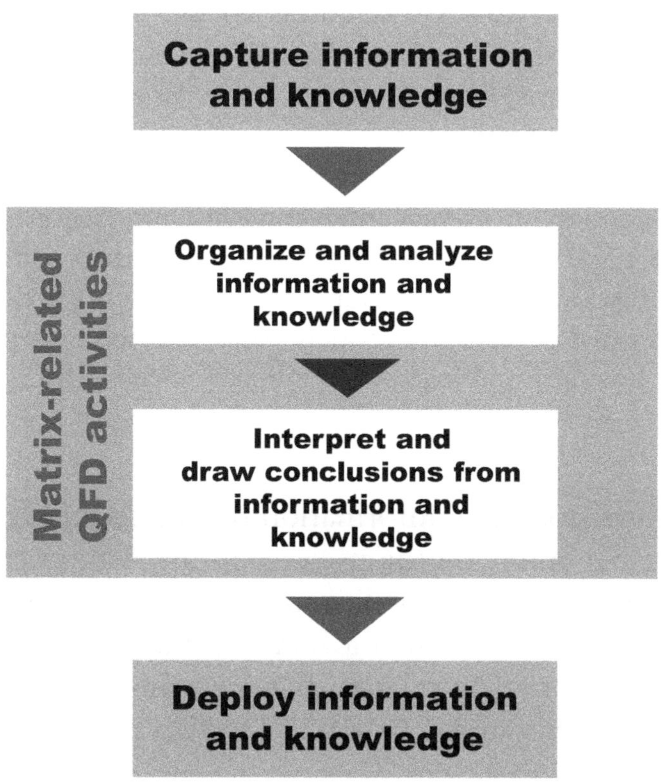

Figure 20.2 QFD in the perspective of an information system.

the spirit of Zahra and George (2002), provides an outstanding instrument to analyse, process, interpret and understand knowledge related to company consumers and customers.

However, as Fig. 20.2 points out, the production of information using more or less advanced marketing research tools, focus groups and clinics, as well as the subsequent deployment of this information in product innovation, should not be considered the "core QFD methodology".

20.3 "Knowledge transformation" in the refinement, analysis and revision of the QFD matrices

> **Denotes a firm's capability to develop and refine the routines that facilitate combining existing knowledge and the newly acquired and assimilated knowledge.**
>
> **(Zahra and George, 2002)**

20.3.1 *Refinement of the QFD matrices*

All QFD matrices, if properly developed, should be created through the collection of both explicit and tacit knowledge, the latter of which unfortunately is often ignored. Matrix structures are often created in an iterative fashion and filling in the relationship matrices are exercises when experience tells that extraordinary opportunities exist for the capturing of team members' tacit knowledge. In the development of the matrices, and during the discussions of individual matrix rooms and relations in each room, tacit knowledge is articulated within the QFD teams and must be saved as explicit information for later retrieval. For those who have not experienced this fascinating articulation of tacit knowledge during such discussions while members of a QFD team, this kind of organizational learning is difficult to comprehend. It has been reported as:

> The matrix proved to develop into a very interesting instrument for organizational learning and collection of tacit LKAB knowledge that already existed to some extent in the R&D organization. As one researcher pointed out, this matrix will provide a platform for the development of "conceptual explanatory models" that describe how product functionalities are related to more or less measurable product inherent characteristics.
>
> (Tottie *et al.*, 2016)

If the QFD software allows storage of information related to different areas of the matrices and can be made retrievable online by the use of pop-up "flag notes" (IDEACore, 2003), it is possible to capture information from all team members and present it as slices of information with individual signatures. In such a fashion, the information under a "flag note" can spell out, for example, that "The relation between this customer requirement and the design requirement is strong because … . T.L.". Or "lab report No XX validates that this technical property is not competitive … B.A." In this way, all of this kind of matrix information constitutes a "living document" that may contain even contradictory statements.

> It was further discovered that this knowledge-building capability of the methodology was closely related to the functionality of the software and its ability to relate and save information to all matrix areas and rooms and even relate information to individual symbols in the matrices. The use of pop-ups displaying this information proved to be very valuable during the working sessions.
>
> (Tottie *et al.*, 2016)

The experience of the development of individual matrices is generally that it stimulates individual team members to express tacit information. However, when one team member presents an idea or information related to a specific part of the matrix, this will only be a starting point for group discussions in the perspective of Nonaka and Takeuchi's *socializing* mode (tacit to tacit knowledge). The matrices can thus be regarded as containers of "synthesized tacit knowledge" combined with the explicit knowledge that, in reality, is the major part of information within the matrices.

> The tacit knowledge of the multifunctional QFD team, when converted into explicit knowledge and organized into matrices showing cause-effect relationship, enabled the construction of a system of standards as a means of transmitting information to shop floor and thus contributing to assure product quality.
>
> (Cheng and de Melo Filho, 2008)

When used properly, explicit and tacit information can be merged, and the information can be made available as organizational knowledge, since the matrices can be studied, communicated and used by selected company organizational departments and functions. ***The company accumulation and subsequent distribution of such information is one of the most important aspects in the use of the methodology; however, if not given proper attention and captured during matrix development and analysis, such information will unfortunately be lost.***

20.3.2 *Analysis of the QFD matrices*

After the matrices have been developed, each individual matrix must be thoroughly analysed during more or less formal events or workshops, as previously discussed in Parts III and IV of this book. Such a matrix analysis should always be carried out after the initial development of each matrix, but this activity is unfortunately generally forgotten or not well explained in QFD textbooks. During a proper matrix analysis, such a holistic perspective in combining a multitude of facts related to product development is a vitally important activity in product innovation.

Cohen (1995) noted that: "*the matrices can be viewed as a top-level view or directory to all the rest of the information. In fact, QFD has sometimes been described as a "visible memory of the corporation"*. Proving the Chinese saying that a picture is worth a thousand words, the easily accessible information contained in a well-built *House of Quality* is tremendous. The visual strength of the matrices is remarkable and gives a unique opportunity to analyse and interpret large amounts of information. The use of coloured symbols and other graphical tools facilitates the review of information surprisingly well, and the interaction between individual rooms is easily grasped. Furthermore, in the use of advanced flexible computer programs, large matrices can be split up into sub-matrices to facilitate more focused discussions of individual parts of the matrices, either in internal company forums or together with select customers.

Experience tells that a proper use of the QFD system can support a more "fact-based" innovation approach; from art to science in product development. ***In the process industries, the easiest paths of development have usually already been trodden, and it is thus likely that future successful product development must be based on a more factual knowledge base — a knowledge base that is more related to findings from applied research.*** In this respect, the matrices can thus serve as a guiding instrument for the identification of interesting areas for R&D. If some areas of a matrix are not well covered (low "fill"), such "white spaces on the map" are often interesting for further research and development (Tottie and Lager, 1995). That is to say one has acquired information that the required information is not available. This can stimulate exploratory research and development activities in order to sort out important or questionable individual relations or even larger matrix areas where present company understanding is insufficient. The use of the *Product Matrices* in the mpQFD system (Lager, 2005b), is a development instrument especially designed and adapted to capture this kind of information.

20.3.3 *Revision of the QFD matrices*

Matrices are never completed; they are and should be in constant flux. A similar point of view was expressed by Franceschini and Rosetto (1998): "*the big investment in QFD is therefore for the first project. For subsequent projects it is necessary only to update and adapt the procedure and documentation arranged the first time*". There should be a never-ending iterative development of all matrices and a *House of Quality* in a continuous dialogue with customers and other stakeholders. The input from customer interactions and new ideas from the development team are merged during the Matrix Analysis, and the results are afterward used as input for a matrix revision.

20.4 "Knowledge exploitation" in the use of QFD matrices in product innovation

> **An organizational capability based on the routines that allow firms to refine, extend, and leverage existing competences or create new ones by acquired and transformed knowledge into operations.**
>
> **(Zahra and George, 2002)**

A company's absorptive capacity was characterized by Zahra and George (2002) as: "*a set of organizational routines and processes by which firms acquire, assimilate, transform and exploit knowledge to produce a dynamic organizational capability*". Assuming that company dynamic capabilities are related to change, the product innovation work process could be viewed as such a "routine and process." In a similar way, the QFD methodology could also be seen as a "routine and process" to be deployed as a methodological tool in the work process. Raper (2000) thus remarked that: "*QFD is a team/communication approach, and its ability is to serve as a mechanism for translating knowledge to new teams*".

20.4.1 *QFD is not the matrices but what you do with the matrices*

In the development of the upstream matrices in a paper mill and during project execution, Scheurell (1992) gained a new insight, which has been noted earlier:

> QFD is not the charts. It's what you do with the charts. If all you do is to make matrices, you are not making full use of the tool. Developing the culture.

It was also demonstrated by mill employees that the matrices provided them with an "excellent learning tool", giving them a clear reference to customers' WANTs (Scheurell, 1993). Day (1993) also recognized that the purpose of the use of the QFD methodology is not to build matrices but to get in touch with the customer and to use this knowledge to develop products that satisfy the customer. For a person unfamiliar with the use of the QFD methodology, this may sound self-evident, but experience unfortunately tells that this is a rather common attitude in the use of the methodology. After the time- and resource-consuming development of the QFD matrices, the QFD teams are often exhausted but also rather satisfied that their "mission is completed". However, after the matrices have been developed, analysed and saved, they must be deployed in the further development of new or improved product concepts during the pre-product development phase.

20.4.2 *Use of the QFD matrices in the pre-development phase of product innovation*

Özgener (2003) points out that:

> The QFD model allows team members to review and recall exact details months, and even years, after a meeting. Thus the QFD documentation process is a highly effective way for getting new members up to speed on what to do and why.

In the deployment of the QFD methodology in product innovation, the target setting after the matrix analysis is an important activity. The knowledge contained totally in the matrices must then be integrated and captured in a multidimensional fashion. Setting provisional and final Target Figures for individual product *Design Requirements* creates a "screen" through which future concepts can be sifted. One reason that such Target Figures are superior to normal product specifications derived from more "ad hoc" work processes is that the product specifications are interactive and traceable in the existing *House of Quality*. That is to say, each product specification can be supported by matrix data and should be motivated by information related to each specification figure (flag notes), which is an important improvement compared to other commonly used procedures in product innovation.

20.5 Renaming the "House of Quality" as the "House of Knowledge"

In view of the presented information and the methodology's strong capability as an instrument for knowledge creation and utilization in product innovation, one could suggest that the "House of Quality" today could more properly be named the "House of Knowledge".

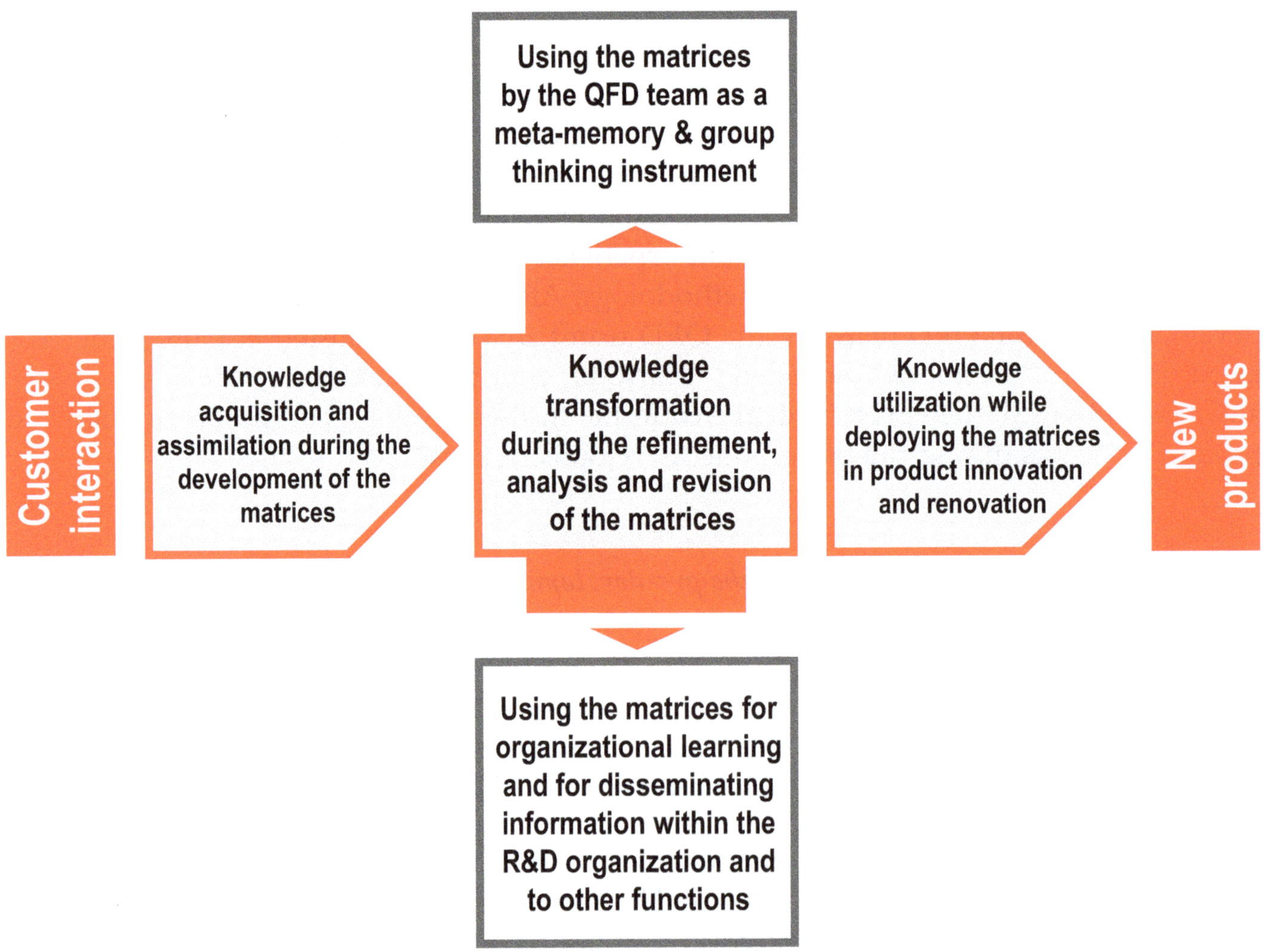

Figure 20.3 A conceptual explanatory model for using QFD methodology as an instrument for knowledge creation and utilization in product innovation (Lager, 2017b).

The overall conclusion from the preceding presentation is that the QFD methodology is an important (and possibly outstanding) instrument for "knowledge creation and utilization" in product innovation. The strong scientific evidence from the surveys and academic publications, coupled with the experiences from industry professionals, indicate that this capability (which, unfortunately, is often unrecognised) is a "hidden gem" in methodology use. It is thus argued that the QFD methodology can serve as an instrument in knowledge creation and utilization during three different phases of methodology use, as shown in Fig. 20.3.

Chapter 21

Consolidating the Knowledge and QFD Matrix Information into Coherent and Reusable Integrated Knowledge Platforms (IKPs)

Product innovation projects in the process industries sometimes have a time-span that lasts many years and are often also very costly. For these reasons, it is vitally important for a company to take advantage of learning from previous innovation experiences and to continually improve its knowledge base. An *Integrated Knowledge Platform* (IKP) amalgamates "explicit" and "tacit" knowledge about customers, related process technology and raw materials for a select product or product family into a solid and reusable framework through the use of the mpQFD system, supporting the overall product innovation work process.

Part III of this book introduced the QFD matrix *House of Quality*, and Part IV introduced the theory and use of the upstream and downstream matrices for the process industries. In the use of the QFD methodology in other manufacturing industries, the matrices can naturally also be used; however, for this part of the book, the process-industrial environment has been selected. Nonetheless, the above definition of an *Integrated Knowledge Platform* is generally applicable to all manufacturing industries.

It seems that today's presenters of the methodology have sometimes been "barking up the wrong tree" in their praise of the methodology's benefits, as they often have disregarded or forgotten the important capability of the QFD methodology as a tool for "knowledge building" in product innovation; a fact which was very early recognized by Eureka:

> After the efforts of a QFD project have been deemed successful, the knowledge base created for that project serves as a repository for engineering knowledge. ... The matrices and charts prepared during the QFD procedure create a working document that can be easily

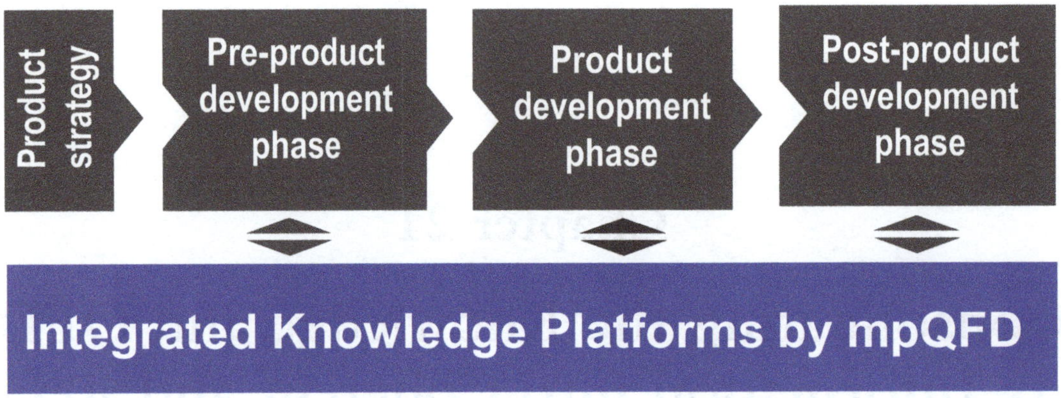

Figure 21.1 Integrated Knowledge Platforms related to select products or product families as supportive foundations for a product development work process.

> referenced and learned from. ... By reviewing the results from successful QFD projects, entry-level engineers begin higher up on the learning curve.
>
> (Eureka and Ryan, 1988)

As such, the matrices will together provide an accumulated and easily retrievable knowledge base for company product innovation in the spirit of the previously quoted Toyota's former director.

The company development and use of individual QFD matrices and related knowledge are thus essential instruments in product innovation, but the importance of an integrated use of all matrices has seldom been considered in previous publications and has usually been disregarded in industrial use. ***However, the dynamic use of all matrix information is certainly an opportunity not to be overlooked in company product innovation — a situation in which the total value of the individual matrix use is less than the user value of the integrated whole.***

Because of that, and in view of the previously advocated use of the QFD methodology as a prime instrument for knowledge creation and utilization in product innovation, such a complimentary structure of all matrices is identified and denominated in this book as an *Integrated Knowledge Platform* (IKP). Moreover, it is argued that such platforms of knowledge should be developed and deployed as platform structures for individual products or product families, supporting an overall company work process. A fundamentally new approach in the use of the QFD methodology is thus advocated, as illustrated in Fig. 21.1.

21.1 Deploying the QFD methodology to build Integrated Knowledge Platforms

> **'Pockets' of specialized knowledge exist, but they tend to be spatially and temporally dispersed across firms because they are constituted and generated in the course of practice and reflection by individuals working on specific projects, products or processes in different locations and times. Therefore, organizational effectiveness is linked**

with the firm's ability to integrate these dispersed pockets of knowledge such that they become generally available for ongoing work throughout the firm.

(Purvis *et al.*, 2001)

In the perspective of R&D organizations as architectures for integrating knowledge, company knowledge platforms are built through the acquisition, storage and dissemination of knowledge, and the importance of generating such platforms within company organizations has been well-articulated by Purvis *et al.* (2001).

Following this line of thought, and recognizing the capability of the QFD methodology as a tool for "knowledge creation and utilization", it is proposed that the methodology is well-suited for the development of company *Integrated Knowledge Platforms.* Such matrix platforms should then be intimately associated with a single product, a product family (product line) or, most preferably, a product platform serving a number of derivative products.

21.1.1 *Integrated Knowledge Platforms as launch-pads for product innovation projects*

The general idea with *Integrated Knowledge Platforms* is that the QFD matrices should already be in place when project-based pre-product development activities start but must initially be revised and updated for each new generation of new or improved products. Starting product innovation in each new development cycle from such accumulated and easily retrievable knowledge bases, in the spirit of the previous citation from Toyota Auto Body, the QFD methodology has a strong potential to speed up related product development activities. The disadvantage of the present time-consuming use of the QFD methodology should thus incentivize companies to change their project-by-project use of the methodology to instead deploy the methodology as an instrument for continuous knowledge building related to selected products or product platforms.

In the use of the QFD matrices as *Integrated Knowledge Platforms*, they should thus initially be dislocated from project-based product innovation activities and instead should be made available as supportive instruments for the overall product innovation work process. The interaction with customers and consumers should naturally be an on-going company process within such platforms that certainly also favours the abandonment of a project-by-project approach in the use of the QFD methodology, a fact well-articulated by Day (1993):

> Keeping in touch with the customers' voice must, therefore, be an on-going process. The voice will change. Its importance level will change. The perception of competitive performance will change. Only by continual surveys can a company keep in touch with its customers' wants and needs.

Deploying the QFD methodology as an instrument to create such *Integrated Knowledge Platforms* will also make company investments in QFD methodology more attractive, since good returns can be expected from a more long-term use of the matrix information in many development projects related to a select product or product family.

21.1.2 *Deploying Integrated Knowledge Platforms for a dynamic integration of all matrix information contained in the mpQFD system*

In the earlier presentation of the Matrix Analysis of individual matrices of the mpQFD system, and in the setting of provisional and final Target Figures for individual matrices, the use of the combined matrix information has not yet been discussed and highlighted. In the use of the full mpQFD system, the combined matrix information in an *Integrated Knowledge Platform* for an individual product or product family can now be deployed and analysed as illustrated in Fig. 21.2. Initially, the information from all individual matrices can first be reviewed in a holistic fashion; moreover, in the light of this new combined stock of knowledge, previously set provisional Target Figures can now be revised into final Target Figures for product innovation. However, this new information may also require a revision of the information within individual matrices in an iterative, dynamic way.

Moreover, the information from individual* Integrated Knowledge Platforms *can also be used in a knowledge-transfer fashion between a number of different but related platforms. One must distinguish between necessary personal QFD capabilities for the development of the matrices and the necessary personal capabilities in the use of the matrices. The latter activity is thus less demanding, which will promote sharing of information and use of the *Integrated Knowledge Platforms* in different organizational settings.

21.1.3 *Using the QFD methodology and Integrated Knowledge Platforms in the development of innovation concepts*

This book focuses on the theory and use of the QFD methodology for the creation of *Integrated Knowledge Platforms* in the use of the mpQFD system. However, it is certainly acknowledged that such platforms not only can support the total product innovation process but also are particularly important instruments in the pre-product development part of the work process.

The final Target Figures from the analysis of the total matrix information in the *Integrated Knowledge Platforms* will guide company development of innovation concepts. This is further illustrated in Fig. 21.3. In the previous review of necessary QFD capabilities for process-industrial use, it was recognized that both product concept development and related process concept development must be contained in such a QFD system. The often-neglected development of process concepts is thus supported by the information generated in the development of the upstream matrices in the *Integrated Knowledge Platforms.*

21.2 Perspectives on the future use of the QFD methodology and Integrated Knowledge Platforms (IKPs)

In the previous review of the industrial use of QFD, some weaknesses in today's general use of the QFD methodology in product innovation were recognized — a situation that is not uncommon among many methodological tools and best practices. Proper guidance will thus be provided in the following how to overcome the three most important of these

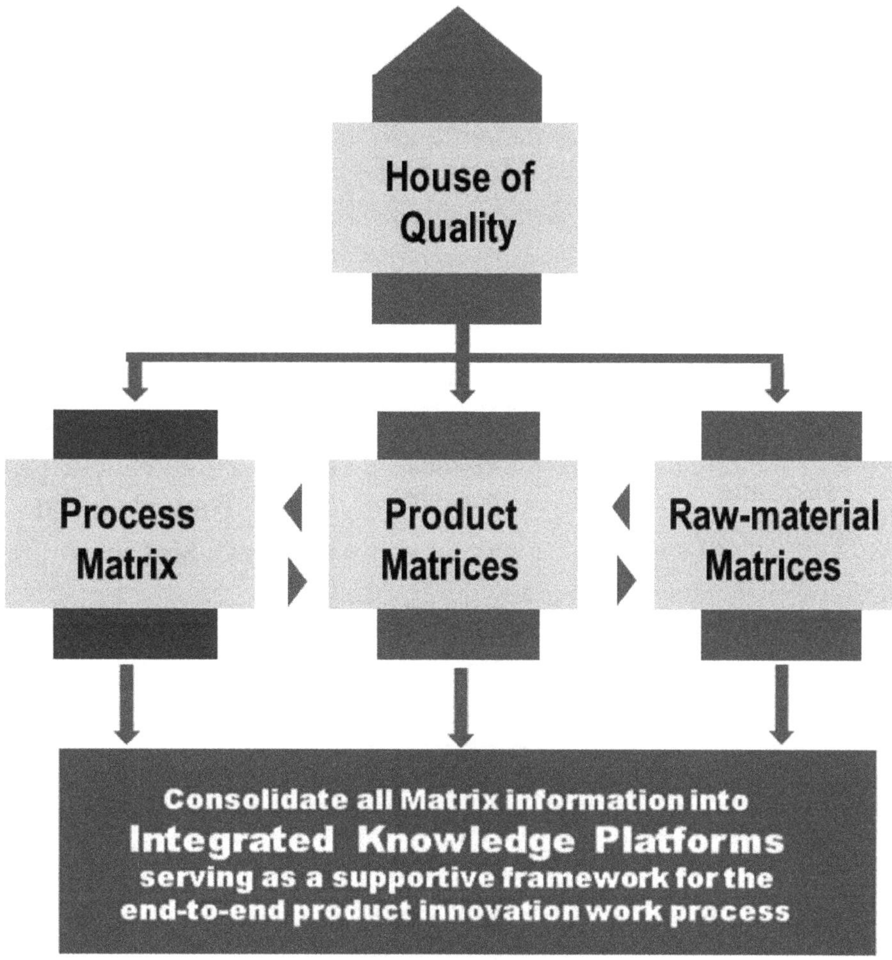

Figure 21.2 The QFD matrices and the related information are consolidated and contained in the Integrated Knowledge Platform. A dynamic use of total matrix information within each Integrated Knowledge Platform is recommended.

disadvantages. It is a general consensus among industry professionals today that the biggest problems in the industrial use of the QFD methodology unfortunately are the complexity and time-consuming nature of the methodology as such.

(1) *Using Integrated Knowledge Platforms to overcome methodology complexities and speed up the company's overall product innovation work process.*

In an early paper about the philosophy of QFD, Hunter and Landingham (1994) urged readers: "*don't let the simplicity of Quality Function Deployment methodology scare you away*". In the light of the previously presented information on the industrial use of the methodology, we might rephrase that recommendation as don't let the complexity of Quality Function Deployment scare you away. One previously recognized disadvantage of the QFD methodology is thus related to the perceived complexity of the methodology and the difficulty involved in learning, understanding, and applying the methodology in an industrial context (Ekdahl and Gustafsson, 1997; Miguel, 2013).

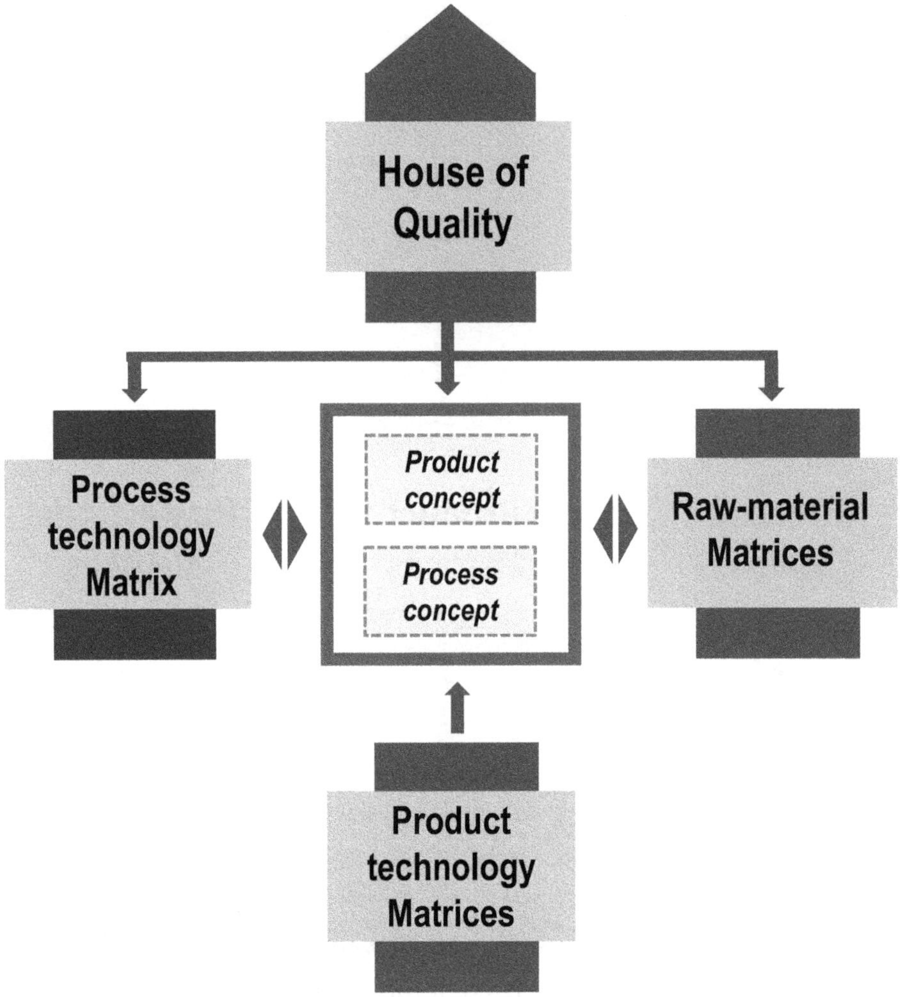

Figure 21.3 Using the consolidated matrix information in Integrated Knowledge Platforms for the development of product and process concepts (Lager, 2005b).

If the decision is made to deploy QFD as a supporting tool in a product innovation project, the project team usually does not have a prior knowledge about the methodology. When the QFD methodology is used as a tool on a project-by-project basis by a team that is unaccustomed to the methodology, it can be expected that this will not speed up project work and could instead make the development even more time-consuming. Even if the time lost is somewhat compensated for later on in a faster product innovation work process, this is not a sufficiently good reason for the use of the methodology in this manner. For this reason, it is generally recommended that expert external facilitators should be used for the first QFD project and that more effort should be invested in training the QFD teams. However, if the methodology will continue to be used on a project-by-project basis, this disadvantage in the use of the methodology will certainly persist. In this context, one should consider Griffin's (1992) comments that *the benefit from QFD will not be realized in the first project* and *the use of the methodology must be seen in the perspective of an improved innovation work process.*

The proposed use of previously developed *Integrated Knowledge Platforms* will overcome this disadvantage, since the matrices will be used in ongoing day-to-day development work and in many product development cycles. Such an approach will not only speed up the development work process and individual product development projects but also overcome the disadvantage of the "time and again" introduction of a complex methodology in the R&D organization. ***Consequently, the industrial deployment of the QFD methodology must be approached in a radically different manner than as an "ad hoc" tool for single development projects.***

Moreover, it must be initially recognized that the development of well-built QFD matrices is a rather time-consuming and complex exercise. There is strong scientific evidence that the most important industrial disadvantage in the use of the methodology is the large amount of necessary resources needed and the time-consuming activity involved in building individual matrices (Griffin, 1992; Miguel, 2013; Norell, 1992). Not only is today's use of the QFD methodology time-consuming, and not only are QFD project time frames difficult to estimate, but also it is clearly indicated that using QFD in its traditional approach can actually prolong the total development cycle and certainly will not reduce time to market.

In addressing this methodological weakness, companies must disregard short-cuts like the "Blitz QFD" approach, and leave other such methodological simplifications to their fate; the misuse of the QFD methodology in such ways not only will discredit an important development tool but also, and most importantly, can seriously harm company innovation activities. Özgener (2003) also noticed that the process of QFD is costly and difficult and requires a much longer view to achieve pay-back. Bouchereau and Rowlands (2000) also concluded that s*ignificant educational and organizational changes will be needed* [...] *Therefore for the full benefits of QFD to be realized, a long-term view is needed and possibly a cultural change.*

In conclusion, in order to reap the fruit of the methodology, the short-termism of earlier QFD approaches must be abandoned, and the QFD methodology should be used in the construction of **Integrated Knowledge Platforms.** In order to speed up innovation activities, this knowledge base must be made available before product innovation activities commence. Moreover, such knowledge bases can be used to capture innovation knowledge from previous development cycles.

The present, and most common, use of the QFD methodology is in the ad-hoc, project-by-project manner described above. In such use, after a project is finished, the results from the QFD activities are often not properly saved, and the information gained through its use — sometimes in vast quantities — is soon forgotten.

For these reasons, companies should abandon the project-by-project use of the QFD methodology in favour of using the methodology in a continuous development of "knowledge platforms" that are intimately related to and associated with product platforms or single products. Starting product development in each new development cycle from such accumulated coherent and easily retrievable knowledge bases has a strong potential to speed up the overall product innovation work process. The general idea is thus that the related QFD matrices must already be in place when pre-product development activities start, and they need only to be revised and updated for each generation of new or improved products. Finally from another perspective and in an overall defence of the industrial use of development methodologies like QFD, which are criticized for being too complex and time-consuming, one must consider whether they not only are imaging today's complex industrial development environment?

(2) *Revitalizing company use of the QFD phase progression systems in order to secure product manufacturability (processability) in innovation.*

The scientific evidence tells that company use of QFD phase progression systems is still scarce. The reason for such neglect can often be attributed to problems with the handling of the multitude of product components and the unfamiliarity with the somewhat complex "Four Phases of Matrices" system for use in other manufacturing industries. However, future product innovation must not only develop new or improved products that delight the consumers and customers but also secure product manufacturability in today's cost-competitive market environment. While seeking technical solutions is the major concern in design, it is in the production stage that product costs are actually committed and product quality and product development lead times are determined.

In terms of the need to integrate production process technology and raw material specifications for the early phases of product innovation, the use of phase progression systems is instrumental in the transfer of product functionalities into process capabilities. In light of the importance of developing products with a high degree of manufacturability, and in order to recognize the importance of production costs in product development, one must conclude that company product innovation must progress *Product Design Requirements* further into the production process and that it is preferable that the development and use of such matrices should already be in in place before product innovation projects start. Relying solely on the *House of Quality* or even solely on the *Voice of the Customer* is thus equivalent to taking a risky shortcut in the industrial use of the QFD methodology.

(3) *Instituting a new "end-to-end" work process perspective in the use of QFD in product innovation.*

Today, the QFD methodology is generally acknowledged as an excellent and possibly outstanding tool, which is primarily to be used in the pre-development phase of the product development work process — that is, the "fuzzy front end." Its usability in this difficult part of the product innovation work process is unquestionable and well-proven. However, this book argues that the use of the QFD methodology should be expanded further, and that QFD also could be deployed in the following product development activities and assist in the marketing of future improved or new products, which is underscored in the following quote:

> The overall impression of those involved in this project, and the general conclusion of this study, is that the QFD system has a strong potential to be deployed as an instrument for knowledge building in product and process development. [...] The systematic use of the matrices and the way the QFD methodology has been used in the pre-development phase — for gathering, valuing and analyzing new knowledge about customer requirements — is

> believed to contribute to the potential absorptive capacity of the case company, and further usage of the method throughout the development and post-development phases will enhance the transformation and exploration of new knowledge, reinforcing the realized potential capacity.
>
> (Tottie *et al.*, 2016)

Figure 21.4 illustrates how today's normal industrial use of QFD focuses on "upfront product innovation". The use of the QFD methodology to build *Integrated Knowledge Platforms* available in advance of the commencement of a product innovation project will, however, "kick-start" a very early QFD use, as illustrated in the supplementary part of the figure. Furthermore, the use of the upstream & downstream matrices will integrate the necessary production system aspects in early phases of product innovation especially during the development of product and process concepts. The industrial experiences in the use of the QFD methodology in advance of and during production plant start-ups and the use of the methodology in subsequent sales and marketing of products confirm the usability of the methodology as an overall supportive instrument in a company's total end-to-end work process product innovation perspective.

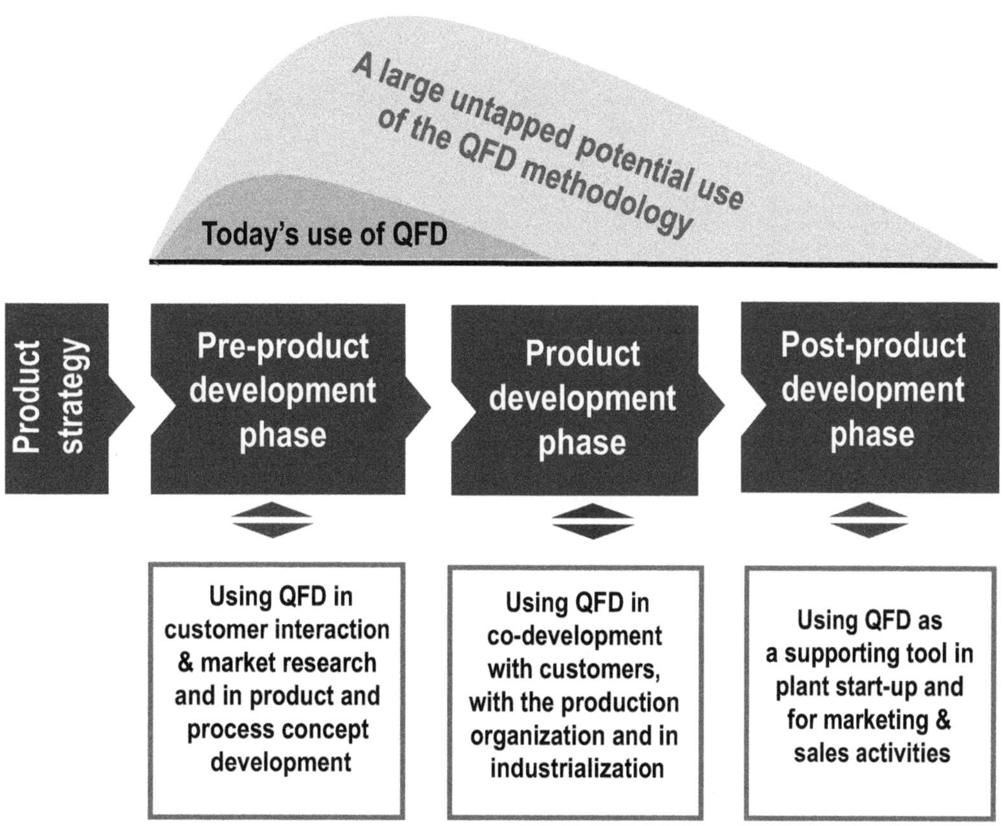

Figure 21.4 Using the QFD methodology in a total end-to-end work process product innovation perspective.

Chapter 22

Summing Up and Some Issues to Reflect Upon

22.1 Expected company outcomes from company use of QFD

In this part of the book, you have learned that expected outcomes from the industrial use of the QFD methodology can be summarized as follows:

Using a "retrievable and accumulated company knowledge base" coupled with "improved company communication", you will get "better decision support", assisting in the development of "better product specifications", ending up with "better products" and ultimately leading to "better overall customer satisfaction".

From the perspective of company R&D, these outcomes are quite impressive and attractive for achieving excellence in product innovation.

22.2 QFD as a system for knowledge management in product innovation

You have now become familiar with the outstanding ability of the QFD methodology as an instrument for knowledge management and for improved company product innovation behaviour during the three following phases.

22.2.1 *QFD for knowledge acquisition and assimilation*

Use of the QFD methodology as an instrument for linking the customer to the company and the product innovation process (Vintage QFD) has long since been acknowledged, and is generally considered to be, the hallmark of the methodology. In a business-to-consumer (B2C) context and in the development of consumer products, the growing importance of customer interaction, co-creation and co-development with consumers is today clearly acknowledged, and these are areas where the use of the QFD methodology fits in very well.

Moreover, experience shows that the structured information in a well-developed *Voice of the Customer* is an efficient platform for further discussions with the marketing function and external marketing research organizations. In the development of inquiries in surveys or guiding frameworks for focus groups or clinics, the *Voice of the Customer* thus serves to pinpoint the questions for which product developers need answers.

22.2.2 *QFD for knowledge transformation*

QFD teams can, during the development, refinement, analysis, and revision of the matrices, use the matrices as transaction memories, serving as meta-memories for "group think". In a dynamic perspective, the QFD sessions can thus be looked upon as endless tacit–explicit, explicit–tacit, tacit–tacit and explicit–explicit sources of knowledge transformations and transfer if the methodology is properly used. Matrices not only merge existing knowledge with new knowledge but also convert the teams' tacit information into explicit information. The matrices also serve as proper instruments for organizational learning, since this explicit knowledge can easily be communicated within the R&D function and to marketing and production.

22.2.3 *QFD for knowledge utilization*

The purpose of company use of the QFD methodology is not to build matrices but rather to get in touch with the customer and to use this knowledge to develop products that satisfy the customer. After the matrices are developed, analysed and saved, they are to be deployed in the further development of new or improved product concepts during the pre-product development phase. The difference of this approach to the development of normal product specifications, is that QFD Target Figures can be interactively traced backwards and related to the knowledge contained in the different matrices. However, you have further learned that, in the use of the QFD methodology, the upstream & downstream matrices are seldom used, and the phase progression ability of the methodology is thus unfortunately overlooked despite its proven effectiveness.

22.2.4 *The House of Knowledge*

The matrix named *House of Quality* by Hauser and Clausing (1988) emerged in an era when the QFD methodology was regarded as one of many quality tools in a Total Quality Management (TQM) overall context (Tottie and Lager, 1995). However, even if one can regard the QFD as a tool that brings quality to the product innovation work process, the methodology is regarded today primarily as a product innovation tool. In the perspective of the presented information and the methodology's strong capability as an instrument for knowledge creation and utilization in product innovation, one could suggest that the *House of Quality* today could more properly be named the *House of Knowledge.*

22.3 Integrated Knowledge Platforms by mpQFD

You have further discovered that there is a need for a fundamentally different industrial approach in the use of the QFD methodology. It is recommended that the system be used in the creation of *Integrated Knowledge Platforms (IKPs)*, supporting end-to-end work processes product innovation:

> An Integrated Knowledge Platform (IKP) amalgamates "explicit" and "tacit" knowledge about customers, related process technology and raw materials for a select product or product family into a solid and reusable framework through the use of the mpQFD system, supporting the overall product innovation work process.

It is concluded that the methodology should not be introduced and used in companies as a development tool for "ad hoc" singular product development projects, since it must be recognized that the development of well-built QFD matrices is a time-consuming and complex exercise. An *Integrated Knowledge Platform* for a product or product family must instead be developed well in advance of the commencement of product development project activities to support the overall product innovation process in the development of new or improved products. The integrated knowledge from all matrices will provide a strong framework for concept generation and product innovation, facilitate production start-up of a new or improved production process, and support related new product marketing and sales.

After reading this part of the book, you should now try to relate the presented information in the perspective of your company's present product innovation work process and philosophy. In order to facilitate this activity, reflect on the following questions.

- When you review the scientific evidence from the industrial use and outcomes of the QFD methodology, does it spark an interest to investigate how the methodology could benefit your company's product innovation activities?
- If you agree that R&D and product innovation in essence are all about knowledge acquisition, knowledge transformation and knowledge utilization, and if you review the explanatory conceptual model in Fig. 20.3, could the model be applicable not only for the use of QFD but also for organizational learning in general?
- What kind of system or "knowledge management approach" does your company use in order to support and promote the capturing of product innovation information and knowledge from individual product development cycles?
- Do you agree with the Japanese point of view that knowledge expressed in words and numbers is only the "tip of the iceberg" of the total product innovation knowledge base? If so, how does your company R&D capture and use tacit information?
- Could Transaction Memory Theory be a useful concept for stimulating questions on how knowledge is shared among product innovation team members through "group

think"? Could this theory, apart from its use in QFD, serve as a mental model in your company's product innovation knowledge management?

- Have you sometimes experienced difficulties locating company expertise in different areas of product innovation in your company? Has your company delineated any kind of support system for the location of internal or external expertise in select areas of R&D?
- If your company conducts product innovation in a project-by-project fashion, how do you capture new knowledge and arrange knowledge transfer from one project into forthcoming generations of innovation projects?
- Since the importance of shortening product innovation cycle time is a major issue in the future product innovation management of excellence, could the concept of *Integrated Knowledge Platforms* serve as a supportive system for the speed-up of the company product innovation work process in your company?
- Do you believe that well-functioning *Integrated Knowledge Platforms* only need to contain customer information for product design, or do you think that they also should include information on related process technology and raw materials?
- Does your company organize the process of "listening to consumers and customers" as an on-going structured work process for individual products or product families?
- Do you agree that the "total product innovation cycle time" includes start-up of the production of new products? If so, do you believe that the use of well-developed QFD matrices could be used in the training of related plant operators?
- Do you agree that knowledge created in product innovation and the *House of Quality* could be used to benefit marketing and sales of new or improved products?

Part VI

Guidelines for Company Implementation of the Quality Function Deployment Methodology and Integrated Knowledge Platforms

Knowing about a methodology as such is not the same thing as knowing how to introduce it and use it well in an industrial environment. One will get good results from a methodology that is used properly and poor results from one that is misused. One sometimes hears unjustified complaints about a well-tried methodology, when the real culprit is the organisation's own inability to introduce and use the methodology in its specific industrial environment according to best practices.

Success factors for company use of QFD

In this part of the book, general success factors for company use of the QFD methodology will initially be reviewed, and you will learn from the scientific evidence that, in order to succeed in the industrial use of the QFD methodology, one must secure necessary management support, select committed team members, provide good training and collect new customer information.

Implementation of the mpQFD system and IKP in the R&D organization

This part of the book will present a comprehensive guide for company implementation and use, with regard to not only the mpQFD system but also the use of the methodology to develop and construct *Integrated Knowledge Platforms* (IKPs) for product innovation. In the process industries, new process technology is generally tested in pilot-planting before being introduced in full-scale industrial operations. Likewise, new methodologies should also be pilot-tested and assessed before they are internalized for the company's general use and a full company roll-out. You will be successively guided on how to introduce and implement the mpQFD system for the development of *Integrated Knowledge Platforms* through a Diagnostic/Pilot Project (DPP). A DPP should include three phases: plan, execute and adapt. In order to speed up the development of the different matrices within the QFD system and to secure their sustainability, the appointment of individual "matrix owners" is recommended. Moreover, it is strongly advisable to use a professional QFD facilitator not only to execute the DPP project but also to select and train a company "super-user" for the QFD methodology. Finally, you will also be introduced to a new organizational approach for matrix development, denominated "lean QFD", which is intended to make the use of the QFD methodology less demanding upon company development resources.

Inter- and intra-organizational integration by QFD

The increasing complexity of innovation in today's industrial environment and the necessity to reinforce "inter- and intra-organizational integration" make this methodology of interest to review, assess and use. Here, it will be pointed out that product innovation in the process industries and other manufacturing industries is no longer a "company internal affair" but rather involves a complex interplay with actors outside company demarcations and borders. Thus, product developers of excellence must, in their everyday work, manage and bridge organizational interfaces of more or less complex characters. It is concluded that the use of the QFD methodology is an unparalleled instrument for improved organizational knowledge integration and cross-functional product innovation. Since the ability to manage this domain is of utmost importance for improved R&D operational and product innovation efficiency, it is concluded that company skills in this area should be carefully considered, monitored and enhanced. However, it is advocated that each company should, from the perspective of its unique product innovation activities, assess what interfaces are important to bridge, to what extent, and what bridging mechanisms to deploy.

Putting the QFD methodology "on trial"

Finally, in the last chapter you will find a number of selected questions that are sometimes posed during different phases of the implementation of the QFD methodology. Some of these questions, related to different aspects of the implementation and use of the QFD methodology, are certain to arise when seeking management support and later on in the introduction of the methodology together with QFD team members.

Chapter 23

General Success Factors for Company Use of the QFD Methodology

Only a fool learns from his own mistakes. The wise man learns from the mistakes of others.

(Otto von Bismarck)

The user-friendliness of a methodology is related to how easy it is to learn and to apply in practice. It is influenced by the complexity of the methodology as well as how well it is explained, introduced, and supported by associated software. However, to reap the fruits of a methodology, it is also necessary to use it properly. Behaving in a way that has previously proved to give good results is usually called a "best practice," which usually can be formulated and condensed into a number of "success factors."

The general idea is that if the user of a methodology behaves according to such well-proven success factors, good outcomes will result. An opposite approach to the development of success factors is to identify general barriers to successful use of a methodology. It is often rewarding for methodology users to identify such potential barriers as well as how to overcome company-specific barriers in the planning of a QFD project. In this book, reported barriers from different studies of the use of the QFD methodology have been translated into success factors by simply posing the question of how to behave in order to overcome these barriers and thus identifying such a behaviour as a success factor.

23.1 A review on success factors for the use of the QFD methodology — lessons from the past

Generally speaking, the best method of QFD use is not to play it by the book. Since each company's conditions are unique, it is important to be imaginative in applying the rules and to find a method that is suitable to your company.

(Yoji Akao)

In a survey and review on Quality Function Deployment in Japan (Akao *et al.*, 1987), it was noted that the information on demanded quality mainly came from customer complaints and field sales rather than from new market research. It was also discovered that the use of QFD was carried out in teams and that department managers took an active part in the development work. In evaluating the use of QFD in US firms as a product development tool (Griffin, 1992), the following success factors for the use of the methodology were identified: "management treats QFD as an investment in methodology and not only as a cost, selection of smaller and not overly complex projects, selection of development projects of the incremental type, secure management support, and building a committed team that believes in the methodology." It was concluded that the QFD project will be even more difficult for companies that are not used to working in cross-functional teams; in other words, cross-functional teamwork is likely to be beneficial. In the study "The Swedish Experience" (Ekdahl & Gustafsson, 1997), problems encountered in using QFD were presented as follows in ranking order: "lack of management support, lack of group member commitment, lack of resources and lack of QFD experience." However, the majority of the respondents considered the use of QFD to be a successful undertaking.

In the study "Is Quality Function Deployment Good For Product Development? Forty Companies Say Yes" (Vonderembse *et al.*, 1997), it was shown, most convincingly, that all companies concluded that QFD had improved their products. How to behave in order to achieve successful use of the methodology, or factors that influence the outcomes of QFD, were grouped in the categories of technical dimensions, organizational dimensions and personal dimensions. The most important factors that correlated to success were: *management support, good development teams* and *strong commitment of the project team.* The methodological aspects of the use of QFD were: *good knowledge of the methodology* and *contact with the customers during the development of the Voice of the Customer.* A positive correlation was noted between large matrix sizes and good products.

The two case studies included in "A Tale of Two QFDs" (Comstock and Dooley, 1998) focused on the implementation of the methodology and the selection of the right application. The results from the study can be summarized as the following recommendations:

- *Adapt the methodology to the needs of the project (do not get stuck in methodological rigidities, but use a pragmatic approach).*
- *Use a well composed cross-functional project team (QFD work cannot be contracted, but there is a need for a QFD expert).*
- *Develop matrices of a manageable and appropriate size (do not let the matrices be too small and "banal"; use of sub-matrices is recommended).*
- *Determine the objectives well and be careful to select a project that fits the use of QFD methodology.*

In the study "Customer-Driven Product Development Through Quality Function Deployment in the U.S. and Japan" (Cristiano *et al.*, 2000, 2001), a majority of the American companies surveyed reported that the work was carried out in cross-functional groups, whereas in Japan the number was significantly lower. The Japanese companies relied to a large extent on customer complaints, while the American ones made extensive use of market research and interviews for their *House of Quality*. The second publication from this study (Cristiano *et al.*, 2001) listed success factors for the application of QFD. Very thorough statistical analyses were carried out, but more descriptive data on individual success factors were unfortunately not explicitly presented. Necessary management support and collection of new customer information were the only factors that could be strongly correlated to better product and process performance in the use of QFD.

The study "Quality function deployment: an empirical study in the UK" (Martins and Aspinwall, 2001), which focused on difficulties and barriers in the implementation of the QFD methodology, listed the following problem areas in ranking order: "time-consuming, lack of good methodological knowledge, lack of group commitment, lack of management support, difficulty in retrieving information and data for the Voice of the Customer and lack of resources." In the study results from "How to balance QFD and the efficiency of the New Product Development process: experiences and practices from the field" (Baglieri *et al.*, 2002), which takes an organizational point of view, the main conditions for making QFD more effective were found to be: "top management's strong commitment, short-term orientation, and strong investment in training." A number of recent studies of QFD best practices (Al-Mashari *et al.*, 2005; Miguel, 2005) largely confirm the previously reported success factors and barriers for successful introduction and use of the QFD methodology.

23.2 A meta-analysis on success factors for the use of the QFD methodology — the overall scientific evidence

In company implementation of the QFD methodology, the scientific evidence is coherent and clear: secure management support, establish a strong QFD team of committed members, provide expert guidance and training, and collect new customer information. Initially following these general recommendations for the QFD project is likely to secure a successful outcome.

Using the previously mentioned sources, the importance of success factors was studied in a meta-analysis of the industrial usability of the QFD methodology (Lager, 2005a). The success factors were structured in the following five different areas (those with strong scientific support are presented in bold font; see Table 23.1). The total list of success factors is

Table 23.1 Success factors for the use of the QFD methodology.

Define the project objectives well

- Define the objectives for QFD well

Select a good application for the use of the methodology

- Select an existing product for the first project
- Select a product that needs improvements
- Give the first project ample time for completion (QFD is time-consuming)

Secure management support

- **Secure necessary management support**
- Secure necessary resources (financial and personnel)
- Sell in the investment perspective on methodologies

Build a QFD team of excellence

- Appoint a strong project manager (a champion for the methodology)
- Create a well-balanced cross-functional team
- **Select committed team members**

Make proper use of the methodology

- **Provide good training of the team in the QFD methodology**
- Use an experienced facilitator throughout the project
- Use the QFD methodology in a flexible and creative manner
- Find clever solutions to handle large matrices
- **Collect new information from customers**

however recommended for use as a general "check-list" in company planning and implementation of the QFD methodology, as well as in companies' general use of the methodology.

After reviewing the presented "scientific evidence" regarding the industrial use of the QFD methodology, the following chapter will describe the author's hands-on practical experiences in his capacity as an experienced facilitator of a large number of industrial QFD projects. In this analysis, you will see part of the above success factors operationalized and discussed as actionable guidelines and advice for company use.

Chapter 24

Implementation of the mpQFD System and IKPs in the Company R&D Organization — Best Practice Recommendations from a Seasoned QFD Facilitator

Success in any human endeavour is never a sure thing and is also true of the use of methodologies. Even if a methodology has proved to be useful for other companies in a similar application, this does not necessarily imply that its use in one's own application will be successful. It is thus recommended that all new methodologies should be pilot-tested and assessed before they are adapted and internalized for general company use and roll-out.

The resources required to implement a new methodology in a company and, later on, to use it in the first development project will vary according to methodological complexity and user-friendliness. In this perspective, the QFD methodology must be considered as a rather complex methodology; however, given proper guidance and supportive tools, it can be regarded as a fairly user-friendly development tool. Nonetheless, it is advisable to initially assess the company usability of the methodology in a diagnostic/pilot project (DPP) and afterward, depending on the results, to adapt, apply and roll out the methodology in company R&D according to Fig. 24.1.

In this context, a diagnostic/pilot project is defined as a project used for the assessment and adaption of the QFD methodology for a company select R&D environment; this is rather different from using the QFD methodology in a "crash company product innovation project." A suitable point of departure for product selection for a DPP project is to choose a single product from a product family of which the overall company knowledge base needs to improve for future product innovation activities. As interesting and innovative ideas for an improved product come forward in such a DPP project (which generally is the case) they

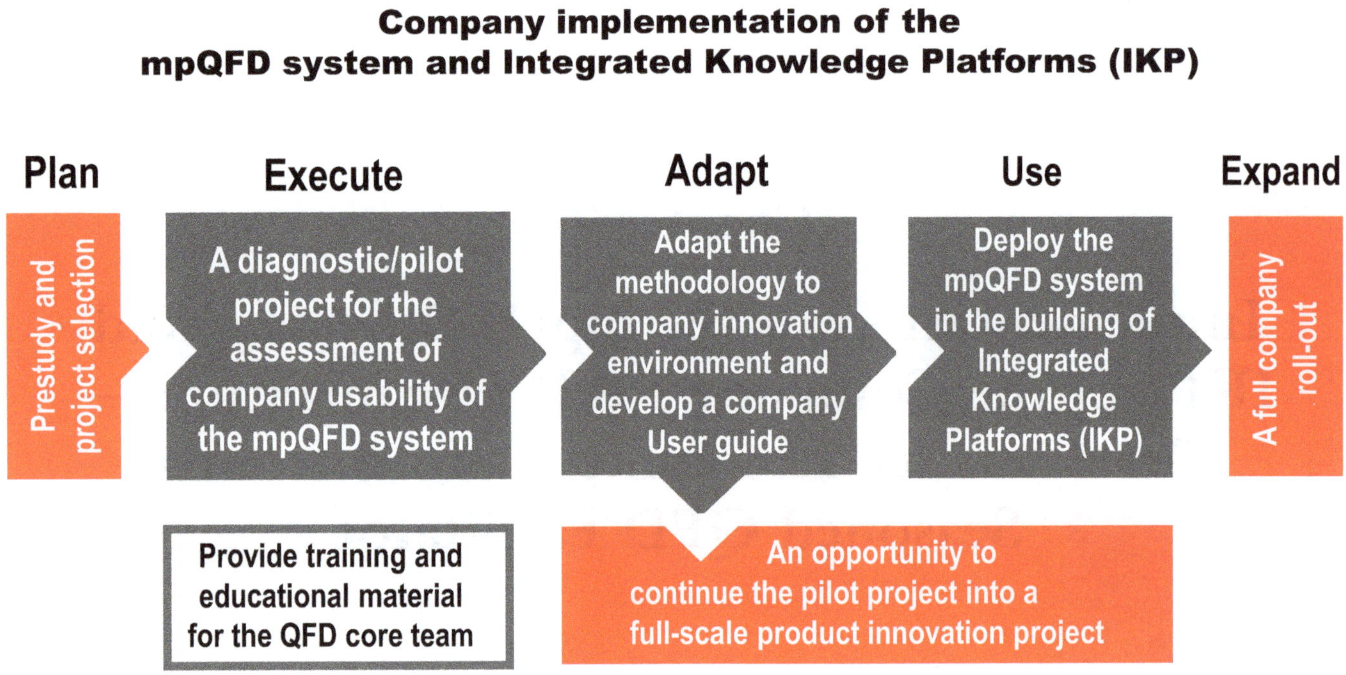

Figure 24.1 A template for company introduction and implementation of the mpQFD product innovation methodology and *Integrated Knowledge Platforms.*

should naturally be captured and afterwards used in a full-scale product innovation project according to Fig. 24.1. The term "implementation of a methodology" is sometimes confused with "use of a methodology." In the present context, implementation means the introduction of a methodology in terms of carrying out a first diagnostic/pilot project (DPP), whereas use of the methodology means its continued application in subsequent innovation projects.

In the following sub-section, you will be introduced to how a DPP project should be executed; however, the presentation also serves as a company guide for further use of the QFD methodology in new or improved product innovation projects.

24.1 Plan the DPP project in a pre-study

Every company and its R&D organization are unique, and the successful implementation and use of methodologies are thus dependant on how well a selected methodology is adapted, implemented and used in the company development environment in order to meet its specific needs. However, the use of the QFD methodology should preferrably not be advertised as a "QFD project"; rather, the methodology is used to create *Integrated Knowledge Platforms* and, as such, is always only one part of product innovation. The development of an IKP where the QFD methodology is deployed for the overall support of the product innovation work process.

As previously discussed in Part II, most companies rely, or ought to rely, on a formalized method and procedure for product development. This is called a company product innovation work process. However, one must not confuse a "Stage Gate system" for a comprehensive work process, as the latter must outline not only associated organizational matters but also related supporting tools and methodologies.

Review and assess the company product innovation work process in the perspective of its need for supporting development methodologies — answering the WHY question

This books advocates that the QFD methodology not only is the preferred tool in the pre-development part of a work process but also should be used to build Integrated Knowledge Platforms (IKPs) as an overall company work process supportive instrument — a platform constructed in advance of project-based product innovation as a prime instrument for customer interaction and target setting. In answering the question of WHY a company should start to use the QFD methodology, one must begin with a review and assessment of present and future needs for the improvement of its product innovation work process, as well as how to improve company customer interaction and how to speed up the delivery of new and improved products.

Define the overall objectives and scope of the DPP project — answering the HOW question

Depending on different companies' R&D organization and size, the implementation of the QFD methodology with a DPP project may differ substantially. In order to evaluate the full benefits of the methodology and exploit its cross-functional advantages, it is certainly recommended that not only the *Voice of the Customer* and the *House of Quality* but also the upstream matrices should be developed. However, in a small organization lacking personal and financial resources, it could be advisable to start with only the development of a *House of Quality* or even to initially explore the usability of only the *Voice of the Customer*.

In Fig. 24.2, the three Phases of a full implementation of the mpQFD methodology are illustrated. If only the *Voice of the Customer* and the *House of Quality* are selected for a DPP project, they can start simultaneously after related organizational issues have been sorted out. Along the same line, if the full mpQFD system has been selected for a DPP project, it should incentivize the company to organize the full team in a simultaneous matrix development approach, which will be further explained in Section 24.2. The overall ambition and objectives for a DPP project will generally be as follows (though it is advisable to identify objectives related to individual company needs):

- Investigate and assess the usability of the QFD methodology in company product innovation; explore company advantages and possible disadvantages to be attended to.
- Adapt the use of the methodology to the overall company business context, product development environment and product characteristics.

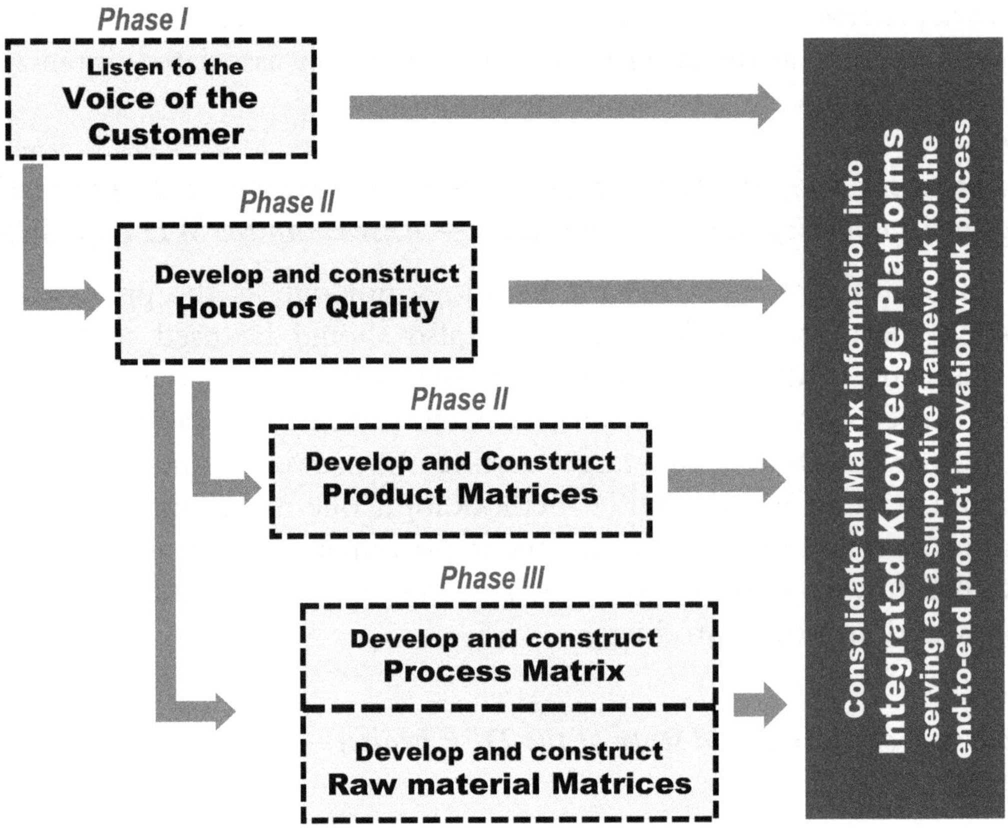

Figure 24.2 Development of the matrices in a diagnostic/pilot project can start in a consecutive fashion but should rely on the initial solid development of the Customer Requirements in the *Voice of the Customer*.

- Secure a "company illustrative case" that can be used in the further communication of the methodology and to be further used in the development of an internal company QFD guide.
- Provide an *Integrated Knowledge Platform* for a select product or product family.
- An opportunity to kick-start and launch a "full scale product innovation project" as a continuation to the DPP project.

Select the product for the DPP — answering the WHAT question

Selection of a proper product for the DPP project is of utmost importance for securing a successful outcome, and there are a number of aspects that generally should be considered and taken into account:

- Experience tells that it is not advisable to choose a completely new product idea or development concept for a DPP project; rather, it is preferable to rely on an already well-established product in the existing company product portfolio.
- The selected product should, in an overall business perspective, preferably belong to an important company product family that, in the long term, will require substantial product innovation efforts in order to stay competitive in the marketplace.

- An alternative kind of product that is also suitable for a DPP project is a product in need of a supportive platform of improved knowledge base, which will benefit from an enhanced integration and input from related applied research.
- The product could be either a B2C or a B2B product. However, the latter kind of product, which often has a limited and well-defined customer structure, will generally facilitate the collection of new customer and market information in the development of the *Voice of the Customer.*
- If there is a consensus in the company R&D organization that a product has "upfront development potential" that would particularly benefit from the use of an advanced product innovation methodology like QFD, such a product is generally an excellent choice for a DPP project and will often simplify the securing of company management support.
- Since the results from the DPP project must be easy to communicate within the company organization and R&D in particular, a well-known product should preferably be selected, which also will benefit in its use as a company illustrative case in an internal QFD guiding manual.
- Organizational and geographical aspects must be taken into account, since the ease of customer interaction and gathering of competitive product information represent a substantial part of QFD general use and thus also for a DPP project.
- If a full mpQFD system is selected for the DPP project, it is important that interaction with the production organization and raw material suppliers can be secured.
- Product and packaging integration could be of interest to explore but will make a DPP project more complex.

Secure necessary company resources for the execution of the DPP project — setting the stage and answering the WHO question

Depending on the selected scope for the DPP project, necessary internal company resources, an experienced QFD facilitator, and support from an external marketing research organization must be estimated. Later in this chapter, you will be introduced to a new "lean QFD" approach for matrix development, and you will also learn that it is recommended to appoint a company Matrix-owner for each QFD matrix. If the DPP project includes the development of the full mpQFD system, it is advisable to select a Project manager and Sub-project managers for each matrix development. If the company project manual requires a steering committee, such a committee should be cross-functional and should include representatives from R&D, Marketing & Sales and production. In a simplified set-up, the person authorized to approve the DPP project could take the sole responsibility as Project-owner. You will later learn that it is also recommended to appoint a Super-user of the QFD methodology; if possible, such a person could also be well suited to serve as the DPP Project manager. The Project manager (Super-user) should generally be an R&D representative which also could be the Matrix-owner of the *House of Quality*. However, the owner of the *Voice of the Customer* should naturally come from Sales & Marketing. The owner of the Process Matrix and the Raw material Matrix should preferably be a representative from the production organization or from R&D.

The external QFD facilitator and the Super-user constitute the "Core QFD team" that should be heavily involved in all matrix development. Each Matrix-owner should affiliate necessary expertise for the development of its matrix and should, depending on matrix complexity and size, organize such expertise in small "Matrix teams."

Outline a plan for the "internal marketing" of QFD methodology use in product innovation — securing management support and answering the WHEN question

Securing management support for the implementation and use of the QFD methodology has previously been recognized and acknowledged as an important success factor. A number of areas related to company introduction of the QFD methodology have already been discussed, and it is important that all of them are properly addressed during the initial planning of the DPP project. For this purpose, it is recommended that a small company pre-study should be conducted, during which necessary contacts with collaborating company functions and securing necessary company resources should be resolved. The pre-study could be finalized in a company PM with the following tentative headings:

- A review of the company need for the use of the QFD methodology as a supportive tool in the improvement of the company product innovation work process.
- A short introduction of the QFD methodology and its use for the creation of *Integrated Knowledge Platforms.*
- A presentation of the overall objectives and scope for the demonstration/pilot project (DPP).
- Selection of the product for the DPP project and an assessment of how necessary new customer and market information will be secured and how competitive product information can be obtained.
- A tentative estimate of necessary company resources for the execution of the DPP project and delineation of a realistic time-line.

Such preparatory work is usually the key for management approval and is in the spirit of the general QFD philosophy, which focuses on upfront development activities.

24.2 Organize and execute the Diagnostic/Pilot Project (DPP)

The development of a well-built "House of Quality" is generally a cumbersome affair that may take from three months to one year. This process involves not only gathering new customer information (e.g., through surveys) but also doing extensive laboratory technical competitive benchmarking and sometimes even experiments. Since these important and

necessary complementary project activities usually take the lion's share of the total resources for matrix development, they should not as such be associated with the idea that the "time-consuming" nature of QFD is a weakness of the methodology!

The advice presented in the following amounts to not only recommendations for the execution of a DPP project but is also an overall guide for a general use of the QFD methodology in an industrial context. Some of these pieces of advice are already common QFD knowledge and generally acknowledged best practices. However, the presented core organisational model is an entirely new approach for matrix development that will help you to overcome some well-known disadvantages with the methodology use and achieve a resource-efficient deployment, here called "lean QFD."

Use an experienced facilitator to guide the planning of the DPP project and to assist in the implementation of the methodology

In company introduction of the QFD methodology, and in particular in the development of the total mpQFD system including the upstream & downstream matrices, it is usually necessary to seek the assistance of an experienced external facilitator of the QFD methodology. Experience tells that reading a book about QFD, attending a two-day seminar or being trained through a "belt-training approach" is generally not a sufficient substitute for the services of an QFD expert.

The guidance of an experienced facilitator is usually most important during the DPP planning phase, but it is also beneficial afterwards in the set-up and design of individual matrices and matrix structures. Later on, during the preparation and implementation of the methodology, the support of a QFD facilitator is generally advisable in order to secure a smooth execution for a "first-time user" of the methodology. Consequently, and referring to the previous review of success factors, it was clearly recognized that the use of an external facilitator is generally a necessity for a company's successful implementation of the QFD methodology. ***Using advanced QFD software and proper communication tools for collaborative and simultaneous development and revisions of individual matrices (using programs like Dropbox), the interaction with an expert QFD facilitator can easily be arranged in a geographically dispersed global matrix development.***

Train an internal company Super-user for the QFD methodology

The use of an external QFD facilitator in the introduction and use of the QFD methodology is only a substitute for general knowledge of QFD theory and methodological best practices in a company. In the planning of the implementation of the QFD methodology, it is thus essential to locate a company representative who shows a particular and genuine interest to learn and test the methodology and for whom the QFD approach has a strong appeal. After the recruitment of such a potential bearer of the methodology within the company organization, the expert QFD facilitator must provide this company representative with necessary training, background information and professional advice in advance of project initiation and especially during project execution. This person shall assist and accompany the facilitator during all matrix development activities and meetings and will then gradually develop into a proper company Super-user and champion of the QFD methodology.

All methodologies need an appointed, well-advertised, internally responsible expert in methodology use and development in order to provide support to managers of new product innovation projects in the use the QFD methodology. Such a Super-user should then attend QFD conferences and acquire new publications about the methodology. The external QFD facilitator is now made redundant and can stay only in a supportive capacity to the Super-user. If the Super-user is also available as a project coordinator for the total DPP project, this is a most advantageous and a preferred organizational solution.

Appoint a company Matrix-owner for each individual QFD matrix

It is further strongly recommended to appoint company Matrix-owners for each of the QFD matrices. Adhering to such an organizational philosophy involves letting each Matrix-owner (sub-project manager) carry out the matrix development in a fairly independent manner. Each Matrix-owner will thus take the responsibility of developing and collection necessary information for his/her matrix, to be used later in the development of the individual QFD matrices. Each draft matrix can be, in essence, developed by each Matrix-owner, with the assistance of the QFD facilitator. Combining the expert knowledge of the subject matter from each sub-matrix owner with the QFD experience of the facilitator in designing tailor-made matrices makes matrix development reliable and agile.

These selected Matrix-owners not only must possess an outstanding expertise in the technical domain of the matrix but also must be interested in trying the methodology in product innovation. The Matrix-owner must take full responsibility for the development of the matrix, from the first embryonic design of the matrix structure to the completion of the matrix. During matrix development the draft matrices must be reviewed, improved, and assessed together with the Matrix-teams, as well as in larger cross-functional meetings and, if necessary, during small workshops. Moreover, the Matrix-owner must afterward ensure that the matrix information is continually updated and that the matrix is properly located in the company's *Integrated Knowledge Platform* and, later on, is properly deployed in the development of new product generations. The owner of an individual matrix should be the only person with the authority to change its content. This is quite an important issue, since it is essential to keep track of matrix versions and to avoid "matrix anarchy" and the use of outdated matrix versions.

Methodology transfer — learning by doing

Here is another quotation from Professor Akao regarding how to learn and use the methodology:

> In learning the methodology of Quality Function Deployment, the most efficient way is not to read one of my books, but to run a practical QFD project.

In this respect, he is underpinning the Toyota philosophy and recommendation of "learning by doing." However, the present author must also confess that, while he strongly agrees with Professor Akao's advice to run a QFD project, he generally also recommends the reading of a textbook on QFD methodology. Despite the fact that the selected and preferred transfer of methodology know how in this book is "learning by doing," it is suggested that the facilitator should introduce the methodology in a workshop together with all selected QFD team members in the start-up of a DPP project.

Nonetheless, the Super-user, all Matrix-owners and the Project manager must be individually trained by the facilitator during the creation of each QFD matrix. Each matrix can be designed in advance by the facilitator and the Super-user, after previous discussions with each Matrix-owner about necessary matrix rooms and other detailed aspects of the individual design of each room. In a second session with each Matrix-owner, the methodology shall be further explained, together with the use of selected software; thereafter, each Matrix-owner can start to develop the matrix together with company representatives. Necessary adjustments, matrix reviews and questions regarding QFD methodology that arise during matrix development can usually be managed via the Internet through collaboration with the Super-user and/or the QFD facilitator. However, at suitable time intervals, new sessions must be individually planned by each Matrix-owner together with the facilitator and the Super-user, when new rooms or new features of the methodology are to be introduced.

Provide the QFD team with supportive tools and equipment

The QFD teams not only must learn to use the methodology well but also must be provided with necessary supportive tools for its deployment. Two of the most important and necessary tools are good software and a reliable printer for printing the matrices in the A1 size.

Software selection

The software for QFD must be available for all Matrix-owners and, preferably, should later be made available to the whole R&D organization. The software should sometimes also be supplied to external collaborating partners in order to facilitate the dialogue with selected company "lead users," which necessitate access to user-friendly QFD software. It is important that the software be flexible and supported by an informative guiding manual. The software must also have an ability to "tailor-make" matrices according to project needs, as well as the capability to store supporting information related to each room and each individual matrix area. There are alternative QFD software on the market and each user of the QFD methodology must select a software that appeals to his/her taste and needs. The author's experiences in using the software **QFD Designer™** in over 15 large QFD projects have confirmed its industrial usability and use-friendliness.

A printer for large matrices

While software often facilitates the use of presentations on screens, the ability to print matrices in A1 format, or even in larger sizes, is generally required for efficient QFD work. There are a few rationales for this recommendation:

- When a matrix is ready, the team members often like to print the matrix in A1 and put it on the wall in their offices in order to easily look at the configuration and individual details and sometimes mark out selected items with colour pens.
- For employees that do not have access to QFD software, a matrix can be handed out in paper form. The objective of doing so can be simply to provide information or to solicit written comments or suggestions for improvements.
- It is generally experienced that a "true innovative atmosphere" is established when the QFD team members congregate around a table or against a wall with a large printed matrix (double A1).
- In discussions with selected customers on their premises, it is convenient to roll out a large printout of a full matrix or selected parts (sub-charts) of the total matrix for discussions, thus avoiding the need to share confidential matrix data via electronic versions.

Deploy a lean QFD approach using the philosophy of decentralized QFD teams for "distributed matrix development" — summing up

It has always been preached by consultants, by academic scholars and in QFD books and publications that a QFD project must be cross-functional from the very beginning. On the other hand, one of the most serious complaints against the methodology is that it is time-consuming and demanding on company resources. The author's early experiences with matrix development in very large cross-functional organizational teams were certainly rewarding but also served to underscore and verify this severe methodological disadvantage.

The previously presented use of the QFD methodology for the development of *Integrated Knowledge Platforms* is securing an efficient use of the QFD methodology in the overall support of the product innovation work process for product innovation projects. However, this recommended approach is not sufficient for overcoming the time- and resource-demanding development of the individual QFD matrices. Thus, it is good news in this book that it is not necessary to sacrifice the methodology's excellent cross-functional product innovation capabilities in order to make the deployment of the methodology less demanding on company resources! ***Obtaining company cross-functional development objectives can certainly be accomplished by the previously presented differentiation of the select organizational solutions and operating practices during the QFD total project execution.*** It is thus essential to select different kinds of organizational set-ups and operating modes for the retrieval and development of new matrix information, the analysis of matrix information and the subsequent dissemination and use of matrix results. Such different operating practices must apply within the QFD team, within the R&D organization and across company functions.

The core idea behind the concept of "distributed matrix development" is thus to decentralize the main part of the matrix development of the individual matrices to appointed Matrix-owners. However, the recommended "lean QFD" approach, using independent Matrix-owners for the main part of matrix development, requires a strong and reliable organizational structure for sharing information within and between the different matrix teams. It is recommended that interactive discussions and frequent meetings should be organized between individual Matrix-owners and their teams in order to share experiences and "glue" the distributed QFD activities together. Moreover, the matrix development will certainly also benefit from frequent discussions within the R&D organization and between different functional department organizations. Needless to say, the development of the individual matrices is favoured by interactions with customers and suppliers and other supportive organizations in an open innovation perspective.

The compromise of letting only one individual develop a matrix must thus be counterbalanced by letting expert teams review these "draft matrices" in frequent review sessions. In selecting the members of the individual matrix teams, the complexity of each matrix must be the overall guiding factor. Furthermore, it is important to select team members who are interested to learn and test the methodology, who are open-minded and who are not afraid of change. Moreover, in the set-up of matrix teams, one should recognize that "knowledge" is more important than "position." It has been generally experienced by the author in several QFD introductions that this distributed simultaneous matrix development model usually works out very well.

The disadvantage of not allowing all participants in the project attend and contribute to all discussions is generally tolerable in light of the experienced swiftness and efficiency of matrix development. Such matrix development and implementation of the QFD methodology may appear somewhat unorthodox to some QFD fundamentalists, but it is in many respects similar to the successful Toyota practice (Liker and Meier, 2006) of "learning by doing — training comes later," "using experts to get fast results," "running a pilot project to show what it is all about and to serve as a model," and so on. It is thus proper to remember the late Professor Akao's advice: "*copy the QFD spirit but not necessarily the form.*"

However, it is most important in a DPP project, or during the use of the QFD methodology in product innovation projects, to keep the teams together and avoid team-member transfer during project execution. It is generally experienced that personnel exchange, both within the QFD teams and in the cross-functional supportive organizational structures, is one of the most detrimental factors to a successful project outcome. Consequently, in the selection of team members, it is important to secure stable and sustainable team organizations.

Compress the QFD project time schedule through a simultaneous matrix development

Since it is generally acknowledged that phase progression is a very rewarding exercise for the process industries but certainly could extend the necessary time frame for the use of the methodology (de Oliveira *et al.*, 1996, Stitt and York, 1993), it is no surprise that company representatives sometimes find a successively matrix development time-consuming. It is thus recommended that all matrices should be developed roughly simultaneously, which will drastically compress the overall lead time for the implementation of the total mpQFD system. In the previous Fig. 24.2, a proper sequence for matrix development is illustrated and outlined.

The total estimated time to create and develop one "draft matrix" can be approximately 2-4 weeks of working time, which usually will translate into four months of company calendar time, depending on the complexity of each matrix (test work and specific information gathering not included). The overall lead time for the development of the mpQFD system, including the collection of necessary market information, is approximately 12–18 calendar months, during which time all four matrices in the Multiple Progression QFD system can generally be completed.

24.3 Adapt the QFD methodology to the company specific R&D environment

During the execution of the DPP project, the QFD facilitator will assist in the creation of the matrices and give advice on the design of individual matrix structures and rooms. This is a difficult activity, since the design of the individual matrices must relate well to the product specifics, and the creative development of new or tailor-made rooms is often of the utmost importance for a successful company use of the methodology. A general point of departure, which has been previously discussed, is the separation of the different kinds of Customer Requirements, including Logistics Requirements, Service Requirements, Core Product Requirements and overall System Requirements.

Develop an internal company guide for the use of the methodology — "The Company Way"

Many previous case studies and surveys on the implementation and use of the QFD methodology show that an important success factor in company implementation of the methodology is the development of a "company internal guide" to support its introduction and use.

The lack of updated recent books on the use of the QFD methodology is a serious problem that this book is intended to overcome. However, it has been argued in this book that each company's product innovation strategy and work process must be adapted to its innovation and business environment. As a consequence, the use of supporting methodologies must also be adapted and streamlined to the needs of the company's product innovation activities.

It is, however, rather difficult to write a manual for a company's use of QFD, if the methodology has not been previously used in a DPP project! Referring to the previously presented Fig. 24.1, it is thus recommended that the development and design of such a company guide and manual should rely on the experiences and results from a previously executed DPP project. If the selection of the product for the DPP project has been made with care, the matrices from the project will then serve as excellent illustrative cases in such a manual, and company organizations will easily comprehend how the methodology should be used.

24.4 Deploy the mpQFD system matrices in the creation of Integrated Knowledge Platforms (IKPs)

> **From a management perspective, the usage of the QFD matrices and the creation of an "integrated knowledge-based platform" contributes to a continuous visualization of knowledge throughout the product and process development processes; an important dynamic capability in order to generate competitive advantage, since it reinforces and may even alter and improve the entire knowledge-base of the company.**
>
> **(Tottie *et al.*, 2016)**

After the completion of a DPP project, the company should assess the outcome and decide whether, how, where and when the methodology should be further implemented and used. At this point, the methodology should be further disseminated across company functions and upwards and downwards in each organization. Seminars and workshops with product development managers and teams are essential to increase the awareness of the methodology as a supportive tool for the product innovation work process.

Starting with the select product or product family for the DPP project, the use of the QFD methodology and IKPs can now be introduced and implemented in the company R&D organization. Initially, it is important to discuss and decide, together with the "owner" of the company's overall product innovation work process, how this supporting methodology should be introduced and used for the work process. In a small to medium-sized company, the DPP team and organizational set-up is usually sufficient for the continuous use of the methodology. Starting with a review of company project portfolios, related production sites and R&D organizational structures, the strategic deployment of *Integrated Knowledge Platforms* can be decided on by the R&D Management.

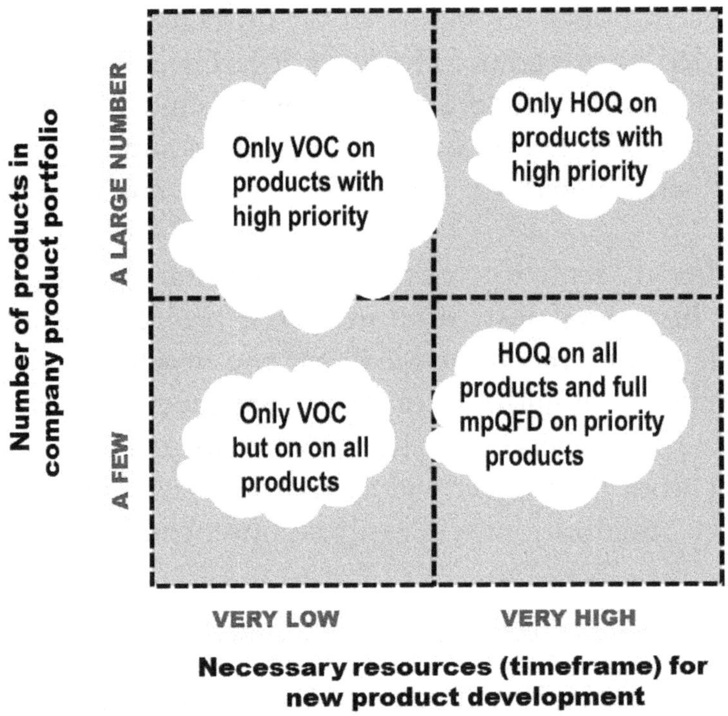

Figure 24.3 A simplified matrix to assist in the selection of the use of mpQFD and different parts of a platform.

Deploy the Integrated Knowledge Platforms selectively and adapt their use to individual project needs

In the use of the QFD methodology, company product portfolios must be analyzed and assessed. In Fig. 24.3, a simplified matrix is presented to assist in the use of the mpQFD system and for individual *Integrated Knowledge Platforms.* Depending on the results of such an analysis, it can be decided when and to what extent the full mpQFD system should be deployed and when only the Voice of the Customer should be used. A careful selection of the use of the QFD instrument in a company's product innovation is an important aspect of an efficient and effective deployment of this supportive tool.

Ensure that the Integrated Knowledge Platforms and related matrices are kept alive and used well

Like all houses, the *House of Quality* and the other matrices must be maintained properly. The most important areas that should be updated at regular intervals are related to customer information. There is often room for improvement in the hierarchic structure of customer requirements as well as the refinement of individual requirements and sometimes the addition of new ones. The importance ratings of individual requirements may change

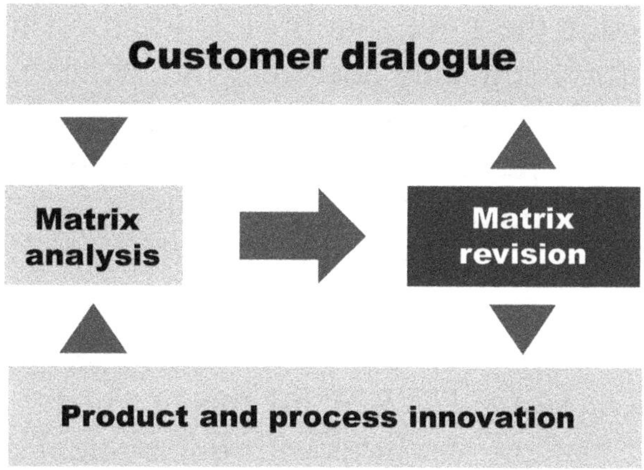

Figure 24.4 The upgrade cycle for a QFD matrix. A matrix should never be "static" or finally completed but must be kept "alive" through regular revisions.

on mature products even on a yearly basis. How well the company product is performing must also be investigated on a yearly basis in benchmarking exercises. Sometimes it can be of interest to add a new kind of room in a matrix, and sometimes old rooms become obsolete. The relationship room in each matrix should be reviewed on a regular basis, and new knowledge should be inserted. The responsibility of keeping the matrices "alive" rests with the Matrix-owner of the methodology in collaboration with each Super-user. The process is illustrated in Fig. 24.4.

Referring to the above figure, the dialogue with customers is providing new information for the *House of Quality* and so does also new knowledge about the product from company product innovation activities. It is also important that new information on the Technical Benchmarking of company products is used in the matrix revisions. Reviewing this information in matrix analyses will incentivize new matrix revisions. This new information not only will give input to new product development but also can be communicated with the customers in order to retrieve new and updated customer information.

24.5 Take an opportunity to continue the DPP into a full-scale product innovation project using the already created information as a point of departure

Using the QFD methodology as an instrument for the creation of *Integrated Knowledge Platforms (IKPs)* will make company investments in QFD methodology more attractive, since they can be expected to give good returns on improved innovation efficiency (ROIE) in a more long-term use of the methodology. Such an approach will not only accelerate the product development work process and speed up individual product development projects but also overcome the disadvantage of the "time and again" implementation of a complex methodology in the R&D organization.

It has previously been argued in this book that the introduction and use of supportive development methodologies is not a company cost item but rather an investment in more efficient product innovation with an expected substantial return. Moreover, it is recommended that an existing product, with a recognized need and potential for product improvements, should be selected for the DPP project. When the DPP project has served its purpose as a pilot product for the QFD methodology, and if carried out according to the previously suggested guidelines, there is a most interesting opportunity to take advantage of and deploy this information in a full-scale product innovation project as well as to introduce the matrices as an *Integrated Knowledge Platform* for this product.

Following the notion of "killing two birds with one stone," the company then not only will introduce the methodology but, at the same time, also will get an opportunity to kick-start product innovation. In companies that have scarce resources for product innovation and for trying new product innovation methodologies, such an approach should be attractive to use in the securing of management support for a DPP project.

24.6 A full company roll-out of the QFD methodology

In larger organizations, and in a full company "roll-out" of the methodology, it is recommended that the DPP team should be used as an embryo for the further dissemination of the methodology. In larger geographically dispersed organizations, the original DPP team, supported initially by the external facilitator, should locate, select and train Super-users at company research centers. After such an initial training period, using the newly developed company "user guide" for QFD and IKP, each new Super-user should select and identify a suitable new product innovation project and decide with the project manager how the QFD methodology should be used in support of such a project.

Chapter 25

Using QFD as the Company's Select Methodology to Reinforce "Inter- and Intra-Organizational Integration" in Product Innovation

> **Industrial innovation processes involve (or should involve) all activities required to make a product available and used on the market, e.g., technology development, market planning, product development, process planning, production, distribution, sales and service.**
>
> **(Säfsten *et al.*, 2014)**

The complexity of today's product development and the necessary adaption of a holistic perspective on company work processes for innovation is well articulated by Säfsten *et al.* (2014) above. The creation of successful new products is thus fundamentally a multidisciplinary, multi-functionality process. Establishing a common "thought world" by bridging intra- and inter-organizational interfaces is generally considered a necessity for excellence in product innovation. From an open innovation perspective, bridging a company's external organizational interfaces and securing customer and consumer collaboration in a co-development approach are activities of foremost importance in product development today (Chesbrough and Appleyard, 2007; Chesbrough, 2007). Since traditional industrial borders have been obliterated, traditional organizations and work processes for product innovation and innovation of manufacturing technology must be reviewed and reconsidered (Lager and Rennard, 2014):

> Because the development of new products often requires a corresponding development of the company's manufacturing system, more interactive innovation work processes are needed, from product design to manufacturing to delivery, bridging the R&D-manufacturing interface and integrating product and process development work processes.

Figure 25.1 Intra- and inter-organizational interfaces in product innovation. (Red arrows illustrate the R&D interfaces, and black arrows illustrate other company functional interfaces).

Not only is there a need for more "interactive work processes" to bridge the R&D-manufacturing interface, but also, in product innovation of excellence, there are a number of supplementary external and internal organizational interfaces to bridge, as illustrated in Fig. 25.1. In the introduction of improved company work processes for product innovation, it is important to ensure not only that both internal and external interfaces are bridged (Gamal and Lager, 2015) but also that supportive organizational bridging mechanisms and methodologies that can facilitate such organizational behaviour are made available (Gamal, 2015).

For more than a century, beginning when companies in the spirit of Taylor (1916) started to separate into departmental organizations to improve their efficiency, functional interfacing has been a topic of interest. Today, most studies focus on how to bridge functional barriers by using different supporting mechanisms and "key business processes" and to further revitalize such processes (Hammer, 1990). However, the study of how companies could improve innovation capabilities through an improved integration of functional silos still remains an interesting area for company improvement. In spite of numerous academic studies and publications related to the methodological use and improvements of QFD, the structural and procedural organizational aspects of the methodology use are surprisingly often ignored. When subordinate methodological issues (however important) are focused on, minor methodological improvements the fundamental capability of the QFD methodology as an instrument in bridging company external and internal interfaces in industrial use is unfortunately often neglected.

25.1 Bridging the R&D–customer interface

> **The company's size is often a hindrance since large companies tend to divide into functional groups that guard their information and are reluctant to engage in a cooperative effort. The customers' voice does not find its way effectively throughout the organization and therefore cannot affect new concepts as it should.**
>
> **(Day, 1993)**

At the end of the 1980s, a new development concept was introduced by von Hippel (1986), the general idea of which was to integrate users of company products, processes and services into the innovation work process as "lead users." The idea was grounded in the assumption that consumers, including industrial customers, often have limited insight into new product needs and potential solutions since they are constrained by their own real-world experiences.

> Lead users are defined as members of a user population who (1) anticipate obtaining relatively high benefits from obtaining a solution to their needs and so may innovate and (2) are at the leading edge of important trends in the marketplace under study and so are currently experiencing need that will later be experienced by many users in the marketplace.
>
> (von Hippel, 1986)

The concept of lead users suggests that this category of product (or process) users are in a better position to provide accurate data on needs related to future conditions. Additionally, the greater the benefit a given user can obtain from a needed novel product, the greater his effort to obtain a solution will be. In discussing lead users, it is important to note that they can be individuals, groups or companies that have product needs beyond what is currently available in the general market (Eisenberg, 2011).

In the identification of lead users and apart from being a trend leader, experience tells that it is important that the lead user perceives a mismatch between his needs and the functions or performance of existing products (Luthje and Herstatt, 2004). Both components in this definition have been proven to be independent dimensions so neither can be dropped without loss of information (Franke *et al.*, 2006). Hippel also discussed the kinds of special incentives that could be needed to induce lead users to act on a more continuing basis. ***He also noticed that in the area of industrial goods, the needed data on important trends are more clear to those with industrial expertise than to normal consumers, since the former often only have a certain number of major potential customers for a given product type and often know the characteristics of each user quite well.***

The theoretical foundations for the lead user concept and its implementation in company management systems were researched by Luthje and Herstatt (2004). They noticed that lead users by definition do not simply face any new need but rather realize needs that most customers in the market will face in the future. They also pointed out that lead users frequently play an important role in the development of new products, particularly for industrial markets.

The importance of using external information and establishing strong company external collaboration has been stressed in a vast number of publications (Huizingh, 2011) and is generally not questioned. The term "open innovation" is often used to describe an extension of company innovation activities across formal organizational borders and collaboration with external suppliers of knowledge and other intellectual assets and properties (Chesbrough, 2010; Chesbrough and Crowther, 2006; Chiaroni *et al.*, 2010). Thus, even the most capable R&D organization today needs to identify, connect to and leverage external knowledge sources as a core process in innovation. Resources from a company's relational capital (customers, regulators, influencers and community members) have been shown to be important in all stages along the product development process, starting with problem recognition and idea generation, through product concept development, and finally to prototyping and testing (Fuller *et al.*, 2011).

Fuller *et al.*'s (2011) findings showed that it is the perceived autonomous, enjoyable and competent experience that enables lead users to free their potential and come up with superior solutions, thus potentially increasing not only the quantity but also the quality of contributed content. The drivers for consumers to engage in co-creation were further studied by Fuller (2010), and the results indicated that monetary incentives are not as important as intangibles, such as feedback or recognition, as well as the interactive experience itself.

Using the *Voice of the Customer* in collaborative development with customers and consumers, in close internal company integration with marketing and sales, is one important area that traditionally has been presented as the hallmark of the QFD methodology (Day, 1993). ***In a study of food and drink industry, Martinez (2014) concluded that this industry has undergone a process of "chain reversal" to incorporate the* Voice of the Consumer *at the centre of the innovation process. One must have an ability to translate subjective consumer needs into objective product specifications to satisfy consumer needs through the development of new products.*** Bilgram *et al.* (2008) concluded that, in the efforts to alleviate the lack of user-acceptance, customer-centric innovations not only harness the *Voice of the Customer* but also take a further step beyond the traditional market research by integrating users as problem solvers in various phases of the individual innovation process.

A case study of five Italian companies using the QFD methodology (Baglieri *et al.*, 2002) emphasized the importance of understanding the expectations of end users, whose purchasing behaviour creates value along the overall value chain. All respondents identified the contribution of QFD as a powerful tool to spread both technical and competitive marketing information at the organizational level. This underlined the fact that QFD can really improve the relationship between marketing and technical departments, providing a common language and interaction grounds.

25.2 Bridging the R&D–marketing interface

> **Quality function deployment reduces the marketing/R&D barriers of different through-worlds, languages and organizational responsibilities and provides mechanisms to increase information utilization across the functions as well as resolving conflicts between them. The processes used to build the House of Quality lead to reduced market uncertainties.**
>
> **(Griffin, 1992)**

In a proposed model for studying the R&D–marketing interface in product innovation, Gupta *et al.* (1986) presented a number of variables for achieving a greater degree of integration: promoting the need for integration, establishing joint reward systems, balancing long- and short-term company objectives, encouraging risk-taking and providing opportunities for R&D and marketing managers to know and understand each other. To reduce the detrimental effects of sociocultural differences between R&D and marketing managers, they proposed hiring marketing managers with a technical background and providing market training to R&D personnel. In a review of the improvement of the R&D–marketing integration, Griffin and Hauser (1996) recommended relocation and physical facility design, personnel movement, informal social systems, organizational structures, incentives and rewards and formal integrative management processes. The use of formal work processes like the stage-gate process was recommended instead of *ad hoc* practices.

To break down the walls among R&D, manufacturing and marketing, Säfsten *et al.* (2014) recommended that techniques such as concurrent engineering and QFD should be used to pave the way for more effective product development. In their study of this triad, Song *et al.* (1997) concluded that in any examination of cross-functional product development, the manufacturing perspective should always be considered. They advocated that manufacturing should take part in the process from the very beginning. Säfsten *et al.* (2014) concluded it is not sufficient for the phases of technology development, product development or production to be efficient individually; rather, a successful innovation work process requires integration among the different phases and functions involved. They further ascertained that, regardless of functional area, the strongest and most direct influence on cooperation and performance comes from a firm's evaluation criteria, reward structures and management expectations.

The capability of the QFD methodology and its work process from the perspective of bridging the R&D–marketing interface is well presented by Day (1993):

> The QFD process requires a number of inputs and decisions that are best done through teamwork. Because of this, the process tends to remove many of the functional barriers that develop in large organizations, thereby helping to merge marketing's knowledge of the customer with product engineering's need to know the customers' requirements.

In the use of integrative organizational tools, Griffin and Hauser (1996) noticed that few would challenge the notion that QFD eliminates barriers to functional integration and improves the information sharing between functions.

25.3 Bridging the R&D–production interface

> **The use of QFD showed that it was possible to build in quality at the beginning of a product development and introduction project. A cross-functional team including mill personnel never lost the customer focus and the competitive objective during mill trials to develop a preferred product. The specification resulting from this effort reflects both the customer WANTs and the mill's capability to consistently deliver the preferred product.**
>
> **(Stitt and York, 1993)**

Ginn and Rubenstein (1986) confirmed early that severe turbulence and conflict tend to arise at the R&D–manufacturing interface because of uncertainties and goal incompatibilities. They argued that barriers may be overcome through the application of superordinate goals (broad aims affecting several departments), power exercises and imperatives for action. In an exploratory study of R&D–manufacturing in the early phases of the product innovation process, Nihtilä (1999) noted that companies used a portfolio of integrating mechanisms like individual integrator and cross-functional teams. The importance of securing production resources and balancing the recruiting between R&D and production in product innovation was stressed.

Vandevelde and Van Dierdonck (2003) discussed how major barriers across the R&D (design)-manufacturing interface should be overcome. They concluded that, although the complexity and newness of products and technologies hinder a smooth production start-up, their effects seem to vanish by introducing formalization and by striving for a design team that has empathy towards manufacturing. In their study, formalization was defined as a variety of mechanisms that contribute to a structured and clear innovation management approach including standard rules, plans and schedules, formal mutual adjustments and dedicated teams.

Unfortunately, the QFD methodology is often presented today in a rather superficial manner, using simplified descriptions of the *House of Quality* and finishing with a presentation of how customer demands are being progressed into production using four consecutive matrices. However, looking into the industrial use of the methodology (Cristiano *et al.*, 2001; Martins and Aspinwall, 2001), one can see that very few companies have ever tried the phase progression into deeper levels (Baglieri *et al.*, 2002). However, there is strong evidence from the literature supporting the importance of going beyond the *House of Quality*. Cohen (1995) argued early that using the QFD upstream matrices can be of significant importance for product innovation and that company development teams miss a great deal when they disregard this opportunity. Experiences from the forest industries in using the phase progression of customer requirements into the production process are encouraging and confirm Cohen's opinion.

The teambuilding experiences from developing the upstream matrices at Kimberly-Clark and tearing down "old functional walls" illustrate the methodology as an "organizational integrating instrument" with production (Stitt and York, 1993). Experiences from using the QFD methodology in the development of upstream matrices and in the start-up of a new production plant are clearly positive (Scheurell, 1992):

> Given we achieved excellent communication, involved the mill operators to make them responsive to customer needs for a new product, using new technology in a new mill run by new operators ... would we do it again? I have been asked by our division management, business, mill, and QFD teams to tell you YES!

Since product development in the future must consider manufacturability and cost to an even greater extent, it is important not only to translate *Customer Requirements* into *Product*

Design Requirements but to ensure that those design requirements are further well-progressed as requirements on the manufacturing process (Meyer and Dalal, 2002; Jianxin *et al.*, 2007). Disregarding the strong evidence from previous studies advocating the importance of going beyond the *House of Quality* and the lack of recent publications in this area, it is evident that manufacturing companies still have not exploited the methodology to its full extent. Using the QFD methodology as a tool in bridging the R&D–production interface is one area of the QFD that should be further revitalized and improved.

25.4 Bridging the R&D–supplier interface

> The initial idea to engage a technology/equipment supplier to the customers as a lead-user proved to be a sound development approach. Involving the customers' technology/equipment supplier as a lead user in B2B product development is thus one general recommendation.
>
> (Tottie *et al.*, 2016)

In the above study of product development in the mineral industry and the collaboration with industrial lead-users in a B2B context, it was concluded that supplier integration is an important aspect in product development. In light of the growing importance of supplier and production integration with R&D in the perspective of product cost reduction and manufacturability (Boothroyd *et al.*, 1994), the interfacial bridging mechanisms that address this area should certainly be more in focus. One can conclude that in today's product innovation, the R&D function cannot solely rely on interfacing with production and marketing but must also have direct access to and input from suppliers and customers.

However, there is sometimes a long and complex production chain of multiple customers and suppliers, and suppliers and customers are sometimes located in the middle of this chain with no direct contact with the end user (Tottie and Lager, 1995). One approach to this problem is to see each company as both a supplier and a customer in this chain (Bigliardi *et al.*, 2010). The user demands on the functionality of the supplier's product in the customers' production environment are translated in the supplier's *House of Quality* into measurable *Product Design Requirements.* This approach gives a development perspective for the supplier in its desire to dig deeper and understand the causes of the demanded product specifications from the customer and user of the product. In a study of the packaging innovation work process at Nestlé (Gamal, 2015), it was concluded that the alignment of different internal functions with external partners and suppliers through timely and accurate communication was critical:

> The packaging innovation process involves many actors from different functional areas within a company and several external partners across company external borders. All those actors need to work cross-functionally in order to avoid organizational, governance, and communication disturbances in the use of the integrative toolkit. The use of Quality Function Deployment will help to eliminate barriers to functional integration.

The importance of supplier involvement in product innovation is unquestionable today (Handfield *et al.*, 1999, 2000) as well as in a future open innovation perspective (Gassman *et al.*, 2010; Huizingh, 2011). However, the use of the QFD methodology as an instrument for an improved integration of suppliers of components or raw materials still seems to be in its infancy in the process industries. It must also be noted that in the process industries, the lack of complex systems of multiple components will make the involvement of raw material suppliers through the raw material matrix a radical methodological simplification.

25.5 A reassessment of QFD methodology from the perspective of organizational integration

The QFD methodology is mainly recognized today as an instrument for structuring *Customer Requirements* and translating them into *Product Design Requirements* using the well-known *House of Quality*. The capability of the QFD methodology as an "organizational integrating instrument" may thus have been overshadowed by its many other important advantages in product innovation. This book advocates the idea that, apart from other important aspects, the QFD methodology can be beneficial as an "organizational integrating instrument" in bridging all four of the previously reviewed interfaces, as further illustrated in Fig. 25.2.

In conclusion, in spite of the scarce general attention to this capability in the product innovation literature and in discussions of interfacial organizational issues, this unique capability of the QFD methodology should be highlighted and exploited in order to enhance company product innovation performance. In search of improved product innovation capabilities, it is not enough to state that, for example, the use of cross-functional teams is the route to follow. Instead, it is equally important to provide the methodological instrument for such cross-functional organizational work and integration. The QFD methodology is certainly one such instrument.

The importance of the integration of R&D–marketing during the early phase of product innovation is a clear indication that there is a need for efficient functional integrating mechanisms and tools. ***However, the importance of a high integration of R&D–production during the early phase of the product innovation work process does not imply a lesser need for integration during later phases.*** The importance of bridging company organizational interfaces is unquestionable, but how much integration is enough?

The necessity to bridge four organizational interfaces to improve product innovation performance seems to be a formidable task even for experienced product development professionals. Different company positions in the supply/value chain in different industrial sectors may require different degrees of organizational integration in relation to each of the four different interfaces.

Song and Thieme (2006) also concluded, in a study of the R&D–marketing interface, that not all innovation projects require the same degree of integration and that the heart of the matter is to ensure that the gap between necessary and realized integration is small. In

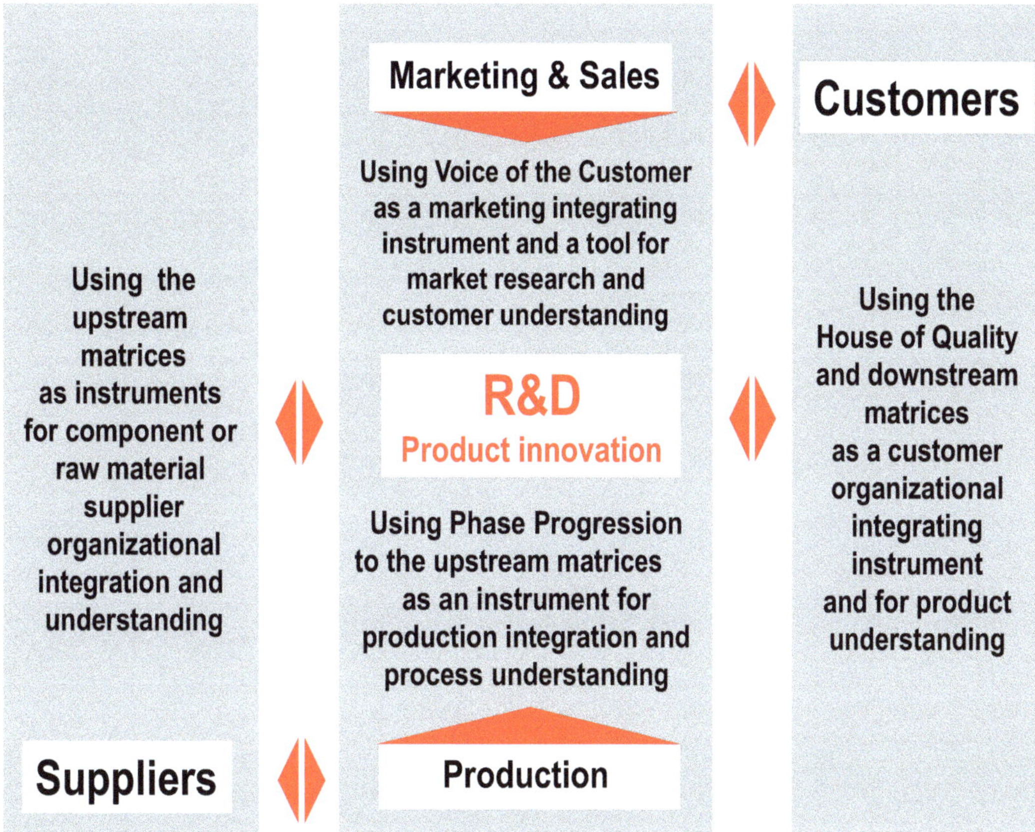

Figure 25.2 A conceptual model of the industrial use of the QFD methodology as an instrument for intra- and inter-organizational integration.

a study of the marketing–R&D–operations triad, Olson *et al.* (2001) found that substantial cooperation between marketing and R&D during the early stages was strongly associated with positive project performance. Counter to their expectations, they also discovered that substantial cooperation between R&D and operations during the early stages was also positively related to project performance, a fact which was later validated by Eng and Ozdemir (2014). From a contingency perspective (Morgan, 1986), each company should, in light of its product innovation activities, assess what interfaces are important to bridge and to what extent, as well as what bridging mechanisms to deploy.

Chapter 26

Summing Up and Selected Questions About Quality Function Deployment (QFD) and Related Integrated Knowledge Platforms (IKPs)

This part of the book has presented a number of success factors that suggest advice on how to behave in order to achieve a successful outcome of the use of the QFD methodology. The importance of selecting "committed team members" and "providing them with training" is highlighted, and the recommended route to follow is "learning by doing." It is then suggested that the methodology, like all methodologies, should be pilot tested and adapted to company innovation environment in advance of a full company roll-out. The opportunity to continue a DPP project as a full-scale product innovation project is advised.

QFD has been recommended for use in the construction of *Integrated Knowledge Platforms (IKPs)* as a novel approach to speed up the product innovation work process. This part of the book has also recommended a "lean QFD" approach to further facilitate and speed up all matrix development in order to make the QFD work less resource-demanding. It is argued that the use of an experienced QFD facilitator in the DPP project is a good investment, not only to organise and set up proper matrix structures but also to provide individual training to the matrix owners and to a company Super user. Finally, it is advocated that the experiences gained from the execution of the DPP project, including the use of the developed matrices as illustrative cases, should be captured in the development of an internal company QFD manual: "This is the company QFD way."

You have now also learned that companies' general organizational set-ups today do not give sufficient support for efficient product innovation, but fortunately there are a number of facilitating organizational bridging instruments available to deploy. It is concluded, apart from other important aspects of the use of the QFD methodology, that QFD can be used as one such instrument and thus beneficial in bridging both internal and external company interfaces. The presented conceptual model in Fig. 25.2 should be used as an "organizational integrating instrument" in the early phase of company implementation and use of the

QFD methodology. Highlighting this capacity of the methodology from the outset will reinforce its capability as an organizational integrating instrument. Recognizing that the use of the QFD methodology is only one of several integrating mechanisms, the model can also serve as a platform for delineating company work processes and organizational structural solutions in bridging organizational interfaces for enhanced product innovation.

Improving collaboration in the triad of R&D-manufacturing-marketing is an important task, but it is also essential to observe and note other related organizational interfaces. Olausson and Berggren (2012) thus noted the structural divide in companies between R&D and manufacturing on one hand and the dispersed network of service activities on the other. ***They advocated not only that R&D needs to interact with manufacturing and service in parallel but also that companies need to create conditions for "integrated knowledge-based approaches" across functions. It is argued in this book that such a knowledge-based approach is achieved in company use of*** **Integrated Knowledge Platforms (IKPs).** After reading the previous chapter, the reader may still have some questions related to methodology use. In this last chapter, the methodology is put "on trial" by the author as being a "devil's advocate." The questions posed are of a kind that a QFD facilitator or company Super-user sometimes must answer during the implementation and further use of the methodology and IKPs.

26.1 Selected questions for companies' general use of the QFD methodology

When a House of Quality is built, one has not developed a new product! Instead, an *IKP* has been created, and, relying on the consensus of the team's best knowledge, interactive explainable Target Figures have been set that will support and guide product concept development and further product innovation.

If one finds the methodology a bit complicated, are there ways to simplify the use of QFD?

There are, as presented in the previous chapters, a number of possible simplifications. The most simplified practice of QFD methodology is to focus on only the development of the *Voice of the Customer*. A less drastic simplification is to build only a *House of Quality*, while neglecting phase progression and construction of the upstream & downstream matrices. Both approaches comply well with the QFD ideology and accepted best practices. However, to simplify each approach further is generally not recommended, since one will lose the "power" and trustworthiness of the QFD system.

Is the QFD methodology a tool that will appeal to everybody?

In spite of the numerous proven company benefits of the use of the methodology, we must accept that systematic product development, and as such the use of the QFD methodology in particular, does not appeal to everybody in an R&D organization. In fact, this can sometimes also apply to some R&D organizations as a whole and to a general resistance to change in a company organisation:

> Many organizations espouse that they desire change and improvements. But almost all organizations actually practice a theory-in-use that is quite hostile to change. In attempting to implement a new method or managing, the scholars face an uphill battle, attempting to instill more order and rationality into processes, which through past experience and inertia, have been running on automatic for many years. Organizations have also developed considerable (if implicit) skills to diffuse the energy and directions from new initiatives.
>
> (Kaplan, 1998)

The author's experiences using the QFD methodology in many different company organisations show that, in the implementation of a new methodology like QFD, it is important to select project team members that have a strong interest in trying the methodology and who are attracted to this approach. It is recommended that non-believers in a methodology should not be forced to join in on QFD implementation and the first DPP project. Instead, it is preferable to let the QFD results speak for themselves after the project is finished, thus winning new QFD devotees in a "bottom-up" fashion. Some people can be even rather hostile to the QFD methodology, even if they only have superficial knowledge about it or have never had an opportunity to try it in a real project application. This is something that experienced facilitators know how to handle, and it is a rather similar situation to driving a car; problems in learning to drive are more often related to how the car is used than to the car itself, and the same is often true of the use of new methodologies.

Can the systematic use of methodologies like QFD be a barrier to teamwork creativity rather than a success factor for innovation?

In development work, there is always a risk of getting stuck in "inside the box" thinking while rejecting ideas from outside. This is not related to the use of the QFD methodology in particular but to all kinds of creative work. The use of the facilitating QFD matrices, however, is intended to make development work more efficient, thus giving more time for creative work and "out of the box" thinking.

Is the QFD methodology only fit for use in incremental product innovation but not really for the development of "new to the world" products?

It is a general misconception that the QFD methodology can only be used in incremental product innovation. On the contrary, in the development of "new to the world" products, when the product developers do not yet have a consumer of customer structure available for interviews and interaction, the need for the structured methodological approach used in QFD is even more important. In such product innovation there is naturally also a need for the development of *Customer Requirements* and their importance ratings, and naturally there are competitive "solutions," if not competitive products, on the market. Since such a product is not yet available on the market, company developers must arrange other substitutes like clinics, focus groups or the use of internal employees in the company as respondents in the development of a *Voice of the Customer.*

The use of the *House of Quality* is rather similar to its use with existing products, but some new, inventive rooms may be developed. Such new products must also be produced in new or modified production processes, and the use of both upstream & downstream matrices is therefore also advisable. If a completely new product will be "test marketed," then these new customers in the marketplace should be used for "on the spot" interviews. In conclusion, even revolutionary new products will finally meet with new customers, and one must not neglect to try to understand their needs in advance.

In the introduction it was mentioned that the QFD methodology could most favourably benefit from an interaction with other methodologies, but this is not further elaborated in this book. Why is this?

In this book, you have become familiar with the use of the "KANO" development methodology, and some supplementary marketing research tools have also been mentioned. In previous QFD books, the supplementary methodologies sometimes take over in the presentations, and the books get stuck on a superficial presentation of the QFD methodology. As one can easily find more information in supplementary QFD readings or other publications, the decision was made at the outset to focus the content of this book on "core QFD methodological areas" and in particular the use of this methodology in the process industries. This is not to say that supplementary methodologies are uninteresting! On the contrary, the advanced KANO methodology, Analytical Hierarchy Process, Pugh concept selection methodologies and Conjoint Analysis are supplementary instruments that are certainly worthwhile to explore and use together with QFD.

How could sensitive product information contained in the QFD matrices be protected?

All information related to company R&D is of a kind that must adhere to a well-articulated information policy. The QFD matrices contain a wealth of sensitive product information that must be handled with care. This is facilitated by the recommendation to appoint a Matrix-owner for each QFD matrix. Such owners will then decide and keep track of who should have access to the individual matrices in paper or software versions. In using matrices in interaction with customers, it is recommended not to present the full matrices but to select parts (sub-chats) of the matrices for use in discussions.

Naturally, for very special industrial B2B customers with whom long-term trustful relations have been developed, sharing total matrix information and letting customers work with the matrices can be a most rewarding experience. The company internal use of the information naturally must be open, and the matrices should be used in day-to-day product innovation work. It is common that product innovation team members put large paper versions on their office walls. Creation of a "QFD team member only access" policy must be avoided in such cases.

Is it very expensive for a company to use the QFD methodology?

One might rephrase this question as: how much does it cost a company not to use product innovation methodologies like QFD? As explained in Chapter 1, methodologies are not cost items; rather, they are instruments to make innovation more efficient and effective. Investments in proper use of the QFD methodology should thus be approached like other company investments; that is, they must produce returns. Proper use of the QFD methodology and the knowledge created in its use must thus be treated as an important and valuable company asset. After the methodology has been implemented and matrices have been created, the update and further development of the matrices is not costly. When a core team has properly learned the methodology and a Super-user has been appointed and is in full swing, the methodology is easy to duplicate and use in many projects.

Could the development of the matrices take too much time away from the truly creative work in product innovation?

The proposed use of the QFD methodology in the creation of *IKPs* to support the overall product innovation work process, developed in advance of the more project-based product innovation projects, will not delay but rather speed up the development process. The initial introduction and implementation of the methodology, however, may require a substantial investment in company resources and time.

The author's experience from facilitating many QFD projects is that it is often not the use of the core methodology that takes a long time. In reality, a fast introduction of the QFD methodology is often not dependent on time-consuming QFD activities as such but rather on company QFD team members' often congested agendas. Related activities like getting product information from competitors are not really a problem with the QFD methodology as such, but indicate a lack of well-functioning company systems for monitoring competitors and to collecting competitive product information. Along the same line, time-consuming collection of customer information is not a QFD-related problem but rather demonstrates a company's lack of a good on-going customer information retrieval system for product innovation. In conclusion, the sometimes time-consuming activities related to QFD use represent not a methodological weakness but rather a methodological strength, since the use of the QFD methodology pin-points neglected areas in a company's ability to retrieve and store vital product innovation information properly.

When a diagnostic/pilot project has been finished, where does the company stand on the "learning curve" in a methodology user perspective?

The author of this book is advocating that excellence in the use of supportive methodologies and instruments for product innovation will, in the perspective of future product innovation efficiency and effectiveness, be an unparalleled success factor. A diagnostic/pilot project (DPP) carried out using proper methodology tools and best practices is only the first step on the road to becoming a "world-class QFD performer." One interesting aspect of methodology use that a company should reflect on after implementation is how often it is necessary for a company to use a methodology in order to give methodology a fair chance of possible success and survival? After a DPP project, the company will gradually start the journey from "QFD general awareness" to "a working knowledge of the QFD methodology" and finally will gradually reach the level of being "a professional QFD performer." After a subsequent roll-out of the methodology into the organization, the company will graduate to the level of "expert QFD performer" when Super-users are available and the general awareness of QFD among company functions is high. Finally, after using the methodology in a number of product innovation projects, the company will reach the level of "a world-class QFD performer."

If this methodology is so successful in other companies, why has it not already been tried in my company?

As stated in the introductory chapter, methodologies and development tools come and go, and company use of methodologies is unfortunately overly tied to the latest fashion and buzzwords. However, the increased focus on company interactive development with customers and consumers is likely to put tools like the QFD methodology in a new spotlight in future industrial product innovation of excellence. The methodology has been developed and practised largely in other manufacturing industries, usually because the methodology has not been adapted to process-industrial use. This book is intended to overcome this deficiency, and hopefully integrate process-industrial use of the methodology.

26.2 Selected questions related to the Voice of the Customer

As pointed out in the previous chapters, the use of QFD is an excellent instrument to support other market research activities. The *Voice of the Customer* should be viewed as a guiding tool for the use of sophisticated marketing research; thus, the QFD methodology and approach is not a substitute for other market research methodologies.

If a company lacks resources for direct customer interaction and market research, can it rely only on its own internal Sales & Marketing information?

In such an unfortunate situation, the best recommendation is not to use the QFD methodology at all; experience tells that such matrices are not really reliable, and in such cases the "QFD spirit" is lost in the execution of the methodology. Interaction with consumers and customers is crucial for successful product innovation and, consequently, for successful QFD use. Please note that "collecting new information from customers" is a success factor that received strong support in the previously presented meta-analysis. In presentations and introductions of the QFD methodology by the author to small- and medium-sized Enterprises (SMEs), the company representatives have sometimes been asked to "play customers" in order to build an artificial *Voice of the Customer* using their own perceived Customer Requirements on a selected product. Experiences from such use in training sessions are that they are very helpful to understand the methodology's approach, but it must always be carefully pointed out that this is not the "real McCoy" of using the methodology in innovation.

Do the "demands" articulated by customers really reflect their "true purchasing behaviour"?

This question relates to the eternal question of how to collect reliable and valid customer information. In order for a respondent to be able to answer certain types of questions, it is always important that the context is correct; for example, in order to answer whether a screwdriver design is good, the respondent must have an opportunity to use the screwdriver himself. Another issue related to this is that customers sometimes give what they consider to be the "proper" answer to questions of a certain kind. The issue of "latent" and "manifest" information arises in all market research, and thus it is also an issue to consider in the use of the QFD methodology.

Focusing on measurable Product Design Requirements and only quantitative information, are we losing the more qualitative information in the use of QFD?

On the contrary, the *House of Quality* is an outstanding instrument to transform the qualitative *Customer Requirements* into quantitative information as *Product Design Requirements*; this is the name of the game! Relying solely on the qualitative information in the *Voice of the Customer* is like a "short cut" in delivering guidelines for the product developers; the information is easy to deliver but difficult to act upon in product design. In the development and use of the matrices, as has previously been stressed, both "tacit information" and qualitative "explicit information" should be carefully captured and stored as background information using "flag-notes" in the QFD software.

26.3 Selected questions related to the House of Quality

Is the use of standard matrices and rooms a somewhat "mechanistic" approach in product innovation?

The QFD methodology is, in general, a very flexible tool for company product innovation, and the selection of which matrices to develop or what kinds of rooms to include should always follow individual R&D and project needs. However, when starting a QFD project, one of the most important tasks for the experienced QFD facilitator is not only to be the "architect for the House of Quality" but also, during project work, to recommend which rooms to deploy in order to facilitate the knowledge-building process and the final matrix analysis.

Are the data in the matrices really reliable, since the relationships in many of the rooms are sometimes based on unverified assumptions or even "guesswork"?

When introducing the QFD methodology in an R&D organization, persons with a strong scientific background may question the validity of relationship matrices and reject them with scorn as simple "guesswork." It must initially be recognized that data in the individual relationship matrices are sometimes based on customers' reliable opinions, sometimes on truly verified experiments, and sometimes on expert judgments from members of the project team. In a relationship matrix that often contains more than 1000 relations, it is clear that not only initially after the creation of the matrix but also after a few revisions, the individual relations in such a matrix are of a mixed quality.

However, if such a matrix had not been built, not only the different individual relationships but also the character and quality of each individual relation would have remained unknown. Thus, the advantage of creating relationship matrices is not only that a number of previously unknown individual relations are put forward to the QFD team but also that each relationship has been questioned and reviewed in a cross-functional approach, despite the fact that this process is initially somewhat superficial. The expert judgement of the team has tried to pinpoint the strength of each relation, and in discussions of the validity of the content of relationships matrices. It was once concluded by a seasoned industry QFD professional:

It is better to be roughly right than precisely wrong!

In the further use of the relationship matrices, the team can focus reviews on selected relations and discuss how this information has been validated. Afterward, the team can choose important and strong relationships that are not well validated for further test work and other kinds of knowledge-building activities; in some cases, this can also be a good incentive to seek external expertise elsewhere.

Is there a risk of getting stuck in "single parameter thinking" in the review of individual requirements?

On the contrary, it has often been expressed by experienced QFD users that one of the most important benefits of the QFD matrices is that they incentivize a more multi-parameter approach and thinking. However, in the development of the room *Directions for improvements*, one wants to encourage a truly single-parameter perspective on each *Product Design Requirement*, since one does not want to start "compromising" at this stage! The matrices are not tools for experimental design but can be important guiding instrument for such an activity. How do the multiple process parameters of the Process Matrix affect the multiple Product Design Requirements? This is a multivariate relationship, which, in many sectors of the process industries, certainly often contains many still-unanswered questions.

Is it necessary to strictly follow the calculated Technical Importance Ratings in the House of Quality when judging the importance of individual product attributes?

There is no such thing as an absolute "truth" in science or in the use of the QFD methodology. In all kind of social sciences, data must be reviewed from many perspectives, and those reviewing it must ask whether the results are reasonable and whether any strange outcomes or errors are identified. Some QFD users prefer not to use the relationship matrices for such calculations but rest with the overall general information in this room. However, the author's own experience with the development of a multitude of such matrices generally confirms the usability of relationship matrices for transforming importance ratings of *Customer Requirements* into importance ratings of *Product Design Requirements.*

26.4 Selected questions related to phase progression and the use of the downstream matrices

If my company is producing assembled products, can we use the mpQFD system for phase progression?

In this book phase progression is only presented and explained for process-industrial innovation and the full mpQFD system cannot be used in other manufacturing industries. However, The Voice of the Customer and the House of Quality part of the system will work quite well also for assembled products. Fortunately, there are many other textbooks on QFD that present phase progression for assembled products in detail.

26.5 A select question related to IKPs

In a knowledge management perspective of IKPs, should a named person be appointed for each platform?

If your company product portfolio is managed by Product Managers, this is the person that should be the owner of an IKP. Product managers are responsible for guiding the success of a product and leading the cross-functional team that is responsible for improving it. The role spans many activities from strategic to tactical and provides important cross-functional leadership — most notably between engineering, marketing, sales, and support teams. The product manager is the person responsible for defining the why, when, and what of the product that the engineering team builds. This means they lead cross-functional teams from a product's conception all the way through to its launch.

Appendix A

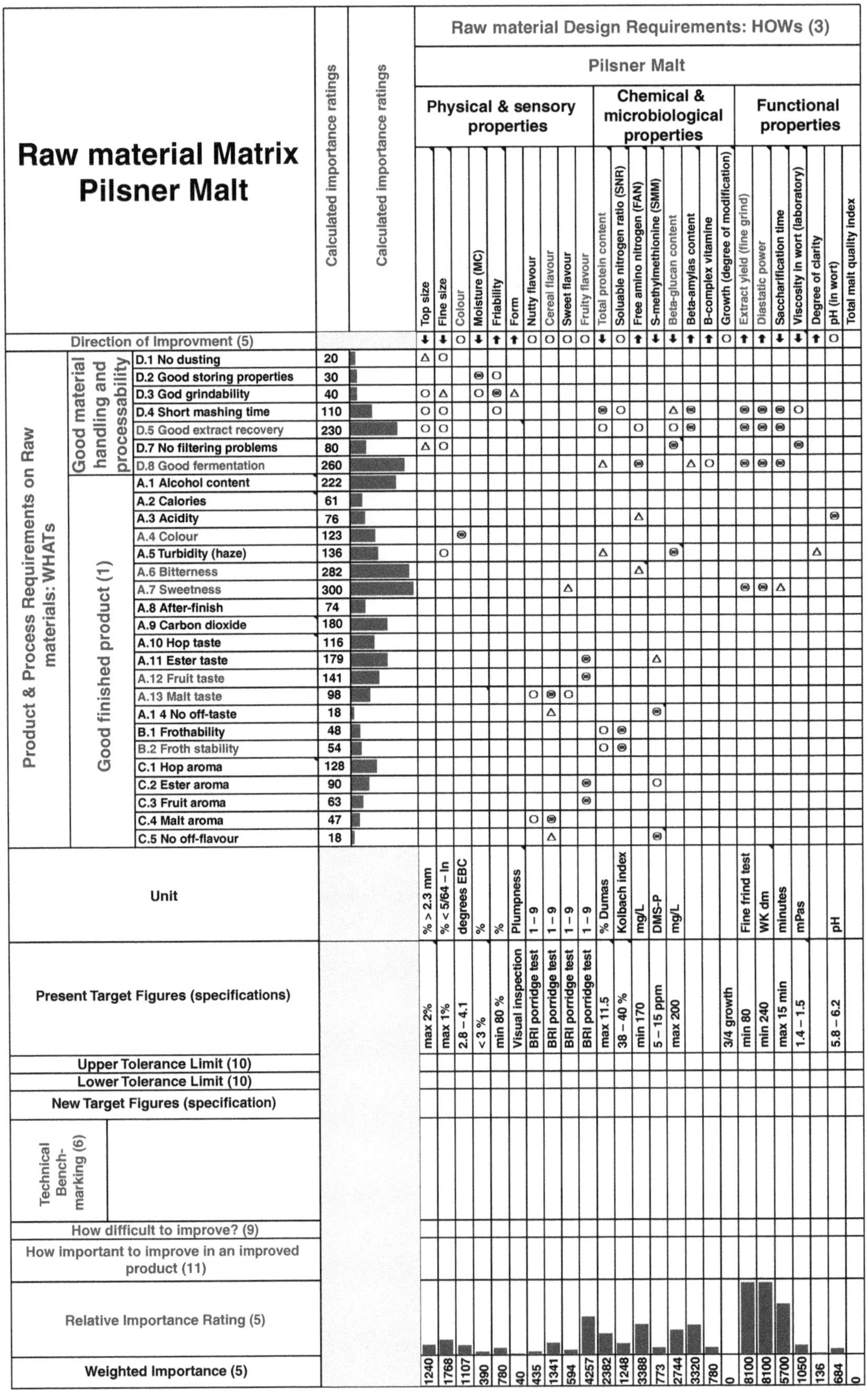

Figure A.1 The Raw material Matrix Pilsner malt.

Appendix B

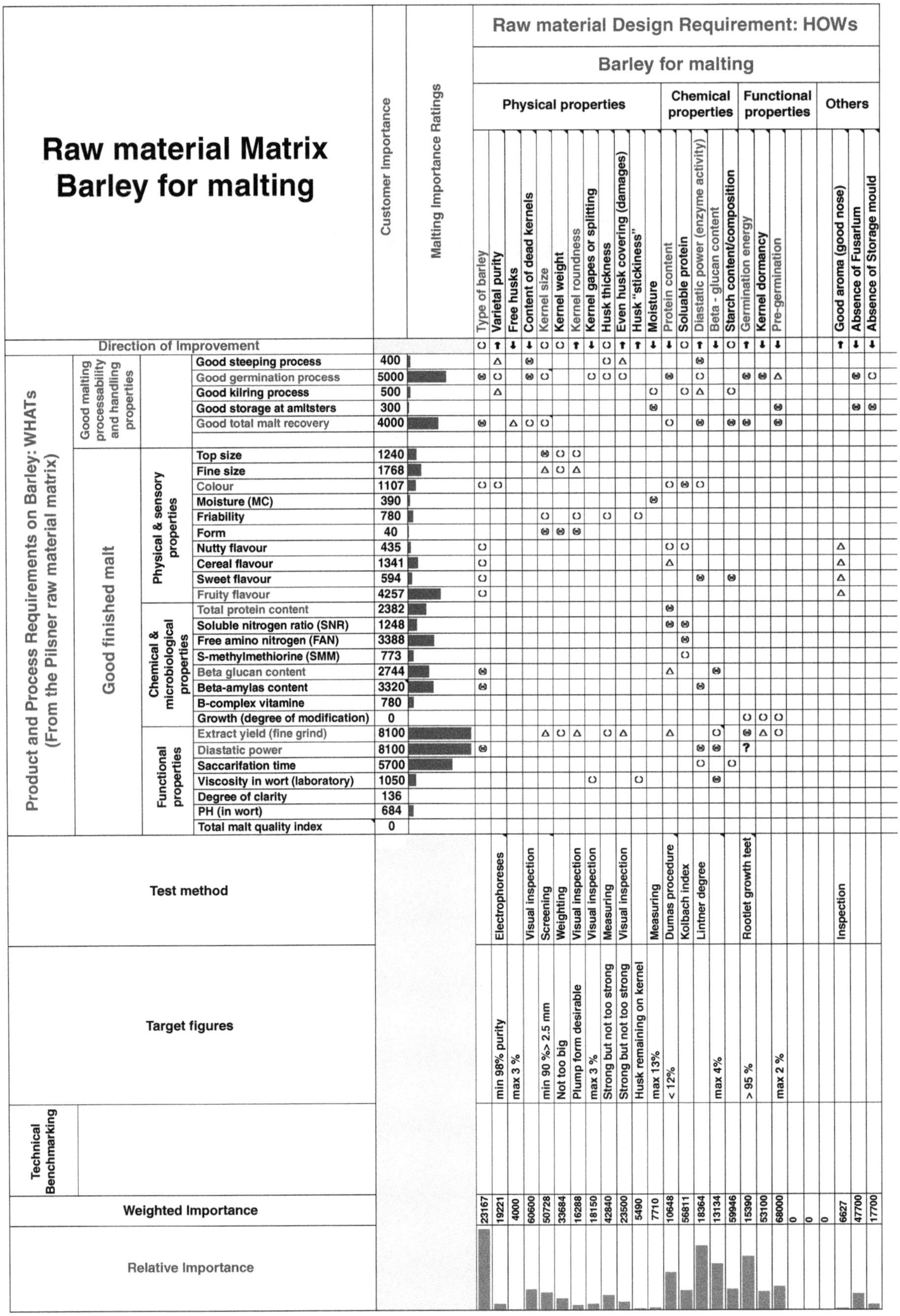

Figure B.1 The Raw material Matrix Barley for malting.

Appendix C

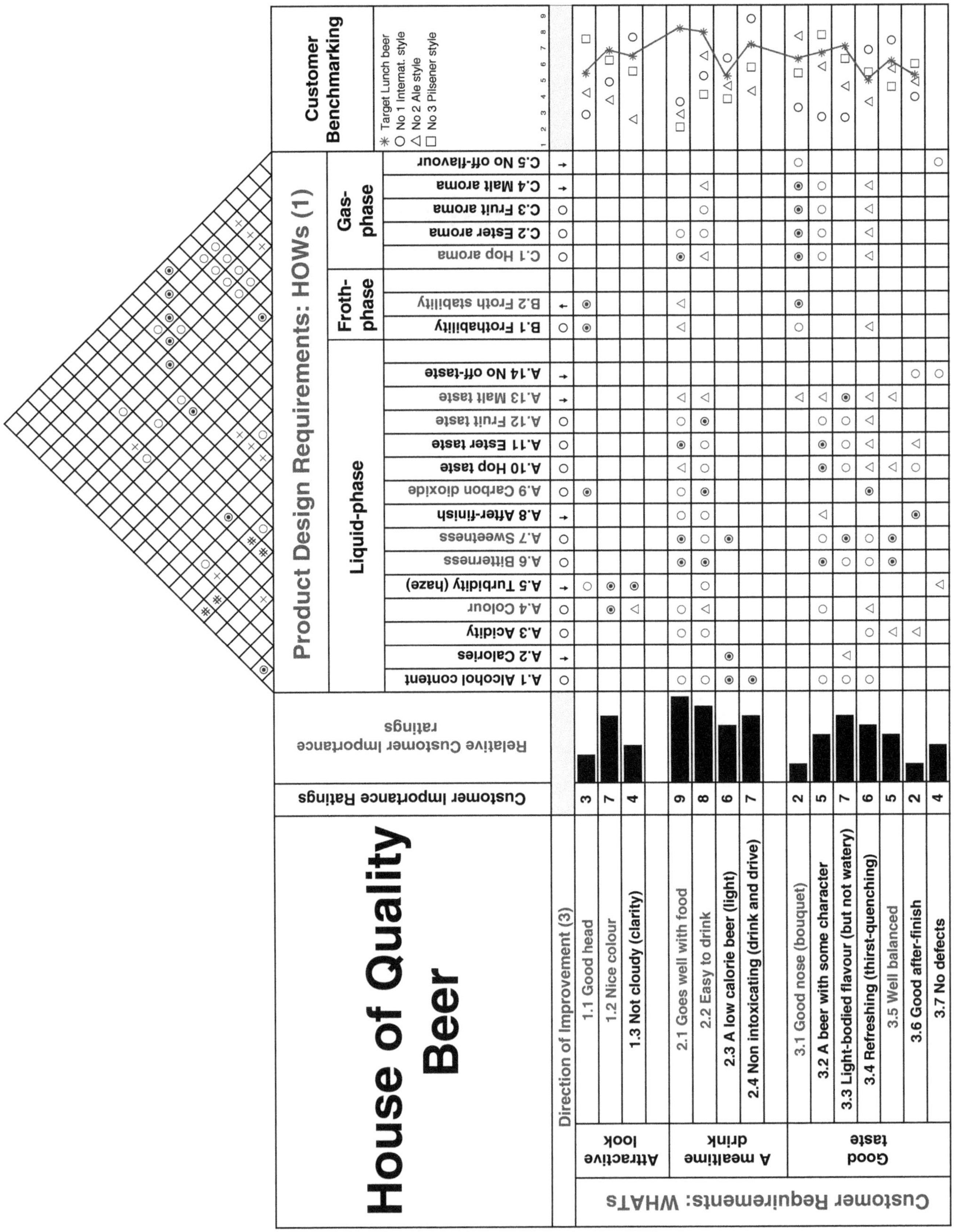
House of Quality
Beer
Customer Importance Ratings
Relative Customer Importance ratings
Product Design Requirements: HOWs (1)
Liquid-phase
Froth-phase
Gas-phase
A.1 Alcohol content
A.2 Calories
A.3 Acidity
A.4 Colour
A.5 Turbidity (haze)
A.6 Bitterness
A.7 Sweetness
A.8 After-finish
A.9 Carbon dioxide
A.10 Hop taste
A.11 Ester taste
A.12 Fruit taste
A.13 Malt taste
A.14 No off-taste
B.1 Frothability
B.2 Froth stability
C.1 Hop aroma
C.2 Ester aroma
C.3 Fruit aroma
C.4 Malt aroma
C.5 No off-flavour
Customer Benchmarking
Target Lunch beer
No 1 Internat. style
No 2 Ale style
No 3 Pilsener style
Direction of Improvement (3)
Customer Requirements: WHATs
Attractive look
1.1 Good head
1.2 Nice colour
1.3 Not cloudy (clarity)
A mealtime drink
2.1 Goes well with food
2.2 Easy to drink
2.3 A low calorie beer (light)
2.4 Non intoxicating (drink and drive)
Good taste
3.1 Good nose (bouquet)
3.2 A beer with some character
3.3 Light-bodied flavour (but not watery)
3.4 Refreshing (thirst-quenching)
3.5 Well balanced
3.6 Good after-finish
3.7 No defects

Relative Technical Importance Ratings (2)																							
Weighted Importance (2)	222	61	76	123	136	282	300	74	180	116	179	141	98	18	0	48	54	0	128	90	63	47	18
Test Unit (1)	Volume (%)	Calories/100 ml	pH	EBC Colour unit	EBC Turbidity unit	EBU Bitterness unit	Degree Plato	1–9	gram/liter CO_2	1–9	1–9	1–9	1–9	1–9		1–9	Seconds		1–9	1–9	1–9	1–9	1–9
Span of beer characteristics (general figures in parenthesis)	0–14 (4,5)	10–70 (40)	4–6 (4.5)	0–80 (10)	0–6 (0.5)	0–100 (20)	0–12 (5)		0–10 (5)								0–200 (>90)						
Test Method (1)	Chemical analysis	Calculation	pH measurement	Absorbance (430 nm)	Turbidity measurement	Absorbance (275 nm)	Analysis (apperent extract)	Test panel	Chemical analysis	Test panel	Test panel	Test panel	Test panel	Test panel		Test panel	Rudin test method		Test panel	Test panel	Test panel	Test panel	Test panel
(4) Technical Benchmarking (value scale) 1–9 ✳ Target Lunch beer ○ No 1 Internat. style △ No 2 Ale style □ No 3 Pilsener style																							
Beer No 1 (Int.) Technical benchmarking figures (intensity scale)	0.6	30	4.6	3.5	0.5	9	4		7	1.5	2.5	3.5	1.0			8	70		2.5	1.5	2.5	1.5	
Beer No 2 (Ale) Technical benchmarking figures (intensity scale)	5.0	45	5.0	30	1.5	20	5		3.0	1.0	5.5	5.0	5.0			4	110		2.0	3.0	5.0	4.0	
Beer No 3 (Pilsner) Technical benchmarking figures (intensity scale)	3.5	35	4.5	15	0.8	30	3.0		5	5.5	2.0	1.5	2.5			6	145		4.6	2.5	1.5	2.4	
Target figures for a "Perfect lunch beer" (7)	2.5	35	4.3	13	0.7	20	5.0		4	3	3.5	4.0	1.0			5	130		3.5	4.5	3.0	1.0	
Upper tolerance level (7)																							
Lower tolerance level (7)																							
Customer complaints (6)																							
Sustainability room (8)																							
How difficult to improve (9)																							
How important to improve (10)																							

Figure C.1 House of Quality for the Perfect Lunchtime Beer.

Appendix D

Process Matrix Brewing				Product Influencing Process Parameters: HOWs (2) — Process control parameters (simplified process)																												
				Mashing										Wort-boiling						Fermentation					Finishing							
Product Design Requirements: WHATs (1)		Importance ratings from House of Quality	Importance ratings from House of Quality	Pilsner malt type/dosage	Caramel malt type/dosage	Chocolate or black malt type/dosage	Particle size after crushing	Adjunct dosage	Water hardness (Ca^{2+} and Mg^{2+})	Mash thickness	Mash pH	Mashing scheme (temp. & time)	Filtering temperature (sparging)	"Bitter hop" dosage	"Aroma hop" dosage	Hop addition scheme	Original gravity	Wort boiling time	Separation	Type of yeast	Yeast dosage	Fermentation temperature	Fermentation scheme	Bunging pressure	Maturation temperature	Maturation time	Filtering	Adjustment of colour	Addition of hop	Sugar priming	CO_2 addition (carbonisation)	Pasteurisation temperature
Liquid-phase	A.1 Alcohol content	222		◉	○	○		◉				○					◉			◉	◉	◉	◉									
Liquid-phase	A.2 Calories	61		◉	○	○		○				○					◉			◉	◉	◉	◉							◉		
Liquid-phase	A.3 Acidity	76							○		◉											△		○							○	
Liquid-phase	A.4 Colour	123		◉	◉	◉												○										◉				
Liquid-phase	A.5 Turbidity (haze)	136					△						○						◉								◉					
Liquid-phase	A.6 Bitterness	282												◉	○	◉													◉			
Liquid-phase	A.7 Sweetness	300		△	○	○		△			○	◉					◉			◉	◉	◉	◉							◉		
Liquid-phase	A.8 After-finish	74		△	○	△		○					○	◉	◉					◉	○	◉	◉		○	○			◉	○		
Liquid-phase	A.9 Carbon dioxide	180																		△	△	△	○	◉							◉	
Liquid-phase	A.10 Hop taste	116												◉	◉	◉		△											◉			
Liquid-phase	A.11 Ester taste	178																		◉		◉	◉	△	◉	○						
Liquid-phase	A.12 Fruit taste	141																		◉		◉	◉	△	◉	○						
Liquid-phase	A.13 Malt taste	98		◉	◉	○	△	○			○	◉										○										
Liquid-phase	A. 14 No off-taste	18											○												◉	◉						
Froth-phase	B.1 Frothability	48		○	○	○						◉					◉	○		△			△	○		△	△					
Froth-phase	B.2 Froth stability	54		○	○	○						◉					◉	○		△			△	○								
Froth-phase																																
Gas-phase	C.1 Hop aroma	128												◉	◉	◉		◉											◉			
Gas-phase	C.2 Ester aroma	90																		◉	○	◉	◉		◉	◉						
Gas-phase	C.3 Fruit aroma	63																		◉	○	◉	◉		○	○						
Gas-phase	C.4 Malt aroma	47		◉	◉	△	△	△			△	○																				
Gas-phase	C. No off-flavour	18											△												◉	◉						

Calculated Relative Importance Ratings of Parameters (5)																														
Calculated Importance		5639	4689	3577	281	3044	210	0	1871	5490	702	5400	3708	4734	6165	1943	1224	10443	6108	10705	10803	2455	4416	2550	1272	1107	5400	3471	1830	0
Measuring instruments and metrics (3)																														
Standard process set-points (3)																														
Upper tolerance level (3)																														
Lower tolerance level (3)																														
Technology Newness (6)																														
Influence on production yield (8)																														
Use of available equipment production capability (9)																														
Relative frequency of equipment outages (10)																														
Display of process interfaces (11)																														
Relative yearly operating cost (12)																														
Relative yearly capital cost (13)																														
Necessary improved production capabilities for a new or improved product (14)																														
How difficult to improve (15)																														
Decided Direction for Improvement (16)																														

Figure D.1 Process Matrix for beer manufacturing.

References

Afshan, N. & Sindhuja, P. N. 2013. Integration of Kano's model into quality function deployment: A review. *Journal of Operations Management*, 12.

Akao, Y. (ed.) 1990. *Quality Function Deployment: Integrating Customer Requirements into Product Design*. Cambridge: Productivity Press.

Akao, Y. 1997. QFD: Past, present, and future. In: Gustafsson, A., Bergman, B. & Ekdahl, F. (eds.) *Proceedings of The Third International QFD Symposium*. Linköping.

Akao, Y. 2002. QFD and Knowledge Management. *7th International Conference on ISO 9000 & TQM*.

Akao, Y. 2003. The leading edge in QFD: past, present and future. *International Journal of Quality & Reliability Management*, 20, 20–35.

Akao, Y., Ohfuji, T. & Naoi, T. 1987. Survey and reviews on quality function deployment in Japan. *Proceedings of ICQC 1987 in Tokyo*.

Al-Mashari, M., Zairi, M. & Ginn, D. 2005. Key enablers for the effective implementation of QFD: A critical analysis. *Industrial Management & Data Systems*, 105, 1245–1260.

American Supplier Institute 1989. *Quality Function Deployment — Three Day Workshop*. Dearborn: American Supplier Institute Inc.

American Supplier Institute 1992. QFD — Kundorienterad produktutveckling hårdvara & process. Dearborn: American Supplier Institute Inc.

Ansari, A. & Modarress, B. 1994. Quality function deployment: The role of suppliers. *International Journal of Purchasing and Materials Management*, Fall, pp. 28–35.

Baglieri, E., Stabilini, G. & Zamboni, S. 2002. How to balance QFD and the efficiency of the New Product Development process; experiences and practices from the field. *Proceedings of the Eighth International Symposium on Quality Function Deployment*. Munich: QFD Institute.

Ballot, G., Fakhfakh, F., Galia, F. & Salter, A. 2015. The fateful triangle: Complementaries in performance between product, process and organizational innovation in France and the UK. *Research Policy*, 44, 217–232.

Bech, A. C., Hansen, M. & Wienberg, L. 1997. Application of house of quality in translation of consumer needs into sensory attributes measurable by descriptive sensory analyses. *Food Quality and Preference*, 8, 329–348.

Bech, A. C., Kristensen, K., Juhl, H. J. & Poulsen, C. S. Development of farmed smoked eel in accordance with consumer demands. In: Luten, J. B. (ed.), *Seafood from Producer to Consumer,*

Integrated Approach to Quality — Proceedings of the International Seafood Conference. Noordwijkerhout: Elsevier Science.

Bettencourt, L. A., Lusch, R. F. & Vargo, S. L. 2014. A service lens on value creation: marketing's role in achieving strategic advantage. *California Management Review*, 57, 44–66.

Bigliardi, B., Bottani, E. & Galati, F. 2010. Open innovation and supply chain management in food machinery supply chain: A case study. *International Journal Of Engineering, Science And Technology*, 2, 244–255.

Bilgram, V., Brem, A. & Voigt, K.-I. 2008. User-centric innovations in new product development — systematic identification of lead users harnessing interactive and collaborative online-tools. *International Journal of Innovation Management*, 12, 419–458.

Boothroyd, G., Dewhurst, P. & Knight, W. 1994. *Product Design for Manufacture and Assembly.* New York, Marcel Dekker Inc.

Bossert, J. L. 1991. *Quality Function Deployment — A Practitioner's Approach.* Milwaukee, ASQC Quality Press.

Bower, D. J. & Keogh, W. 1996. Changing patterns of innovation in a process-dominated industry. *International Journal of Technology Management*, 12, 209–220.

Brockhoff, K. K., Koch, G. & Pearson, A. W. 1997. Business process re-engineering: Experiences in R&D. *Technology Analysis & Strategic Management*, 12, 209–220.

Calo, R. & Kramer, S. Total quality management in the tight tolerance aluminum extrusion process. *5th International Aluminum Extrusion Technology Seminar*, 1992 Chicago. The Aluminum Association, pp. 467–483.

Cantista, I. & Tylecote, A. 2008. Industrial innovation, corporate governance and supplier–customer relationships. *Journal of Manufacturing Technology*, 19, 576–590.

Chakravorty, S. S. 2000. Improving a V-plant operation: A window manufacturing case study. *Production and Inventory Management Journal*, 41.

Charteris, W. 1993. Quality function deployment: A quality engineering technology for food industry. *Journal of the Society of Dairy Technology*, 46, 12–21.

Cheng, L. C. & De Melo Filho, L. D. R. 2008. An application of QFD for production start-up of a generic drug in a government-owned pharmaceutical institution. *The 14th International Symposium on Quality Function Deployment*, 2008, Beijing.

Chesbrough, H. & Appleyard, M. M. 2007. Open innovation and strategy. *California Management Review*, 50, 57–76.

Chesbrough, H. & Crowther, A. K. 2006. Beyond high tech: Early adopters of open innovation in other industries. R&D Management, 36, 229–236.

Chesbrough, H. W. 2007. Why companies should have open business models. *MIT Sloan Management Review*, 48 Winter, 21–28.

Chesbrough, H. 2010. Business model innovation: Opportunities and barriers. *Long Range Planning*, 43, 354–363.

Chiaroni, D., Chiesa, V. & Frattini, F. 2010. Unravelling the process from closed to open innovation: Evidence from mature, asset-intensive industries. *R&D Management*, 40, 222–245.

Cobbenhagen, J., Den Hertog, F. & Philips, G. (eds.) 1990. *Management of innovation in the processing industry: a theoretical framework.* London: Printer Publishers.

Cohen, L. 1995. *Quality Function Deployment — How to Make QFD Work for You.* Reading, Addison-Wesley Publishing Company.

Cohen, W. M. & Levinthal, D. A. 1990. Absorptive capacity: a new perspective on learning and innovation. *Administrative Science Quarterly*, 35, 128–152.

Comstock, T. N. & Dooley, K. 1998. A tale of two QFDs. *Quality Management Journal*, 5, 32–45.

Contreras, J., Sistemas, I. & Ferrer, J. M. 2005. Dynamic simulation: A case study. *Hydrocarbon Engineering*, May.

Cooper, R. G. 1994. Perspective. Third-generation new product processes. *Journal of Product Innovation Management*, 11, 3–14.

Cooper, R. G. 2008. Perspective: The stage-gate idea-to-launch process — update, What's new, and NexGen Systems. *Journal of Product Innovation Management*, 25, 213–232.

Cooper, R. G. 2012. Best practices in the idea-to-launch process and its governance. *Research-Technology Management*, MarchApril, 43–54.

Cooper, R. G. 2014. What's next: After stage gate. *Research-Technology Management*, January–February.

Cooper, R. G. & Kleinschmidt, E. J. 1986. An investigation into the new product process: Steps, deficiencies and impact. *Journal of Product Innovation Management*, 3, 71–85.

Cooper, R. G. & Kleinschmidt, E. J. 1988. Resource allocation in the new product process. *Industrial Marketing Management*, 17, 249–262.

Corma, F. 1993. QFD in Ceramics. In: Ourén, P. & Fernández, J. F. (eds.) Third Euro-Ceramics, 1993. 1077–1082.

Cristiano, J. J., Liker, J. K. & White, C. C. 2000. Customer-driven product development through quality function deployment in the U.S. and Japan. *Journal of Product Innovation Management*, 17, 286–308.

Cristiano, J. J., Liker, J. K. & White, C. C. 2001. Key factors in the successful application of quality function deployment (QFD). *IEEE Transaction on Engineering Management*, 48, 81–95.

Crossan, M. M., Lane, H. W. & White, R. E. 1999. An organizational learning framework: from intuition to institution. *Academy of Management Review*, 24, 522–537.

Crossan, M. M., Maurer, C. C. & White, R. E. 2011. Reflections on the 2009 AMR decade award: Do we have a theory of organizational learning? *Academy of Management Review*, 36, 446–460.

Davenport, T. H. 2005. *Thinking for a Living*. Boston, Massachusetts, Harvard Business School Press.

Day, R. G. 1993. *Quality Function Deployment — Linking a Company with Its Customers.* Milwaukee, ASQ Quality Press.

De Saint-Exupéry, A. 1984. *The Wisdom Of The Sands.* Chicago, University of Chicago Press.

De Oliveira, C. A., Magri, W. & Torres, I. V. 1996. QFD in a Brazilian Steel Company. *Transactions from The Eighth Symposium on Quality Function Deployment.* Novi: QFD Institute.

De Waal, G. A. & Knott, P. 2013. Innovation tool adoption and adaption in small technology-based firms. *International Journal of Innovation Management*, 17.

Dussauge, P., Hart, S. & Ramanantsoa. 1987. *Strategic Technology Management,* New York, John Wiley & Sons.

Duxbury, D. 1991. "High Tech" QFD and TQM programs produce all — natural ingredients. *Food Processing*, 52, 46–48.

Eisenberg, I. 2011. Lead-user research for breakthrough innovation. *Research-Technology Management*, January–February, pp. 50–58.

Ekdahl, F. & Gustafsson, A. 1997. QFD: The Swedish Experience. *Transactions from The Ninth Symposium on Quality Function Deployment.* Novi: QFD Institute.

Eng, T.-Y. & Ozdemir, S. 2014. International R&D partnerships and intrafirm R&D-marketing-production integration of manufacturing firms in emerging economies. *Industrial Marketing Management*, 43, 32–44.

Eureka, W. E. & Ryan, N. E. 1988. *The Customer Driven Company.* Dearborn, American Supplier Institute Inc.

Fix, G. 1999. *Principles of Brewing Science.* Boulder: Brewers Publications.

Flapper, S. D. P., Fransoo, J. C., Broekmeulen, R. A. C. M. & Inderfurth, K. 2002. Planning and control of rework in the process industries: a review. *Production Planning & Control*, 13, 26–34.

Floyd, R. C. 2010. *Liquid Lean: Developing Lean Culture in the Process Industries.* New York, Productivity Press.

Franceschini, F. & Rosetto, S. 1998. Quality function deployment: How to improve its use. *Total Quality Management*, 9, 491–500.

Franke, N., Von Hippel, E. & Schreier, M. 2006. Finding commercially attractive user innovations: A test of lead-user theory. *Journal Of Product Innovation Management*, 23, 301–315.

Frishammar, J., Lichtenthaler, U. & Kurkkio, M. 2012. The front-end in non-assembled product development: A multiple case study of mineral- and metal-firms. *Journal of Engineering and Technology Management (JET-M)*, 29, 468–488.

Frishammar, J., Söderhlm, P., Bäckström, K., Hellsmark, H. & Ylinenpää.Håkan 2014. The role of pilot and demonstration plants in technological development: synthesis and directions for future research. *Technology Analysis & Strategic Management.*

Foster, R. N. 1986. *Innovation — the attackers advantage.* New York, Summit Books

Fuller, J. 2010. Refining virtual co-creation from a consumer perspective. *California Management Review*, 52, 98–122.

Fuller, J., Hutter, K. & Faullant, R. 2011. Why co-creation experience matters? Creative experience and its impact on the quantity and quality of creative contributions. *R&D Management*, 41, 259–273.

Gamal, R. 2015. *Packaging Innovation Work Process Improvement In The Food Industries: Towards A New Integrated Approach.* Dr Of Business Administration, Grenoble Ecole De Management.

Gamal, R. & Lager, T. 2015. Bridging the R&D — manufacturing interface: the case of packaging innovation At Nestlé. 22nd Euroma Conference. Neuchatel, Switzerland: European Operations Management Association.

Gassman, O., Enkel, E. & Chesbrough, H. 2010. The future of open innovation. *R&D Management*, 40, 213–221.

Ginn, M. E. & Rubenstein, A. H. 1986. The R&D/production interface: A case study of new product commercialization. *Journal of Product Innovation Management*, 3, 158–170.

Granger, R. J. Value for money in R&D. Arthur D. Little International Inc.

Green, P. E., Krieger, A. M. & Wind, Y. 2001. Thirty years of conjoint analysis: reflections and prospects. *Interfaces,* 31 May–June, pp. 56–73.

Griffin, A. & Hauser, J. R. 1991. *The Voice of The Customer.* Cambridge: Massachusetts Institute of Technology.

Griffin, A. & Hauser, J. R. 1996. Integrating R&D and marketing: A review and analysis of the literature. *Journal of Product Innovation Management*, 13, 191–215.

Griffin, A. 1992. Evaluating QFD's use in US firms as a process. *Journal of Product Innovation Management*, 9, 171–187.

Guedes, L. B. R., Antoni, I., Andrade, J. C., Galvao, M. T., Olivera, L. C., Matsunaga, P., Saraiva, R., Cremaschi, M., Soffiatti, T., Watanabe, R. & Santos, W. 1999. Obtaining countrywide success through QFD implementation in the development process of a popular Brazilian food product. *Fifth International Symposium on Quality Function Deployment.*

Gupta, A. K., Raj, S. P. & Wilemon, D. 1986. A model for studying R&D — marketing interface in the product innovation process. *Journal of Marketing*, 50, 7–17.

Gustafsson, A. 1996. *Customer Focused Product Development by Conjoint Analysis and QFD.* Doctoral, Linköping University.

Hammer, M. 1990. Reengineering work: Don´t automate, obliterate. *Harvard Business Review*, 104–112.

Handfield, R. B., Krause, D. R., Scannell, T. V. & Monczka, R. M. 2000. Avoid the pitfalls in supplier development. *Sloan Management Review*, Winter.

Handfield, R. B., Ragatz, G. L., Petersen, K. J. & Monczka, R. M. 1999. Involving suppliers in new product development. *California Management Review*, 42, 59–82.

Hanson, D. 1993. Quality function deployment for product and service improvement. *Transactions from the Fifth Symposium on Quality Function Deployment.* Novi: QFD Institute.

Harding, J. A. & Popplewell, K. 2000. Simulation: an application of factory design process methodology. *Journal of Operational Research Society*, 440–448.

Hauser, J. R. & Clausing, D. 1988. The house of quality. *Harvard Business Review*, May–June, 63–73.

Hirsch-Kreinsen, H. 2008. "Low-Technology": a forgotten sector in innovation policy. *Journal of Technology Management & Innovation*, 3, 11–20.

Hofmeister, K. R. 1991. Quality function deployment: market success through customer-driven Products. Food Product Development.

Huizingh, E. K. R. E. 2011. Open Innovation: State of the art and future perspectives. *Technovation*, 31, 2–9.

Hullova, D., Trott, P. & Simms, C. D. 2016. Uncovering the reciprocal complementarity between product and process innovation. *Research Policy*, 45, 929–940.

Hullova, D., Simms, C. D., Trott, P. & Laczko, P. 2019. Critical capabilities for effective management of complementarity between product and process innovation: Cases from the food and drink industry. Research Policy, 48, 339-354.

Hunter, M. R. & Van Landingham, R. D. 1994. Listening to the customer using QFD. Quality Progress, April, pp. 55–59.

Ideacore 2003. QFD Designer. 5.0 ed. Birmingham: IDEACore.

Jackson, M. 1991. The New World Guide to Beer, London: The Apple Press.

Jianxin, J., Simpson, T. W. & Siddique, Z. 2007. Product family design and platform-based product development: A state-of-the-art review. *Journal of Intelligent Manufacturing*, 18, 5–29.

Katz, G. M. 2004. Practitioner note: a response to Pullman *et al.*'s (2002) comparison of quality function deployment versus conjoint analysis. *Journal of Product Innovation Management,* 21, 61–63.

Kaplan, R. S. 1998. Innovation action research: Creating new management theory and practice. *Journal of Management Accounting Research*, 10, 89–118.

King, B. 1987. Better Design in Half the Time — Implementing QFD Quality Function Deployment in America, Methuen, GOAL/QPC.

King, P. L. 2009. *Lean for the Process Industries: Dealing with Complexity.* New York, Productivity Press.

King, A. M. & Sivvaloganathan, S. 2010. Development of a Methodology for Concept Selection in Flexible Design Strategies. *Journal of Engineering Design*, 10.

Kline, S. J. 1985. Innovation is not a linear process. *Research Management*, July–August, 36–45.

Krueger, R. A. 1998. *Developing Questions for Focus Groups*, London, SAGE Publications.

Krueger, R. A. & Casey, M. A. 2009. *Focus groups: A Practical Guide for Applied Research*, London, SAGE.

Kubota, I. 1990. Using quality function deployment: the case of Nippon carbon. In: Akao, Y. (ed.) *Quality Function Deployment Integrating Customer Requirements into Product Design.* Cambridge: Productivity Press.

Laganier, F. 1996. Dynamic process simulation — trends and perspectives in an industrial context. Computers & Chemical Engineering, 20, 1595–1600.

Lager, T. 2000. A new conceptual model for the development of process technology in process industry. *International Journal of Innovation Management*, 4, 319–346.

Lager, T. 2002. Product and process development intensity in process industry: A conceptual and empirical analysis of the allocation of company resources for the development of process technology. *International Journal of Innovation Management*, 4, 105–130.

Lager, T. 2005a. The industrial usability of quality function deployment: a literature review and synthesis on a meta-level. *R&D Management*, 35, 409–426.

Lager, T. 2005b. Multiple progression — a proposed new system for the application of quality function deployment in process industry. *International Journal of Innovation Management*, 9, 311–341.

Lager, T. 2010. *Managing Process Innovation — From Idea Generation to Implementation.* London, Imperial College Press.

Lager, T. 2017a. A conceptual analysis of conditions for innovation in the process industries and a guiding framework for industry collaboration and further research. *International Journal of Technology Learning, Innovation and Development*, 9.

Lager, T. 2017b. A reassessment of the QFD product development methodology from the perspective of knowledge creation and utilization. *International Journal of Knowledge Management Studies*, 8.

Lager, T. & Blanco, S. 2010. The commodity battle: A product-market perspective on innovation resource allocation in the process Industries. *International Journal of Technology Intelligence and Planning*, 6, 128–150.

Lager, T. & Kjell, Å. 2007. Multiple progression QFD: A case study of cooking product functionality at Arla Foods. In: MAZUR, G., ed. *13th International & 19th North American Symposium On QFD*, September 7–8, 2007. Williamsburg: QFD Institute, 271–295.

Lager, T. & Rennard, J.-P. 2014. Managing the manufacturing — R&D interface: An extended editorial viewpoint. *Journal Of Manufacturing Technology Management*, 25, 146–154.

Lager, T., Blanco, S. & Frishammar, J. 2013. Managing R&D and Innovation in the process industries. *R&D Management*, 43, 189–195.

Lager, T., Samuelsson, P. & Storm, P. 2017c. Modelling company generic production capabilities in the process industries: A configuration approach. *International Journal of Operations & Production Management*, 37, 126–161.

Leonard-Barton, D. 1992. The factory as a learning laboratory. *Sloan Management Review*, 34, 23–38.

Lewis, K., Lange, D. & Gillis, L. 2005. Transactive memory systems, learning, and learning transfer. *Organizational Science*, 16, 581–598.

Liebeskind, D. 1998. Reengineering R&D work processes. *Research Technology Management*, 45, 43–48.

Liker, J. K. & Meier, D. 2006. *The Toyota Way Fieldbook: A Practical Guide For Implementing Toyota's 4ps*. New York, Mcgraw-Hill.

Linton, J. D. & Walsh, S. T. 2007. A theory of innovation for process-based innovations such as nanotechnology. Technology Forecasting and Social Change.

Luthje, C. & Herstatt, C. 2004. The lead user method: An outline of empirical findings and issues for future research. *R&D Management*, 34, 553–568.

Macdonald, E. K., Wilson, H., Martinez, V. & Toosi, A. 2011. Assessing value-in-use: A conceptual framework and exploratory study. *Industrial Marketing Management*, 40, 671–682.

Martinez, M. G. 2014. Co-creation of value by open innovation: Unlocking new sources of competitive advantage. *Agribusiness*, 30, 132–147.

Martins, A. & Aspinwall, E. M. 2001. Quality function deployment: An empirical study in the UK. *Total Quality Management*, 12, 575–588.

Marx, K. 1867. Das Kapital, Gateway Editions.

Matzler, K. & Hinterhuber, H. H. 1998. How to make product development projects more successful by integrating Kano's model of customer satisfaction Into quality function deployment. *Technovation*, 18, 25–38.

Mclaughlin, C. P. & Stratman, J. 1997. Improving the quality of corporate technical planning: Dynamic analogues of QFD. *R&D Management*, 27, 269–279.

Melan, E. H. 1992. *Process Management — Methods for Improving Products and Service*, New York, McGraw-Hill, Inc.

Meyer, M. H. & Dalal, D. 2002. Managing platform architectures and manufacturing processes for nonassembled products. *Journal of Product Innovation Management*, 19, 277–293.

Miguel, P. A. C. 2005. Evidence of QFD best practices for product development: a multiple case study. *International Journal of Quality & Reliability Management*, 22, 72–82.

Miguel, P. A. C. 2013. Benchmarking QFD application for developing packaging products: A comparison between a company in Italy and one in Brazil. *Benchmarking: An International Journal*, 20, 419–433.

Mintzberg, H. 1990. The Manager's job: Folklore and fact. *Harvard Business Review*.

Mizuno, S. & Akao, Y. (eds.) 1994. *QFD: The Customer-Driven Approach to Quality Planning and Deployment*, Tokyo: Asian Productivity Organization.

Mongeon, D. L. 1996. RE: Bethlehem Steel Experience.

Moore, D. S. 1991. *Statistics — Concepts And Controversies*. New York, W.H. Freeman and Company.

Morgan, G. 1986. *Images of Organization*. London, Sage Publications. Inc.

Nace 2006. Statistical classification of economic activities in the European Community Rev 2. Luxembourg: Office for Official Publications of the European Communities.

Neelamkavil, F. 1987. *Computer Simulation and Modelling*. Chichester: John Wiley & Sons.

Nevo, D., Benbasat, I. & Wand, Y. 2012. Understanding technology support for transactive memory: Requirements, application, and customization. *Journal of Management Information Systems*, 28, 69–97.

Nihtilä, J. 1999. R&D — production integration in the early phases of new product development. *Journal of Engineering and Technology Management (Jet-M)*, 16, 55–81.

Nijssen, E. J. & Frambach, R. T. 2000. Determinants of the adoption of new product development tool by industry firms. *Industrial Marketing Management*, 29, 121–131.

Nijssen, E. J. & Lieshout, K. F. M. 1995. Awareness, use and effectiveness of models and methods for new product development. *European Journal of Marketing*, 29, 27–44.

Nonaka, I. & Takeuchi, H. 1995. *The Knowledge-Creating Company*. Oxford, Oxford University Press.

Norell, M. 1992. Advisory Tools and Co-operation in product development. Doctoral, Royal Institute of Technology.

Norling, P. M. 1997. Structuring and managing R&D work processes — Why bother? CHEMTECH, October, 12–16.

Novitsky, M. P. 1983. Process industry — where are you? *Production and Inventory Management*, 21, 1–26.

Olausson, D. & Berggren, C. 2012. Managing asymmetries in information flows and interaction between R&D, manufacturing, and service in complex product development. *R&D Management*, 42, 342–357.

Olson, E. M., Walker, O. C., Ruekert, R. W. & Bonner, J. M. 2001. Patterns of cooperation during new product development among marketing, operations and R&D: implications for project performance. *The Journal of Product Innovation Management*, 18, 258–271.

Ootaki, A., Oyaizu, M. & Koura, K. 1996. How to connect material and technology seeds to customer needs. *Transactions from the Eighth Symposium on Quality Function Deployment*. Novi: QFD Institute.

Oyaizu, M. 1990. New product development using quality technology deployment: The case of Nippon Zeon. In: Akao, Y. (ed.) *Quality Function Deployment: Integrating Customer Requirements into Product Design*. Cambridge: Productivity Press.

Oyaizu, M. 1994. QFD applied in the development and planning of a material producer. Hinshitsu, p. 24.

Özgener, S. 2003. Quality function deployment: a teamwork approach. *TQM and Business Excellence*, 14, 969–979.

Palluzi, R. P. 1992. *Pilot Plant Design, Construction and Operation*. McGraw Hill, Inc.

Palluzi, R. P. 1997. Succeed at crash pilot–plant construction. *Chemical Engineering Progress*, 93.

Park, S. & Gil, Y. 2006. How Samsung transformed its corporate R&D center. *Research Technology Management*, 24–29.

Pertl, A. 1992. Present Day Quality Assurance for Manufacture of Refractories. Radex-Rundschau.

Pisano, G. P. 1996. Learning before doing in the development of new process technology. *Research Policy*, 25.

Pisano, G. P. 1997. *The development factory: Unlocking the potential of process innovation*. Boston, Mass.: Harvard Business School.

Polanyi, M. 1983. *The Tacit Dimension*. Gloucester, Mass.

Purvis, R. L., Sambamurthy, V. & Zmud, R. W. 2001. The assimilation of knowledge platforms in organizations: An empirical investigation. *Organizational Science*, 12, 117–135.

Reichstein, T. & Salter, A. 2006. Investigating the sources of process innovation among UK manufacturing. *Industrial and Corporate Change*, 15, 653–682.

Rousselle, T. 2002. R&D Management Conference, 2012. Grenoble Ecole de Management.

Säfsten, K., Johansson, G., Lakemond, N. & Magnussson, T. 2014. Interface challenges and managerial issues in the industrial innovation process. *Journal of Manufacturing Technology Management* 25, 218–239.

Sahay, B. S. 2003. Supply chain collaboration: The key to value creation. Work Study, 52, 76–83.

Samuelsson, P., Storm, P. & Lager, T. 2016. Profiling company-generic production capabilities in the process industries and strategic implications. *Journal of Manufacturing Technology Management*, 27, 662–691.

Scheurell, D. M. 1992. Taking QFD through to the production planning matrix: Putting the customer on the line. *Transactions from the Fourth Symposium on Quality Function Deployment.* Novi: QFD Institute.

Scheurell, D. M. 1993. Concurrent engineering and the entire QFD process: One year after start-up of a new mill. *Transactions from the Fifth Symposium on Quality Function Deployment.* Novi: QFD Institute.

Scheurell, D. M. 1994. Beyond the QFD House of quality: Using the downstream matrices. *World Class Design in Manufacture,* 1, 13–20.

Schroeder, R., Van De Ven, A., Scudder, G. & Polley, D. 1986. Managing Innovation and Change Processes: Findings from the Minnesota Innovation Research Program. *Agribusiness*, 2, 501–523.

Shelden, R. A. & Dunn, I. J. 2001. Dynamic Simulation — Modelling Processes. *Chemical Engineering Progress*, Dec 44–48.

Silayoi, P. & Speece, M. 2007. The importance of packaging attributes: A conjoint analysis approach. *European Journal of Marketing,* 41, 1495–1517.

Song, M. & Thieme, R. J. 2006. A Cross-national investigation of the R&D-marketing interface in the product innovation process. *Industrial Marketing Management*, 35, 308–322.

Song, X. M., Montoya-Weiss, M. M. & Schmidt, J. B. 1997. Antecedents An consequences of cross-functional cooperation: A comparison of R&D, manufacturing, and marketing perspectives. *Journal of Product Innovation Management*, 14, 35–47.

Soosay, C. A., Hyland, P. W. & Ferrer, M. 2008. Supply chain collaboration: Capabilities for continuous innovation. *Supply Chain Management: An International Journal*, 13/2, 160–169.

Stewart, D. W., Shamdasani, P. N. & Rook, D. W. 2007. *Focus Groups: Theory and Practice*, LA, SAGE Publications.

Stitt, J. & York, C. 1993. Just do it. *Transactions from the Fifth Symposium on Quality Function Deployment.* Novi: QFD Institute.

Storm, P. & Lager, T. 2014. Application development: A strategic corporate innovation activity in the process industries. *International Journal of Technology Intelligence and Planning,* 10, 129–149.

Storm, P., Lager, T. & Samuelsson, P. 2010. A material conversion perspective on firm innovation in the Process Industries: the production system revisited. *The R&D Management Conference.* Manchester.

Svensson, J.-E. 1995. Ölkunskap, Falköping, Liber utbildning AB.

Taylor, S. G., Seward, S. M. & Bolander, S. F. 1981. Why the process industries are different. *Production and Inventory Management Journal, Fourth Quarter*, 9–24.

Taylor, W. T. 1916. The principles of scientific management. In: Shafritz, J. M. & Ott, J. S. (eds.) *Classics of Organization Theory.* Belmont, CA: Earl Mcpeek.

Tennant, G. 2002. *Design for Six Sigma.* Aldershot, Gower Publishing Limited.

Terninko, J. 1997. *Step-by-Step QFD — Customer-Driven Product Design.* Boca Raton, Florida, St. Lucie Press.

Tottie, M. & Lager, T. 1995. QFD-linking the customer to the product development process as a part of the TQM concept. R&D Management, 25 (July), 257–267.

Tottie, M., Lager, T. & Nordqvist, S. 2016. From customer understanding to product understanding: Collaboration with industrial lead users in a B2B context. *Journal of Business Chemistry*, 13.

Utterback, J. M. 1974. Innovation and the Diffusion of Technology. *Science*, 183, 620–626.

Vandevelde, A. & Van Dierdonck, R. 2003. Managing the design — manufacturing interface. *International Journal of Operations & Production Management*, 23, 1326–1348.

Vargo, S. L. & Lush, R. F. 2011. It's all B2B ... and beyond: Toward a systems perspective of the market. *Industrial Marketing Management*, 40, 181–187.

Viaene, J. & Januszewska, R. 1999. Quality function deployment in the chocolate industry. *Food Quality and Preference*, 10, 377–385.

Von Hippel, E. 1986. Lead users: A source of novel product concepts. *Management Science*, 32, 791–805.

Vonderembse, M., Van Fossen, T. & Raghunathan, T. S. 1997. Is quality function deployment good for product development? Forty companies say yes. *Quality Management Journal*, 4, 65–79.

Wegner, D. M., Giuliano, T. & Hertel, P. (eds.) 1985. Cognitive Interdependence in Close Relationships. New York: Springer-Verlag.

Williams, F. S. 1992. Improving customer/supplier relationships between alumina refining and smelting. In: CUTSHALL, E. R. (ed.) *Light Metals.* The Mineral, Metals & Materials Society.

Witell, L., Löfgren, M. & Dahlgaard, J. J. 2013. Theory of attractive quality and the kano methodology — the past, the present, and the future. *Total Quality Management,* 24, 1241–1252.

Young, B., Monnery, W. & Svrcec, W. 2001. Dynamic simulation improves gas plant. *Oil & Gas Journal*, 99, 54–57.

Zahra, S. A. & George, G. 2002. Absoptive capacity: A review, reconceptualisation, and extension. *Academy of Management Review*, 27, 185–203.

Index

A

Adapting the QFD methodology, 276
Adjuncts, 29
Advanced Multiple Progression, 151, 153
Akao, 18
A lean QFD approach, 274
A medium strong relation, 14
Applied research, 211
Associated test units, 121
A strong relation, 14
A weak relation, 14

B

B2B customer requirements, 102
Beer development, 216
Beer development and manufacturing, 171
Beer quality, 35
Beer styles, 34
Best Practice Recommendations, 265

C

Chemical and Petrochemical Industries, 69
Combination, 234
Company implementation, 259
Company introduction and implementation, 266
Company resources, 269
Company roll-out, 280
Competitor understanding, 110
Complex production chain, 165
Concept development, 17
Conditioning process, 33
Consolidate matrix information, 144, 223
Contextual production, 50
Continuing the DPP, 279
Correlation matrix, 15, 131, 137, 143, 169, 187
Correlation matrix symbols, 133
Cross-functional product innovation, 6
Customer benchmarking, 10, 107
Customer categories, 90
Customer complaints, 132
Customer dimension, 115
Customer importance ratings, 10
Customer relationship matrix, 123
Customer requirements, 9, 94, 100
Customers and suppliers, 74
Customer's shoes, 99

D

Decentralized QFD teams, 274
Demonstration plant, 61
Design requirements, 18
Diagnostic/Pilot Project (DPP), 270
Difficulty of changing, 187
Directions for improvement, 15, 126, 128, 137, 185
Directions for product improvements, 127
Downstream matrices, 151
Downstream multiple progression, 215, 217
Dynamic integration, 248

E
Easy to drink, 114
End-to-end, 252
End-to-end work process product innovation perspective, 253
End-user understanding, 215
Equipment outages, 169
Ethnographic investigations, 99
Evenness of Target Figures, 134
Expected company outcomes, 223, 225
Expected outcomes, 228
Experienced facilitator, 271
Experimental innovation environment, 59
Explanatory customer comments, 111
Explanatory Product Characteristics, 200, 206
Externalization, 234

F
Facilitating methodologies, 37
Fermentation process, 32
Final synthesis, 144
Focus groups and clinics, 98
Food and Beverage Industries, 67
Forest Industries, 68
Formal product innovation, 56
Formal work processes, 55
Four phases of matrices, 19
Fukuhara, 19
Functional design requirements, 74
Fuzzy front end, 5

G
GEMBA, 99
Goes well with food, 113
Good head, 113
Good nose, 114
Group sessions, 98
Group work, 144

H
Hierarchic structure, 100
Hops, 29
House of Knowledge, 242, 256
House of Quality, 11, 81, 115, 118–119, 141

I
Importance ratings, 107
Improved production capabilities, 170
Incoming material, 50
Industrial usability, 226
Information policy, 294
Inherent conditions, 52
Initiate laboratory testing, 202
Innovation concepts, 248
Integrated Knowledge Platforms (IKPs), 145, 223–224, 245–247, 249–250, 277
Intensity scales, 130–131
Interaction with other methodologies, 294
Inter- and intra-organizational integration, 260, 281
Intermediate product interfaces, 170
Internal company guide, 276
Internalization, 234
Internal marketing, 270
Internal process requirements, 182
Interview settings, 97
Intra- and inter-organizational integration, 289
Intra- and inter-organizational interfaces, 282
Investment approach, 37
Investment perspective, 39

J
Judge beer, 36

K
Kano model, 95–96
Knowledge acquisition and assimilation, 236
Knowledge creation and utilization, 231, 235
Knowledge exploitation, 240
Knowledge management, 223, 232
Knowledge transformation, 238

L
Laboratory testing, 59

M
Malt, 28
Malting process, 27
Malt production, 26

Manufacturing, 216
Manufacturing of beer, 23, 112, 135
Market & Customer Understanding, 87
Market segments, 90
Mashing process, 30
Material Delivery, 102
Matrix Analysis, 17, 139, 173, 191, 203, 208
Matrix of Matrices, 18
Matrix-owner, 272
Matrix reviews and revisions, 124
Meta-analysis, 228
Meta-analysis on success factors, 263
Methodology transfer, 273
Metric, 166, 216
Mineral and Metal Industries, 66
mpQFD, 21
Multiple methodologies, 39
Multiple progression, 20, 213

N
New to the world products development, 293
Nice colour, 113
Non-assembled products, 50
Not assembled components, 72

O
Objectives and scope, 267
One-on-one interviews, 98
Ordinal scale, 109
Organizational integration, 288
Organizational learning, 232
Outcomes of methodologies, 38
Outgoing product, 52
Outside Company Demarcations, 213

P
Packaging (bottling) process, 33
Packaging integration, 134
Perfect Lunchtime Beer, 136
Phase progression, 17
Pilot plant testing, 60
Planning the DPP project, 266
Primary (strategic) needs, 100
Printer for large matrices, 274
Processability, 252
Process control status, 168
Process development, 62
Process-Industrial Illustrative Case, 23
Process industries, 46, 49
Process matrix, 157, 159–161, 172
Process parameters, 170
Process relationship matrix, 166–167
Process requirements, 184
Process supply/value chain, 75
Process understanding, 152, 157
Product and process concepts, 75
Product attributes, 122
Product characteristics, 196
Product & Competitor Understanding, 115
Product design requirements, 12, 119, 121, 182
Product development, 62–63
Product development work process, 62
Product influencing process parameters, 164, 171, 206
Product innovation, 1
Production capability, 169
Production plant tests, 61
Production processes, 73
Product matrices, 195
Product Matrix A, 195, 197–198
Product Matrix B, 196, 205, 207
Products, 73
Product target figures, 134
Product understanding, 153
Product yield, 169
Provisional target setting, 193

Q
QFD spirit, 297
QFD system capabilities, 71
Qualitative information, 297

R
Radically new products, 88
Raw material design requirements, 184
Raw material matrices, 177–180, 188, 190, 218

Raw material matrix for malt, 189
Raw material relationship matrix, 185
Raw materials, 73
Raw material understanding, 177
R&D–customer interface, 282
R&D–marketing interface, 284
R&D–production interface, 285
R&D–supplier interface, 287
Relationship matrix, 14, 137, 142
Relationship symbols, 125

S
Samsung, 40
S-curve, 168
Secondary (tactical) needs, 100
Selected questions, 291–292, 296, 298–299
Selecting the product, 268
Service and Support, 102
Side roof, 202
Simulation, 59
Simultaneous matrix development, 276
Single parameter thinking, 299
Socialization, 234
Software selection, 273
Spoken and unspoken, 95
Stakeholder structure, 90–93
Success factor, 259, 261, 293
Super-user, 271
Supplementary rooms, 110, 132, 186, 202, 208
Supply/value chain, 25, 213
Supporting development methodologies, 267
Supportive tools and equipment, 273
Sustainability room, 134, 187
System perspective, 187

T
Tacit and explicit knowledge, 233
Target Figures, 134, 187
Target setting, 139, 144, 191
Technical Benchmarking, 16, 128, 137, 186
Technical dimension, 115, 143
Technical Importance Ratings, 16, 137
Technology Newness, 168
Telephone interviews, 98
Tertiary (operational) needs, 100
Three phase model, 58–59
Tolerance levels, 166
Total quality management, 6
Toyota Auto Body, 8
Transactional memory systems, 232
Transformation process, 51
True purchasing behaviour, 297
Types of rooms, 117

U
Understanding of Product Functionalities, 195
Upfront product innovation, 5
Upgrade cycle, 279
Upstream, 213–214
Upstream & downstream, 154

V
Value scales, 130–131
Voice of the Customer, 8, 81, 94, 111, 142

W
Water, 28
Well balanced, 114
Where to do it, 140
Who should attend, 140
Work process, 46, 55, 61, 89, 119–120, 159, 161, 181, 199, 208
Wort boiling process, 32

Y
Yearly capital cost, 170
Yearly operating cost, 170
Yeast, 29
Yoji Akao, 7

Lightning Source UK Ltd.
Milton Keynes UK
UKHW052211220819
348338UK00004B/51/P

9 789813 279872